Remote Sensing in Mangroves II

Remote Sensing in Mangroves II

Editor

Chandra Giri

MDPI

Basel • Beijing • Wuhan • Barcelona • Belgrade • Novi Sad • Cluj • Manchester

Editor
Chandra Giri
United States Environmental Protection Agency
Research Triangle Park, NC, USA

Editorial Office
MDPI
St. Alban-Anlage 66
4052 Basel, Switzerland

This is a reprint of articles from the Special Issue published online in the open access journal *Remote Sensing* (ISSN 2072-4292) (available at: https://www.mdpi.com/journal/remotesensing/special_issues/remote_sensing_mangroves_Part_II).

For citation purposes, cite each article independently as indicated on the article page online and as indicated below:

Lastname, A.A.; Lastname, B.B. Article Title. *Journal Name* **Year**, *Volume Number*, Page Range.

ISBN 978-3-0365-8886-5 (Hbk)
ISBN 978-3-0365-8887-2 (PDF)
doi.org/10.3390/books978-3-0365-8887-2

Contents

About the Editor

Chandra Giri

Chandra Giri is a senior advisor at the United States Environmental Protection Agency (US EPA) and an adjunct professor at Duke University. He has previously worked in various countries in the government, educational, and private sectors. Some of the institutions that he has been associated with include the Nepal Government, Asian Institute of Technology in Thailand, the United Nations Environment Program (UNEP), Columbia University, Duke University, North Carolina State University, South Dakota State University, and the United States Geological Survey.

He has visited 125 countries through formal and informal visits and collaborated with local politicians and luminaries. Known as the "Pioneer of Mangrove Mapping and Monitoring of the World", he is also a successful writer, editor, and songwriter. To date, he has published 3 books and more than 126 articles. Similarly, some of his literary articles have been published and sixteen songs have also been released.

Dr. Giri lives in Durham, North Carolina, USA, with his wife, and he has a son and a daughter. He likes to play golf in his spare time.

remote sensing MDPI

Editorial
Frontiers in Global Mangrove Forest Monitoring

Chandra Giri

Office of Research and Development, United States Environmental Protection Agency, 109 T.W. Alexander Dr., Durham, NC 27713, USA; giri.chandra@epa.gov; Tel.: +1-919-349-1569

Enhanced spatial, spectral, thematic, and temporal resolution is imperative to optimize the monitoring of mangrove forests, ensuring their effective conservation and management as crucial global resources [1]. These forests play a vital role in providing invaluable ecosystem goods and services to both human society and nature. A prime example of their significance lies in their exceptional ability to sequester carbon, surpassing that of other tropical forests, thus making a substantial contribution towards mitigating the climate change crisis. Furthermore, these forests act as natural protective barriers, safeguarding coastal communities from the destructive forces of hurricanes and tsunamis.

Unfortunately, mangrove forests are under threat due to both natural and anthropogenic forces. Currently, the conversion of mangroves to alternative land uses stands as the dominant factor driving these changes. However, the increasing prominence of sea level rise, global warming, and the intensification of natural disasters, such as hurricanes, is likely to play a more significant role in the future. Thus, observation and monitoring of mangrove distribution and their dynamics are crucial for understanding and addressing the impacts of these changes along with their broader ecological implications for both society and nature. Such information is also needed to manage these forests on a sustainable basis.

The utilization of remote sensing technology is of utmost importance in accurately mapping and monitoring mangrove forests across various scales, ranging from local to global scales. The advancement of this invaluable tool enables us to achieve an enhanced spatial, spectral, thematic, and temporal resolution, thereby significantly improving our scientific comprehension of these expansive ecosystems. Over the past two decades, the availability of new and higher resolution satellite data has played a pivotal role in advancing our knowledge in this field. Moreover, there have been notable advancements in methodologies, computing technologies, and data interpretation expertise, further enhancing our capabilities. As a result, the acceptance and recognition of remote sensing-derived findings by a broader community have also witnessed a substantial improvement.

Remote Sensing is pleased to publish the second volume of a Special Issue dedicated to the observation and monitoring of mangroves using remote sensing techniques. This Special Issue covers a wide range of applications, including the utilization of optical and radar data. Researchers have employed state-of-the-art techniques for data acquisition, management, exploitation, processing, and analysis of remote sensing data in the context of mangrove forest applications. Contributions in the form of eleven research papers and one review paper have been included in this volume.

In the first paper of this Special Issue, Purwanto et al. utilized decision trees and random forest classification algorithms to map and monitor one of the most extensive mangrove forests in Indonesia. The mangrove forests in the Sembilang National Park were facing threats from human activities, necessitating rapid mapping and monitoring as a result. These authors conducted a change analysis from 2002 to 2019 using the decision tree and random forest algorithms. They employed secondary data sources, such as the multi-error-removed improved-terrain digital elevation model (MERIT DEM) accompanied with existing mangrove maps, along with various indices, like the normalized difference moisture index (NDMI) and the normalized difference soil index (NDSI). This

Citation: Giri, C. Frontiers in Global Mangrove Forest Monitoring. *Remote Sens.* **2023**, *15*, 3852. https://doi.org/10.3390/rs15153852

Received: 3 July 2023
Accepted: 2 August 2023
Published: 3 August 2023

study compared the performance of the classification algorithms for the interpretation of the Landsat 7 ETM+ and Landsat-8 OLI data. Their results indicated that the decision tree algorithm with the parameter combination of NDMI, NDSI, and DEM effective for the classification of Landsat-7 ETM+. The random forest classification algorithm outperformed the decision tree algorithm in mapping mangrove forests, as all the parameters of the random forest model exhibited a higher producer accuracy.

Vul et al. conducted a change analysis in three provinces on the northern coast of Vietnam, namely Thai Binh, Nam Dinh, and Hai Phong, using multi-temporal Landsat imagery. This study aimed to monitor the dynamic nature of mangrove forests in this region, focusing on the time period between 1990 and 2022. These researchers employed the Google Earth Engine (GEE) cloud computing platform. The overall accuracy obtained in the year 2022 was 91.98%, while the Kappa coefficient was 0.84. Their results also revealed fluctuations in the mangrove area over time, with periods of decline and elevation. However, since 2005, mangrove forests have exhibited a continuous increase, mainly due to the implementation of restoration programs and policies by the Vietnamese government and local authorities. Notably, this study demonstrated the potential of Landsat time series imagery, pixel-based algorithms, and the GEE platform for the long-term monitoring of mangrove forests.

Ghosh et al. conducted a multiscale diagnosis of mangrove status in a data-poor context using very high spatial resolution (VHSR) satellite images in the Pichavaram mangrove forest, located in Tamil Nadu, India. Pichavaram mangrove forests face threats from cyclones as well as reduced freshwater flow from upstream sources. This study analyzed the changes in the mangrove area spanning from the years 2003 to 2019 at a spatial resolution of 4 m. Specifically, they employed Quickbird (QBD) images in the years of 2003 and 2005 (that were captured shortly after the Indian Ocean tsunami which significantly impacted the region on 26 December 2004), along with GeoEye-1 (GEO) images that were obtained in 2011 and 2016, and a Worldview-3 (WV3) image from 2019. They classified the mangrove and non-mangrove areas using supervised classification available in ERDAS Imagine processing software. Accuracy assessment was performed for each year, achieving an overall accuracy of 85% or higher. Post-classification change analysis was conducted for both natural and planted mangroves between 2003 and 2019, with their results revealing that the mangrove forest area experienced a 28.0% increase (201.2 hectares) from 2003 to 2019. The expansion of these mangrove areas predominantly resulted from the conversion of non-mangrove areas into mangrove habitats. These researchers concluded that despite the limited availability of ground-truth data, VHSR data proved to be valuable in providing a multiscale diagnosis of this ecosystem's condition.

Bunting et al. emphasized the work performed by the Global Mangrove Watch (GMW) in updating the extent of mangrove forests from the year 2010. These researchers highlighted that the latest version of the GMW, namely GMW v2.5, represents a significant improvement over the previous version. They identified 204 regions that were either inaccurately mapped or completely missing from the previous maps, and these areas were updated accordingly. The primary objective of this endeavor was to enhance the existing map through the incorporation of new information. To achieve this, they employed the XGBoost binary classification algorithm for their classification process. As a result of this revision, an additional 2660 square kilometers of mangroves were added, leading to a revised global mangrove extent equivalating to approximately 140,260 square kilometers for the year 2010. However, it is important to note that this study did not examine for any potential forest loss in other areas, which thereby presents a significant limitation of this research.

Rahmandhana et al. conducted mangrove species mapping based on spectral reflectance using extremely high-resolution satellite data from WorldView 2. This study was carried out in the Karimunjawa and Kemujan Islands, located in the central Java province of Indonesia. To create a detailed inventory of mangrove forest biodiversity, particularly in Java, where the diversity of mangrove species is exceptionally high, the utilization of very high-resolution imagery for species mapping in this context was not only possible, but also essential. The study area was found to house approximately 44 species, including 25 true mangroves and 19 mangrove associates. To perform species mapping, these researchers employed the spectral angle mapper (SAM), spectral information divergence (SID), and spectral feature fitting (SFF) algorithms. Field data, such as mangrove species identification, coordinate locations of targeted mangrove species, and spectral reflectance measurements of mangrove species using a field spectrometer, were all collected. Dendrogram analysis utilizing the Ward linkage method was conducted to classify mangrove species based on their distances between clusters of spectral reflectance patterns. This classification process resulted in two, four, and five species groups for Levels 1 to 3, respectively, while individual species were identified under Level 4. The SID algorithm achieved the highest overall accuracy, while the SFF algorithm yielded the most accurate results for mapping individual species. As they expected, these findings indicated that as the number of classes to be mapped increased, the mapping accuracy consequently decreased.

Zhu et al. conducted a spatiotemporal simulation of mangrove forests under different scenarios in Hainan Island, China. Their study utilized a total of 12 drivers, including the elevation, slope, enhanced vegetation index (EVI), EVI change trends, distance to major roads, distance to minor roads, distance to the sea, distance to rivers, distance to aquaculture ponds, distance to building land, distance to suitable land for mangroves, and a spatial autocorrelation factor, for the simulation. Various models, such as logistic regression, support vector regression, and the random forest model, were compared with one another in terms of their ability to capture the spatial characteristics of mangrove forests. Three development scenarios were established: a natural growth scenario (NGS), an economic development scenario (EDS), and a mangrove protection scenario (MPS). The CLUE-S model was employed to predict the spatiotemporal distribution of mangrove forests from the years 2022 to 2037 under these different scenarios. Subsequently, based on the prediction results, the future change trends of mangrove forests from 2017 to 2037 were analyzed. The simulation results of these different models demonstrated that AutoRF (random forest with spatial autocorrelation) performed the best in simulating the spatial characteristics. Importantly, specific factors, such as the enhanced vegetation index (EVI), various location indices, and the spatial autocorrelation factor were all found to significantly improve the accuracy of the mangrove simulations. The prediction results for Hainan Island indicated that under the NGS, the mangrove area would experience slow growth, while under the EDS, it would decrease significantly. Conversely, the mangrove area would increase significantly under the MPS. Based on these findings, the MPS was identified as the most suitable development direction for the future, as it offers a balanced approach that promotes economic development while ensuring mangrove preservation.

Chambarlain et al. utilized Landsat dense time series data to monitor the mangrove forest cover and its phenology in Central Queensland, Australia. Although the exploitation of mangroves by coastal communities is strictly regulated in Australia, these forests are still being threatened with landscape modifications and hydrological alterations upstream. This study employed Landsat reflectance data from the years 2009 to 2019 to track mangrove forest cover changes. To address the persistent cloud cover issues impacting certain areas, mosaics of three-year windows were used to generate cloud-free mosaics. Analysis was conducted using the Google Earth Engine and a random forest classifier. Additionally, secondary information, such as the shuttle radar topography mission (SRTM) and previously established mangrove land cover maps, were incorporated. The overall classification accuracies and Kappa coefficient for the land cover maps in the years 2008–2010 and 2018–2020 were both determined to be 95%. Furthermore, a decrease of 1480 hectares (−2.31%) in mangrove coverage was also observed from 2009 to 2019. Extending from this, an examination of intra- and inter-annual seasonality in mangrove growth phenology was conducted using an NDVI-based time series. Linear and harmonic regression models, as well as TIMESAT metrics, were employed to analyze mangrove forests in three sections of the study region. Together, these findings indicated a correlation between the growth phenology of mangrove forests, precipitation anomalies, and the occurrence of severe tropical cyclones over the time series.

Niu et al. conducted a global sensitivity analysis for the canopy reflectance and vegetation indices of mangroves. This study focused on a one-dimensional canopy reflectance model to systematically analyze the sensitivity of mangroves to various biophysical and environmental factors. Different scenarios were created, including sparse and dense canopies, to assess the impact of these factors on simulated canopy reflectance spectra and selected Sentinel-2A vegetation indices. This study employed a variance-based method and a density-based method to compare the computed sensitivity indices. The results revealed that the fractional cover and leaf-to-total area ratio of mangrove crowns were among the most influential factors across all examined scenarios. These findings highlight the significant role of these factors in shaping the canopy reflectance and vegetation indices of mangroves.

Ghorbanian et al. utilized Sentinel-1 and Sentinel-2 satellite data to map the mangrove ecosystem in the Hara protected area in Qeshm, Iran. The analysis was conducted on the Google Earth Engine at a spatial resolution of 10 m. The data acquired using these Sentinel satellites in 2019 were utilized to generate a composite of optical and synthetic aperture radar (SAR) data. To classify the different components, a pixel-based random forest (RF) classifier was employed. This study successfully generated six distinct classes: mangrove, mudflat, deep water, tidal zone, shallow water, and aerial roots. The resulting mangrove ecosystem map exhibited a high accuracy, with an average overall accuracy of 93.23% and a Kappa coefficient of 0.92. These findings demonstrate the reliability and effectiveness for mapping mangrove ecosystems using Sentinel-1 and Sentinel-2 satellite data.

Kanniah et al. utilized the MODIS-derived leaf area index (LAI) and gross primary productivity (GPP) to investigate the fragmentation of mangrove forests in the southern region of Peninsular Malaysia, specifically in Iskandar Malaysia. This study aimed to assess the impact of land cover changes, including urbanization, plantations, and aquaculture activities, on the mangrove forest ecosystem. The results revealed a decline in the mangrove forest area due to land cover changes. However, the areas that remained undisturbed revealed an increase in both the mean LAI and GPP, signifying the mangrove forest's capacity to absorb CO_2 when left undisturbed. Furthermore, areas that experienced mangrove loss but were replaced with oil palm plantations exhibited a decrease in the mean LAI. The fragmented mangrove patches also showed an increase in GPP, potentially due to the smaller patch sizes (<9 ha) and the edge effects, which promote higher levels of productivity in areas exposed to abundant solar radiation along the patch edges. The impact of fragmentation on the GPP was found to depend on the type of land transformation along with patch characteristics, such as size, edge, and shape complexity.

Muang and Sasaki conducted a study to evaluate the natural recovery of mangroves in the Wunbaik mangrove forests of Myanmar following human disturbance, specifically focusing on the abandoned shrimp ponds. These researchers utilized cloud-free Sentinel-2 images that were captured on 21 January 2020 and 23 December 2015, both during the dry seasons. They employed an artificial neural network (ANN) classification approach with a transfer learning method to classify these two dates' images. Additionally, a post-classification change analysis approach was employed to assess the changes that occurred. To identify the naturally recovering mangroves, three abandoned shrimp ponds were selected based on field investigations and their change detection results were then extracted. The proposed methodology achieved a high level of accuracy, with an overall accuracy of 95.98% and a Kappa coefficient of 0.92 for the 2020 classification. For the 2015 prediction, the transfer learning approach improved the model's performance, resulting in an overall accuracy of 97.20% along with a Kappa coefficient of 0.94. The analysis of the change detection results revealed a slight decrease of mangrove forests within the Wunbaik mangrove forests between the years 2015 and 2020. However, naturally recovering mangroves were identified in approximately 50% of each abandoned site within a relatively short abandonment period. These findings contribute to our understanding of the natural regeneration process of mangroves following human disturbance, specifically in abandoned shrimp ponds.

In the final paper of this Special Issue, Tran et al. presented a comprehensive review paper on spectral indices for mangrove remote sensing. This study covered the period between 1996 and 2021, examining the range of spectral indices that have been developed and utilized in the context of mangrove remote sensing. This review revealed that spectral indices have been predominantly used for various aspects of mangrove analysis, excluding the identification of mangrove species. The aspects that were covered included the mangrove extent, their distribution, and above-ground parameters (such as the carbon density, biomass, canopy height, and leaf area index estimation), as well as changes in these aspects over time. Among the spectral indices, the NDVI emerged as the most widely applied, appearing in 82% of the reviewed studies, followed by the EVI, which was noted as being used in 28% of the studies examined. While the development and application of potential indices for characterizing mangrove cover have shown growth (with six indices having been published so far), the NDVI remains the most popular index for mangrove remote sensing. However, this review also highlights the limitations and gaps that are present within current studies. Considering the digital era, the authors suggested future directions for research, emphasizing the need to explore spectral index applications in connection with time series imagery and the fusion of optical sensors for more comprehensive and accurate mangrove studies.

In summary, the utilization of cloud computing platforms for data analysis in mangrove research is growing. Researchers benefit from the flexibility, cost savings, enhanced security, scalability, and improved collaboration offered by cloud computing. There is a rising demand for mapping mangroves at higher resolutions, specifically at levels higher than 5 m. Compared to the past, supervised classification approaches are now more commonly used instead of unsupervised classification. The most popular application used in the remote sensing of mangroves is the mapping of mangrove and non-mangrove areas and conducting time series change analysis. Both optical and synthetic aperture radar data are currently being employed for mangrove mapping. Overall, researchers are generating higher resolution mangrove data and producing improved results for change analysis, ranging from local to global scales. This progress has been made possible with the increased availability of free and higher resolution remote sensing data, advancements in computing technology (both hardware and software), growing collaborative efforts among researchers, increased availability of expertise, better understandings of mangrove forests, and the development of improved methodologies.

Funding: This research was funded by the U.S. Environmental Protection Agency through its Office of Research and Development (ORD) as a contribution to ORD's Safe and Sustainable Water Resources National Research Program. The APC was funded by the ORD.

Conflicts of Interest: The author declares no conflict of interest. The views expressed in this article are those of the author and do not necessarily represent the views or the policies of the U.S. Environmental Protection Agency.

Reference

1. Giri, C.; Ochieng, E.; Tieszen, L.L.; Zhu, Z.; Singh, A.; Loveland, T.; Masek, J.; Duke, N. Status and distribution of mangrove forests of the world using earth observation satellite data. *Glob. Ecol. Biogeogr.* **2011**, *20*, 154–159.

remote sensing

MDPI

Article

Assessing the Natural Recovery of Mangroves after Human Disturbance Using Neural Network Classification and Sentinel-2 Imagery in Wunbaik Mangrove Forest, Myanmar

Win Sithu Maung [1,2,*] and Jun Sasaki [1]

1 Department of Socio-Cultural Environmental Studies, Graduate School of Frontier Sciences, The University of Tokyo, Kashiwa 277-8561, Japan; jsasaki@k.u-tokyo.ac.jp

2 Forest Department, Ministry of Natural Resources and Environmental Conservation, Naypyitaw 15015, Myanmar

* Correspondence: sithumg1989@gmail.com

Abstract: In this study, we examined the natural recovery of mangroves in abandoned shrimp ponds located in the Wunbaik Mangrove Forest (WMF) in Myanmar using artificial neural network (ANN) classification and a change detection approach with Sentinel-2 satellite images. In 2020, we conducted various experiments related to mangrove classification by tuning input features and hyper-parameters. The selected ANN model was used with a transfer learning approach to predict the mangrove distribution in 2015. Changes were detected using classification results from 2015 and 2020. Naturally recovering mangroves were identified by extracting the change detection results of three abandoned shrimp ponds selected during field investigation. The proposed method yielded an overall accuracy of 95.98%, a kappa coefficient of 0.92, mangrove and non-mangrove precisions of 0.95 and 0.98, respectively, recalls of 0.96, and F1 scores of 0.96 for the 2020 classification. For the 2015 prediction, transfer learning improved model performance, resulting in an overall accuracy of 97.20%, a kappa coefficient of 0.94, mangrove and non-mangrove precisions of 0.98 and 0.96, respectively, recalls of 0.98 and 0.97, and F1 scores of 0.96. The change detection results showed that mangrove forests in the WMF slightly decreased between 2015 and 2020. Naturally recovering mangroves were detected at approximately 50% of each abandoned site within a short abandonment period. This study demonstrates that the ANN method using Sentinel-2 imagery and topographic and canopy height data can produce reliable results for mangrove classification. The natural recovery of mangroves presents a valuable opportunity for mangrove rehabilitation at human-disturbed sites in the WMF.

Keywords: mangrove; natural recovery; artificial neural network; Sentinel-2; transfer learning; change detection

Citation: Maung, W.S.; Sasaki, J. Assessing the Natural Recovery of Mangroves after Human Disturbance Using Neural Network Classification and Sentinel-2 Imagery in Wunbaik Mangrove Forest, Myanmar. *Remote Sens.* **2021**, *13*, 52. https://dx.doi.org/10.3390/rs13010052

Received: 10 November 2020
Accepted: 22 December 2020
Published: 25 December 2020

1. Introduction

Mangrove forests provide highly valuable ecosystem services consisting of tangible goods, such as wood and non-wood products, and intangible services, such as the protection of coastal regions from storms and wave attacks and the sequestration of substantial volumes of carbon dioxide [1,2]. Despite these invaluable services, mangrove forests worldwide are being depleted at an alarming rate every year because of anthropogenic pressures, such as the development of aquaculture, agricultural expansion, oil plantations, and urbanization [3–5]. Mangroves possess a high capacity for self-recovery; this facilitates mangrove rehabilitation. Mangrove self-recovery involves the natural restoration of mangrove species in a deforested area if environmental parameters, such as natural hydrology, salinity, and site elevation, satisfy the preferences of seeds or propagules dispersed with tidal flow [6,7]. However, limited research has been conducted to address the natural recovery of mangroves in rehabilitation programs. This study highlights the natural recovery of mangroves after human disturbance using remote sensing classification and a change detection approach.

The Wunbaik Mangrove Forest (WMF) in Myanmar is one of the largest remaining mangrove forests in the world and provides ecological, environmental, and socio-economic goods and services to the local community. Due to the expansion of aquaculture and agriculture, WMF has experienced severe degradation since the 1990s. According to the 2011 Food and Agricultural Organization (FAO) inventory [8], WMF possesses abundant seed productivity and a high mangrove species germination rate and is suitable for examining the natural recovery of mangroves depleted by human disturbances. Despite the presence of such a large remnant mangrove forest with a diverse range of species, there has been a prominent information gap regarding mangrove distribution and changes since 2014. Current and accurate information is necessary before implementing any restoration efforts for sustainable mangrove management.

Mangroves are under threat from natural and anthropogenic disturbances; however, accessing reliable and accurate information for many mangrove forests worldwide is difficult [9]. Remote sensing classification contributes to the provision of reliable information for different research topics, such as mangrove extent mapping, species identification, and estimation of above-ground biomass, contributing to the sustainable management of mangrove forests [10,11]. However, classification accuracy is highly variable depending on the performance of the selected classifier because each classification method has benefits and drawbacks [12]. Machine learning is an emerging advanced classification methodology in which supervised classifiers, such as maximum likelihood, random forest (RF), support vector machine (SVM), and artificial neural networks (ANNs), can outperform traditional unsupervised classification [13]. In mangrove remote sensing, many studies have reported that supervised classification is superior to unsupervised approaches [14–16]. Supervised machine learning classifiers, such as RF, SVM, classification and regression tree, and regularization in two discriminant analysis (RDA), have been compared and evaluated for mangrove cover changes [17–19]. However, ANNs require extensive training data and are rarely applied in mangrove classification. However, they have been widely used in many fields of remote sensing classification because of promising results and model robustness [20–23].

In terms of the role of remotely sensed data, the effect of input features is a key component in image classification. Many mangrove studies have been conducted using optical images with different resolutions (e.g., IKONOS, QuickBird, WorldView-3, Landsat series, and SPOT), synthetic aperture radar (SAR) data, and airborne hyperspectral data [11,24–28]. The launch of the Sentinel-2 satellite by the European Space Agency in 2015 provided a valuable opportunity for the remote sensing community to develop classification approaches in a wide variety of fields [29]. Owing to the advantage of spectral, temporal, and spatial resolutions of freely accessible multi-spectral bands, Sentinel-2 imagery outperforms that of Landsat 8 by yielding more precise results in remote sensing classification [30–32]. Wang et al. [33] described that the Sentinel-2 sensor bested that of Landsat 8 by offering more accurate information in mapping mangrove extent and species.

Studies have improved results in their targeted fields through the application of various machine learning classifiers, including the ANN method and Sentinel-2 imagery [34–39]. The combination of other machine learning classifiers and Sentinel-2 imagery has been evaluated for mangrove mapping [40,41]. There is a limited number of mangrove studies using the ANN classification with various satellite imagery [42–44], and the potential performance of the ANN classification in conjunction with Sentinel-2 imagery has not been previously investigated for mangrove classification.

The integration of additional features, such as digital elevation models (DEMs), into natural resource assessment using remote sensing approaches, can improve classification accuracy [45]. DEM generally implies two different meanings: A digital surface model (DSM), providing elevation data for the earth's surface and objects on it, and a digital terrain model (DTM), representing true topography [46]. Current widely-used DEMs, such as the Shuttle Radar Topography Mission (SRTM), Advanced Land Observation Satellite (ALOS), and Advanced Spaceborne Thermal Emission and Reflection, provide DSM data; however, these data are associated with high error margins for true topographic information [46]. Mangrove canopy estimation has been performed using SAR datasets, such as TerraSAR X, Tan DEM X, and Sentinel-1 [47–49]. There is a good opportunity for geographic studies to use the new Multi-Error-Removed-Improved-Terrain (MERIT DEM), which was developed to reduce the vertical errors of SRTM3 (90 m resolution) and ALOS World three-dimension (3D)—30 m in 2017 [50]. The MERIT DEM offers a true DTM with a spatial resolution of 90 m in the geoid datum for a global area between 90° N–60° S and provides a significant accuracy improvement in swamp forests. This new methodology was applied in this study to investigate potential improvements in classification accuracy. A canopy height model (CHM) was created from the SRTM and MERIT DEMs in the remote sensing of large spatial mangrove areas, particularly with a mix of mangroves and other types of vegetation. Distinguishing between mangroves and other vegetation types remains challenging in remote sensing classification. A few studies have attempted mangrove species classification using the CHM with hyperspectral data taken by an unmanned aerial vehicle (UAV) as a small-scale approach [28]. However, there is a lack of information regarding the application of the CHM in large-scale mangrove classification.

The aims of this study were to: (1) Explore the ANN classification using Sentinel-2 satellite imagery for mangrove distribution by integrating field information, (2) investigate the effectiveness of the CHM in distinguishing between mangroves and other vegetation types, (3) address information gaps regarding changes in mangroves in the WMF between 2015 and 2020, and (4) assess the natural recovery process of mangroves in different abandoned shrimp ponds. We conducted various experiments on input features and hyper-parameter tuning to propose a promising classification approach for mangrove distribution. Furthermore, transfer learning was introduced to improve the proposed method by applying a different seasonal dataset. It is assumed that the use of this proposed approach can achieve more reliable results for mangrove classification to provide information on the change in mangrove extent in the WMF and to assess the natural recovery of mangroves following human disturbance.

2. Materials and Methods

2.1. Study Site

The study area was located between 19°07′02″ N–19°23′30″ N and 93°51′0″ E–94°02′30″ E in Yambye Township, Kyauk Phyu District, Rakhine State, Myanmar (Figure 1). Mangrove forests in the study area were locally known as the WMF, of which the main portion of mangroves was part of a reserved forest known as the Wunbaik Reserved Mangrove Forest (WRMF), formally protected by the Forest Department since 1931. The WRMF is one of the largest remaining mangrove communities, with an area of 22,919 ha [8,51]. Despite the high altitude of the terrestrial region in the southern part, the topography of the WMF is almost flat. Three types of climate, namely, summer, monsoon, and winter, are experienced here [8]. Annual rainfall recorded for 2019 was 4860 mm, and the maximum and minimum temperatures were 33 °C and 27 °C, respectively (http://themimu.info/). The FAO has identified 34 mangrove species, such as *Rhizophora*, *Avicennia* and *Sonneratia*, in the WRMF since 2011 [52].

2.2. Field Survey

Field surveys were conducted from 15 September to 10 October 2019 to acquire accurate and current information on the study area. The supervised classification required a reliable ground truth image to train the neural network model for mangrove classification. The study area contained abundant natural mangroves and various land-use types (Table 1). Therefore, the latitudes and longitudes of the ground truth points for each land-use type were collected using a handheld global positioning system (GPS) (GARMIN etrex 10), providing a horizontal accuracy of approximately 3 m. Figure 2 shows the locations of the ground truth points with matching photos taken in the field.

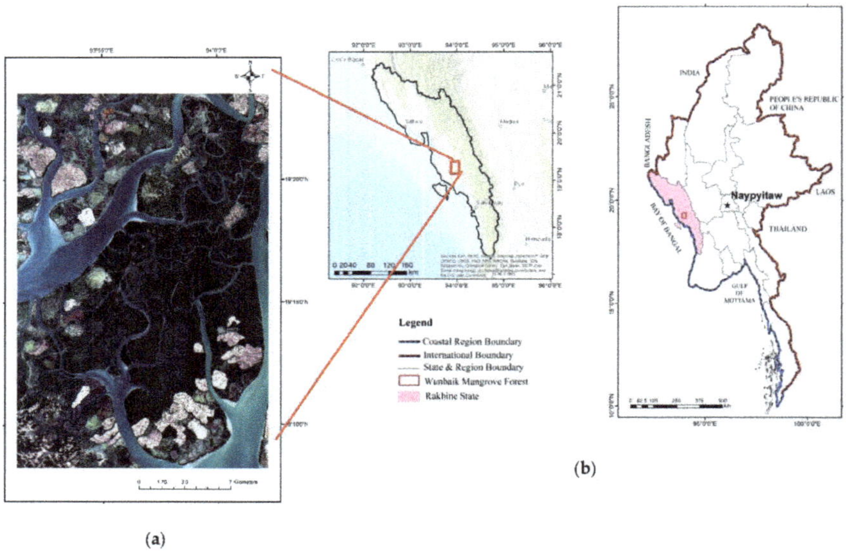

Figure 1. Location of the study area, (**a**) Wunbaik Mangrove Forest (Sentinel-2 true color image), (**b**) Rakhine coastal region in Myanmar.

Table 1. Ground location points of different land uses in the study area.

Sr	Land-Use Types	Number of Points
1	Water	20
2	Paddy field	6
3	Shrimp pond	13
4	Natural mangrove	30
5	Mangrove plantation	5
6	Other vegetation	6
	Total	80

2.3. *Preparation of a Ground Truth Image*

Preparing accurate ground truth data was necessary in supervised classification and was a major challenge for training and evaluating a neural network model. To obtain a reliable ground truth image, location points collected in the field were imported into ArcGIS. Polygons were manually drawn by referring to the same land cover of the ground truth points, using high-resolution Google Earth imagery and different combinations of Sentinel-2 satellite bands. For accurate mangrove recovery analysis, two major classes were assigned as mangrove and non-mangrove (rather than multi-classification, which may lead to massive time consumption for creating ground truth images). Assigned polygons were converted into a raster image with a resolution of 10 m. A ground truth image, which has assigned values of 0 and 255 for every pixel, was obtained. All 53,939,393 pixels, of which the ground truth image had a width and height of 1914 × 2814 pixels with a resolution of 10 m, were counted as the dataset for the ANN prediction.

To assess natural recovering mangroves in abandoned shrimp ponds and to ensure the identification of natural recovery sites, we set 3 criteria: (1) No planted mangroves, (2) same abandonment year, and (3) site accessibility. Three disused ponds, which met these criteria, were selected after discussions with the local staff of the Forest Department and the pond owners (Figure 3). The locations of these abandoned ponds were collected to create polygons of the recovered ponds. Recovering mangrove species at the selected abandoned sites were observed by conducting a field inventory in 50 sample plots with the assistance of knowledgeable local laborers and Forest Department staff (Figure 3).

Figure 2. Ground location points of different land covers of mangrove plantation, natural mangrove, other vegetation, shrimp pond, paddy field, and water body (Photos were taken by Win Sithu Maung during the field survey of this study).

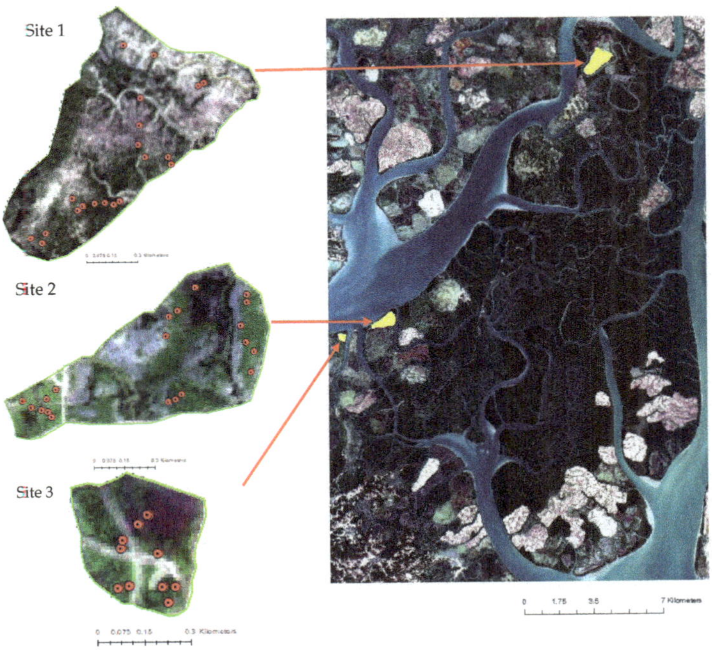

Figure 3. Locations of three selected abandoned sites and sample plots.

2.4. *Earth Observation Data*

2.4.1. Satellite Band Information

To create the required datasets for the ANN classification, sets of multi-spectral bands of Sentinel-2 satellite imagery were collected from the Copernicus Open Access Hub using the semi-automatic classification plugin (SCP) in QGIS [53]. To detect changes in the mangroves and identify recovering mangrove areas within the selected abandoned sites, cloud-free Sentinel-2 images from 21 January 2020, and 23 December 2015 (during the dry seasons) were utilized. Detailed information regarding Sentinel-2 data acquisition is provided in Table 2.

Table 2. Specification of Sentinel-2 products used in this study.

Date of Acquisition	Tile Number	Cloud Coverage	Processing Level	Bands Used	Central Wavelength (nm)	Spatial Resolution (m)
21 January 2020	T46QEG	0	Level-2 A	Band 2	490	10
23 December 2015	T46QEG	0	Level-1 C	Band 3	560	10
				Band 4	665	10
				Band 5	705	20
				Band 6	740	20
				Band 7	783	20
				Band 8	842	10
				Band 8A	865	20
				Band 11	1910	20
				Band 12	2190	20

Currently, 2 major products of Sentinel-2 imagery (Level-1 C and Level-2 A) are disseminated to the public by the European Space Agency [54]. Level-1 C provides orthorectified Top-Of-Atmosphere reflectance, whereas Level-2 C, which was extended globally in December 2018, offers orthorectified Bottom-Of-Atmosphere reflectance. The spatial resolutions of multi-spectral bands provided by Sentinel-2 were 10, 20, and 60 m [55]. We used 10 Sentinel-2 spectral bands, excluding bands 1, 9, and 10, which had a poor spatial resolution of 60 m and were unrelated to the aim of mangrove classification. Sentinel-2 bands with various resolutions were resampled to a resolution of 10 m using the nearest neighbor algorithm for 2015 and 2020. As the multi-spectral bands acquired in 2015 were Level-1 C products, an atmospheric correction was conducted using the dark object subtraction 1 method provided in the SCP QGIS tool [53].

2.4.2. Spectral Indices

In addition to the multi-spectral bands of satellite images, spectral indices were one of the most effective features that offered information on land use and land cover. The normalized difference vegetation index (NDVI), normalized difference water index (NDWI), and combined mangrove recognition index (CMRI) were used as input features to train the neural network model. The NDVI was calculated using band 4 (RED) and band 8 (near infrared (NIR)) (Equation (1)) [56], whereas the NDWI was derived from Equation (2) using band 3 (GREEN) and band 8 (NIR) (Equation (2)) [57]. The CMRI was determined based on the difference between the NDVI and NDWI (Equation (3)) [58].

$$\text{NDVI} = \frac{\text{NIR} - \text{RED}}{\text{NIR} + \text{RED}} \tag{1}$$

$$\text{NDWI} = \frac{\text{NIR} - \text{GREEN}}{\text{NIR} + \text{GREEN}} \tag{2}$$

$$\text{CMRI} = \text{NDVI} - \text{NDWI} \tag{3}$$

2.4.3. Digital Elevation Model

Collecting topographic data over the entire study area during the field survey was problematic because of difficulties in accessing the mangrove area. To overcome this challenge, we applied the SRTM for DSM and the MERIT DEM for DTM. During the SRTM mission in February 2000, the SRTM 1 arc second DEM was created from the C-band radar data acquired by the cooperation of the National Aeronautics and Space Administration and the National Geospatial-Intelligence Agency. The SRTM1 DEM provided user elevation values with a resolution of 30 m, referring to the WGS84 ellipsoid as a horizontal datum and the geoid as a vertical datum. It may be freely downloaded from the USGS EarthExplorer website (https://earthexplorer.usgs.gov/) [59]. The MERIT DEM was downloaded free of charge from the MERIT_DEM website (http://hydro.iis.u-tokyo.ac.jp/~yamadai/MERIT_DEM/).

2.4.4. Canopy Height Model

This study used canopy height information for the large-scale classification of mangrove extent. Applying the SRTM DEM as a DSM data source was reasonably acceptable in estimating forest height [60], and the MERIT DEM provided true topographic information. We ensured the reliability of the two DEMs by comparing field data collected during the post-processed kinematic (PPK) survey (Figure 4). The resulting CHM applied was derived from the difference between the SRTM DEM and MERIT DEM.

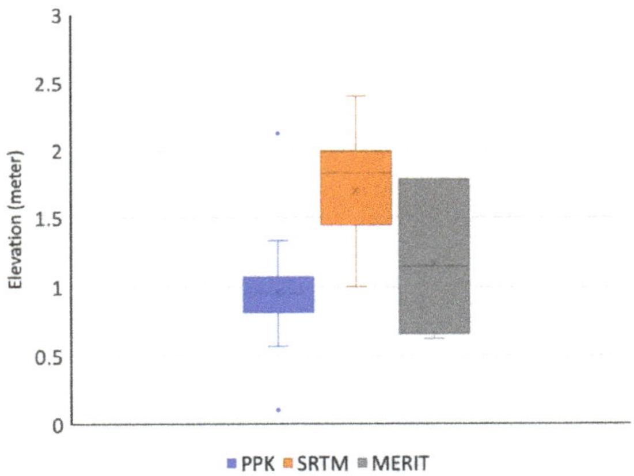

Figure 4. Comparison of Shuttle Radar Topography Mission (SRTM) and Multi-Error-Removed-Improved-Terrains (MERIT DEMs) with post-processed kinematic (PPK) ground truth elevation data.

2.5. Artificial Neural Network Classification

An ANN model is based on the structure of the biological neural network of the human nervous system and primarily consists of a multi-layer perceptron with an input layer, a hidden layer, an output layer, and interconnected operating nodes similar to brain nerve cells in each layer [61]. The nodes in the input layer transfer feature values (x_j), weight information (w_j), and bias (b_i)) from a dataset to the hidden layer. In the hidden layer, data obtained from the input layer are analyzed through an activation function in each node, and the generated data (y_i) are conveyed to the output layer (Equation (4)).

Based on the basic model architecture, the ANN model for this study was first created with 1 hidden layer and 12 neurons as a basic model using a sequential model of the Keras Application Programming Interface (https://keras.io/api/models/) and TensorFlow 2.3.0 (https://www.tensorflow.org/api_docs/python/tf/keras/sequential). We explored input features and hyper-parameter tunings, where multiple experiments were conducted to obtain the most suitable combination of input features for mangrove classification. An optimum ANN model design was established by tuning the number of hidden layers and neurons. In terms of the activation functions that analyze input data within the neural network process, a rectified linear unit (ReLU) was applied to the hidden layers, and the Softmax function for the output layer. The ReLU function produced linear values for input data values larger than 0 and negative values for 0 (Equation (5)). The Softmax function [62,63] normalized the output value ranging from 0–1 and yielded the input value probability belonging to a particular class by dividing the summation (Equation (6)).

$$y_i = \sum(w_j x_j) + b_i \tag{4}$$

$$f(y_i) = \begin{cases} y_i, & y_i > 0 \\ 0, & y_i \leq 0 \end{cases} \tag{5}$$

$$f(y_i) = \frac{\exp(y_i)}{\sum_j \exp(y_j)} \tag{6}$$

2.6. *Workflow of Assessment of Natural Recovering Mangrove*

Figure 5 illustrates the workflow of this research, which contained 3 primary approaches: (1) Input feature selection and hyper-parameter tuning for ANN classification; (2) application of a trained model to a new dataset and improvement of model performance through the transfer learning method; and (3) post-classification change detection. To delineate information on naturally recovering mangroves, the 3 abandoned sites were extracted from the mangrove change detection results. A detailed explanation of each process is provided in the following sections.

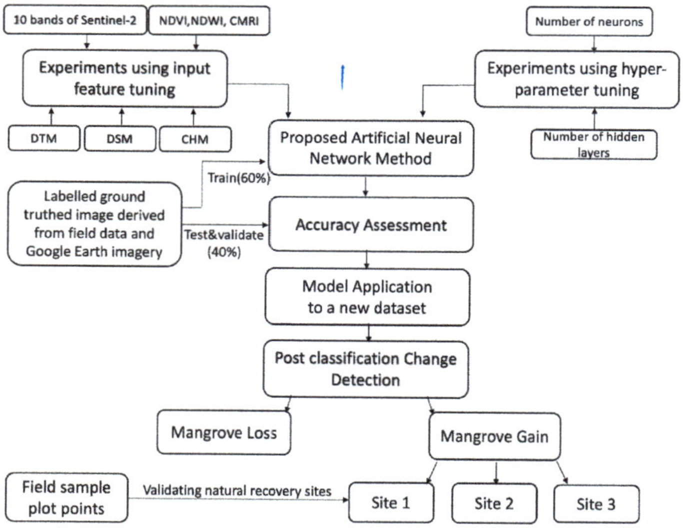

Figure 5. Workflow of assessment of natural recovering mangrove at different abandoned sites.

2.6.1. Experimental Analysis through Input Feature and Hyper-Parameter Tuning

To select the most appropriate input feature combination for mangrove classification, different experimental classifications were first conducted using the basic ANN model. The raster images of the input features, multi-spectral satellite images of 10 bands, NDVI, NDWI, MERIT DEM, SRTM DEM, CHM, and ground truth images were converted into numerical datasets. Fifteen datasets were created and divided into 3 parts: Training data (60%), testing data (20%), and validation data (20%). This division was performed using the pre-processing modules of the scikit-learn data processing library (https://scikit-learn.org/stable/). The training dataset was applied to teach the ANN model by referring to the labeled data of the ground truth image, and the validation dataset was used to observe the behavior of the model to check for overfitting problems in every epoch of the training session simultaneously. The model was trained using the training dataset with 30 epochs for all experiments, and the trained model was applied to predict the testing dataset. Model performance was evaluated through overall accuracy and the kappa index.

After classification was completed using the trained model, the output datasets were exported in a raster format containing geographic information. The output raster images were visually assessed using true color Sentinel-2 and high-resolution Google Earth imagery in ArcGIS. Using the same approach, all datasets were analyzed to select a combination of input features, allowing the highest accuracy and a high-quality output map for mangrove classification.

Following the acquisition of the most suitable combination input features, the basic model was tuned in terms of the number of hidden layers and neurons. By adding one hidden layer to the basic model, the performance of each model was evaluated through accuracy and loss reduction, and a suitable number of hidden layers were selected. The number of neurons was then doubled in the hidden layers of the selected model to optimize the design of the ANN model.

2.6.2. Model Application to a New Dataset Using Transfer Learning

One of the advantages of neural network classification is that once the model is established with the desired accuracy, related new datasets can be predicted without re-training the model. After classifying the 2020 dataset through input features and hyperparameter tuning, the trained ANN model was applied to a new dataset of the same input features for 2015.

Transfer learning is an emerging technique of machine learning in which a model trained for a given problem can theoretically be applied to a different yet related problem [64]. The performance of the pre-trained model may vary depending on the variation of input features of a new dataset, such as the spectral signatures of remotely sensed data acquired in different seasons. Retraining a model whenever a new problem is confronted in future predictions is a very time-consuming practice. To overcome such problems, transfer learning has become a practical and effective method for neural network classification. In this study, different layers of the original model were experimentally frozen and retrained to predict the new dataset.

During transfer learning, a new ground truth image was created by manually digitizing based on Google Earth images in 2015. 80% of the new dataset with the corresponding reference image randomly selected as training data and 20% of the dataset was the validation data. After conducting transfer learning through different experiments, the entire new dataset in 2015 was also predicted by all models.

2.6.3. Accuracy Assessment

Accuracy assessment is one of the major phases in image classification used to evaluate model performance. In this study, 20% of the ground truth image was used for validation after training the model. Model performance was evaluated using the overall accuracy and kappa coefficient, the most widely used measure to assess the accuracy of satellite image classification [65]. For the new prediction in 2015 using the ANN model trained with the 2020 dataset, 1000 reference points with geographic locations and land cover information were randomly generated to represent the entire study area. The predicted points were then extracted from the classified images from 2015, and model performance was assessed by comparing them with the reference points.

A confusion or error matrix representing the ground truth pixels in the column and classified pixels in the row for each class was created to assess the classification accuracy. The overall accuracy was then calculated by dividing the sum of truly classified pixels of each class by the total number of pixels in the diagonal elements of the confusion matrix. The kappa coefficient, precision, recall, and F1 score were also used, as they considered the true positive pixels, the false-negative pixels, and the non-diagonal elements in a confusion matrix [66,67]. These parameters ranged from 0 to 1; values closer to 1 denoted higher classification accuracy.

2.6.4. Post-Classification Change Detection

Acquiring accurate information on changes to the mangroves is critical for mangrove conservation, to determine past occurrences, implement future restoration programs, and develop the necessary regulations [68]. Among the various available change detection methods, pixel-to-pixel-based post-classification change detection was used in this study to provide promising classification accuracy results [69]. The 2 classified images in 2015 and 2020 were differentiated in the post-classification change detection analysis. Areas that underwent extensive changes were highlighted to identify the primary drivers of mangrove dynamics. Using Sentinel-2 and high-resolution Google Earth imagery, the findings were visually clarified to understand the causes of these changes.

2.6.5. Assessment of Natural Recovering Mangroves at Different Abandoned Sites

Based on the post-classification change detection results, 3 selected sites were extracted from the mangrove gain areas to identify naturally recovering mangroves for different abandoned shrimp ponds. Naturally recovering areas at t3 selected sites were quantified in ArcGIS, and the capacity of natural recovery was compared with the extent of the ponds.

3. Results

3.1. Artificial Neural Network Classification

3.1.1. Experimental Results of Input Feature Selection

Using the basic ANN model, 15 experiments using various combinations of input features were conducted to identify the 2020 mangrove distribution of the WMF. Different levels of accuracy were obtained depending on the input feature combination in each experiment (Table 3). Experiment 1 using 10 bands of the Sentinel-2 satellite image produced an unsatisfactory accuracy of 56.39%, revealing that the use of only spectral information from satellite bands was insufficient to classify mangrove distribution using the ANN model. As such, Experiment 1 was not adopted to produce an output image of the classification result because of its poor accuracy.

Table 3. Experimental results of input feature adjustment.

Experiment	Combination of Input Features	Overall Accuracy
1	10 bands (B2, B3, B4, B5, B6, B7, B8, B8A, B11, B12) of Sentinel-2 image	56.39%
2	10 bands, NDVI, NDWI	93.43%
3	10 bands, NDVI, NDWI, CMRI	94.02%
4	10 bands, NDVI, NDWI, MERIT	95.49%
5	10 bands, MERIT	71.05%
6	10 bands, NDVI, NDWI, MERIT, CHM	95.85%
7	10 bands, NDVI, NDWI, MERIT, SRTM	95.79%
8	10 bands, NDVI, NDWI, SRTM, CHM	93.71%
9	10 bands, NDVI, NDWI, MERIT, SRTM, CHM	93.73%
10	10 bands, NDVI, NDWI, CHM	95.33%
11	10 bands, MERIT, CMRI	72.00%
12	10 bands, NDVI, NDWI, CMRI, MERIT	95.70%
13	10 bands, NDVI, NDWI, CMRI, MERIT, CHM	95.81%
14	NDVI, NDWI, MERIT, CHM	95.76%
15	4 selected bands (B2, B3, B4, B8), NDVI, NDWI, MERIT, CHM	95.65%

Based on the outcomes of Experiment 1, more informative input features were required in the classification; two spectral indices of NDVI and NDWI were included in Experiment 2. After implementing the same procedures, a relatively higher accuracy of 93.43% was achieved in Experiment 2. Despite the satisfactory accuracy in Experiment 2, the output image showed that this combination of input features was still incapable of distinguishing mangroves from other types of vegetation (Figure 6c). To undertake this differentiation, a newly developed index, CMRI, was applied in Experiment 3. The resultant accuracy was improved to 94.02%; however, there were no effective changes in terms of the determination of other vegetation types (Figure 6d).

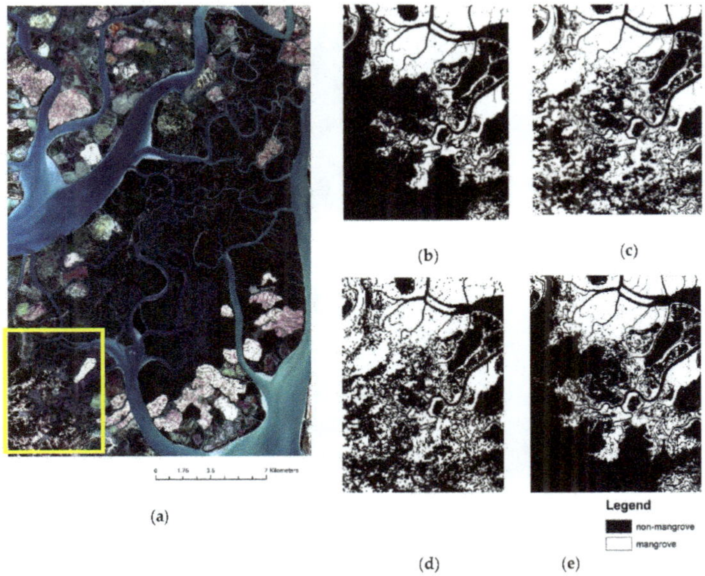

Figure 6. (**a**)The study area (Sentinel-2 true color image) and mixed vegetation area in (**b**) ground truth image, classified images of (**c**) Experiment 2, (**d**) Experiment 3, and (**e**) Experiment 6.

In Experiment 4, topographic information was utilized (instead of CMRI), as most of the other vegetation in the study area thrives in the upper land area. Therefore, the MERIT DEM was applied to Experiment 4 in an effort to remove the misclassified pixels of other vegetation types. Following Experiment 4, the other forested areas misclassified as mangrove vegetation in the previous experiments could be converted into non-mangrove areas with an overall accuracy of 95.49% (Figure 6e). However, noisy and coarse pixels were still present in the mixed vegetation areas of the output image, potentially impacting classification accuracy due to the 90 m spatial resolution of the MERIT DEM.

We attempted to delineate the classification of mangrove distribution by applying the CHM derived from the SRTM DSM and MERIT DTM. After implementing different combinations of the DTM, DSM, and CHM in Experiments 6–9, the resultant accuracy was 95.85% in Experiment 6. Here, the DTM and CHM were integrated with the Sentinel-2 bands, NDVI, and NDWI. Furthermore, Experiment 6 eliminated noisy pixels from the output image in Experiment 4, producing a smooth map of the mangrove distribution. After implementing different classifications by adjusting input features, Experiment 6—containing 10 bands of the Sentinel-2 image, NDVI, NDWI, MERIT DEM, and CHM—produced the highest accuracy and the best delineation for mangrove distribution of the 15 experiments (Figure 7c). Other than experiments 1, 2, 3, 4, and 6, the results of the remaining experiments were excluded from this study because these experiments were tested with minor changes in input features and demonstrated no outperformance in mangrove classification. The accuracies of all the experiments are provided in Table 3.

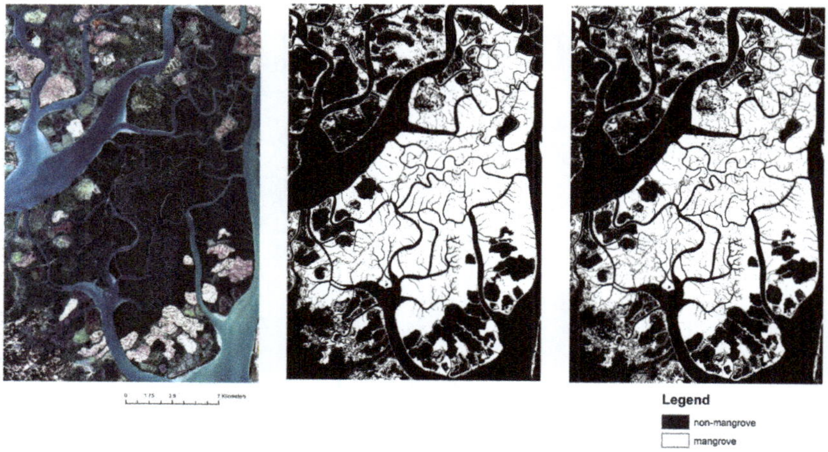

(a) (b) (c)

Figure 7. (**a**) Sentinel-2 true color image, (**b**) ground truthed image, and (**c**) classified image of Experiment 6 of the whole study area.

3.1.2. Experimental Results of Hyper-Parameter Tuning

Using the dataset in Experiment 6 described in the input parameter tuning section, hyper-parameter tuning was evaluated by adjusting the number of hidden layers and neurons. First, one hidden layer with 10 neurons was added to the basic model used in the input feature adjustment. The model trained and evaluated the accuracy while visually checking the behavior of the model in the accuracy learning curves and loss using the validation dataset. Following the addition of up to four hidden layers, the model with two hidden layers (12:10) produced the highest accuracy of 95.89%. The accuracy of the model with three hidden layers (12:10:8) and four hidden layers (12:10:10:8) was the same at 95.76% (Figure 8).

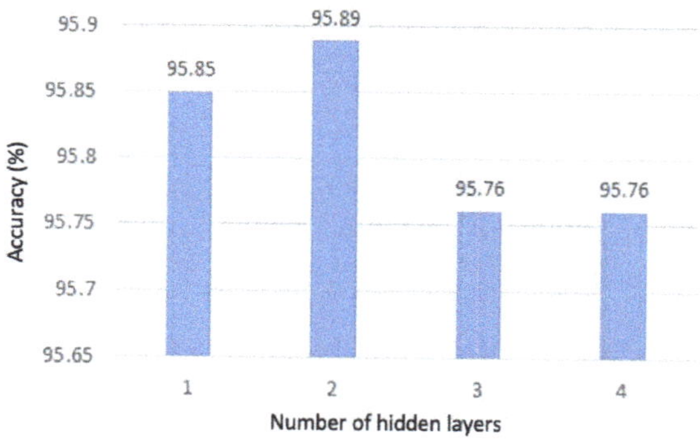

Figure 8. Evaluation results of artificial neural network (ANN) models with different hidden layers.

To evaluate model behavior, the accuracy and loss of training were compared to those of the validation prediction. Figure 9a shows that the loss reduction of the training prediction of the model with two hidden layers was less divergent and had a direction similar to the validation at the end of the learning curve. In contrast, the models with three and four hidden layers showed greater differences between training and validation measures and were overfitted; validation losses significantly increased at the end of learning despite the stable declining trend in training (Figure 9b,c). Thus, the model with two hidden layers was selected based on evaluating the classification accuracy and model behavior during the learning time.

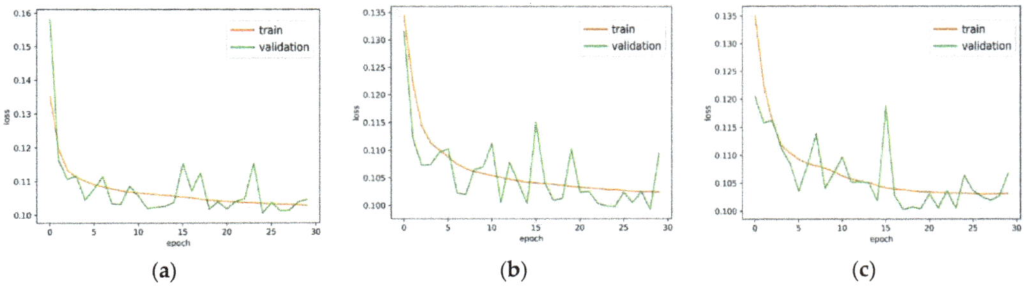

Figure 9. Learning curves of loss reduction of model with (**a**) two hidden layers, (**b**) three hidden layers, and (**c**) four hidden layers during the training phase.

The neurons in the hidden layers played a key role as processing units in analyzing input data and producing output; hence, the number of neurons influences the performance of the ANN model. The number of neurons in the two selected hidden layers was adjusted by halving and doubling the neurons in each layer to test the performance. Comparing the accuracy and loss of training prediction to that of validation in each model, the model with greater neurons (544:320) performed with the highest accuracy (95.98%), least error, and the smallest difference between the training and validation predictions (see Figure 10). After performing input features and hyperparameter tuning, the ANN model with two hidden layers (544:320 neurons) was selected as the model for mangrove classification using the input features of Experiment 6 (Figure 11).

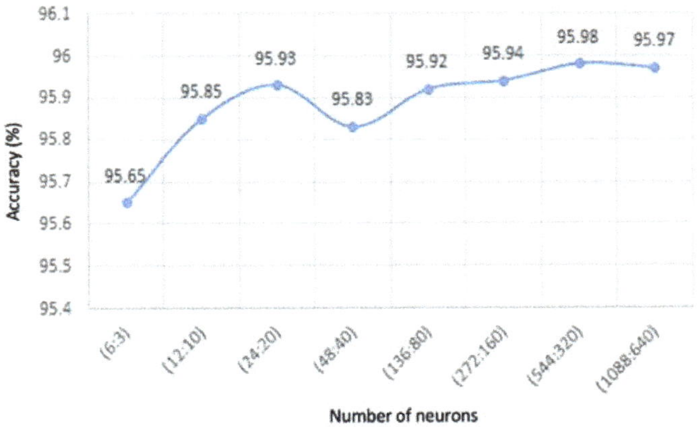

Figure 10. Model accuracy depending on the number of neurons in hidden layers.

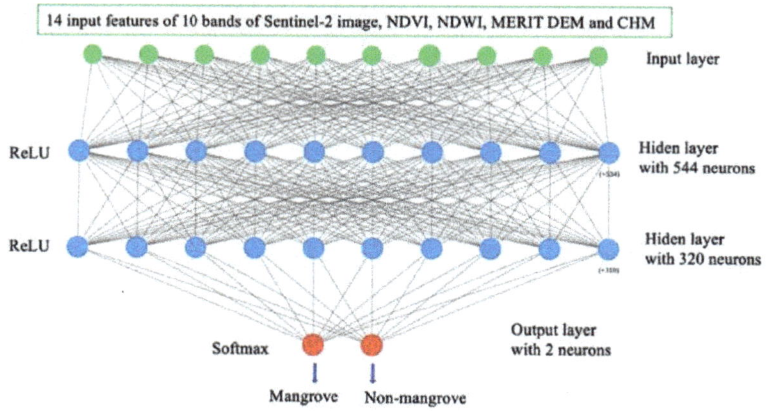

Figure 11. Proposed ANN model and input dataset for classification of mangrove distribution.

3.2. *Classification Results of New Prediction Using Transfer Learning*

The ANN model with two hidden layers (544:320), obtained through hyper-parameter tuning, was applied to a new dataset from 2015, with the same input features. The model produced an overall accuracy of 94.50% and a kappa coefficient of 0.92. Despite the high classification accuracy, the model falsely classified green paddy fields as mangroves in the new prediction due to the spectral variation in different seasons. To overcome this problem, certain parameters of the original model were fixed with existing knowledge, whereas others were retrained with a new dataset that contained information on the green paddy fields (Figure 12).

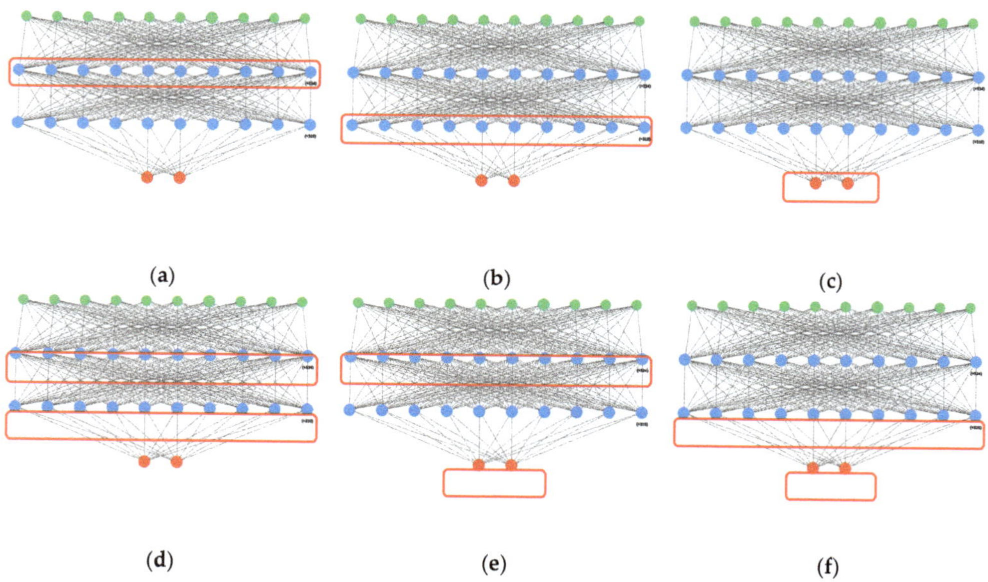

Figure 12. Six different types of models in transfer learning process: (**a**) T0 model fixed layer 0, (**b**) T1 model fixed layer 1, (**c**) T2 model fixed layer 2, (**d**) T01 model fixed layer 0 and 1, (**e**) T02 model fixed layer 0 and 2, and (**f**) T12 model fixed layer 1 and 2.

The classification results of the experiments are listed in Table 4. The original model produced a low accuracy of 72.59% for the transfer learning dataset owing to the spectral variation problem in the paddy fields. The original model was tested with different layer fixing and re-training settings. Of the six different models re-trained in transfer learning, the T12 model, which fixed the second hidden layer (layer 1) and output layer (layer 2), could be trained within the shortest training time of 32 s per epoch (using one node on the supercomputer ITO-B at Kyushu University). The T12 model also outperformed the other models, obtaining the highest accuracy of 97.2% and a kappa coefficient of 0.94 for the whole dataset, and a high accuracy of 95.77% for the transfer learning dataset. As a result, the T12 model could eliminate misclassified paddy field pixels predicted by the original model (see Figure 13c). After conducting mangrove classifications using the ANN model in 2020 and 2015, the validation parameters obtained in this study are described in Table 5; the final mangrove distribution maps of the two periods are shown in Figure 14.

Table 4. Classification results of different models in transfer learning.

Models	Transfer Learning Dataset		Accuracy for Whole Dataset (%)
	Accuracy (%)	Training Time per Epoch	
Original model	72.59	-	94.5
T0 (Model fixed layer 0)	96.09	42 s	93.8
T1 (Model fixed layer 1)	95.83	33 s	95.10
T2 (Model fixed layer 2)	96.07	40 s	93.00
T01 (Model fixed layer 0 and 1)	95.57	38 s	94.00
T02 (Model fixed layer 0 and 2)	96.04	38 s	96.20
T12 (Model fixed layer 1 and 2)	95.77	32 s	97.20

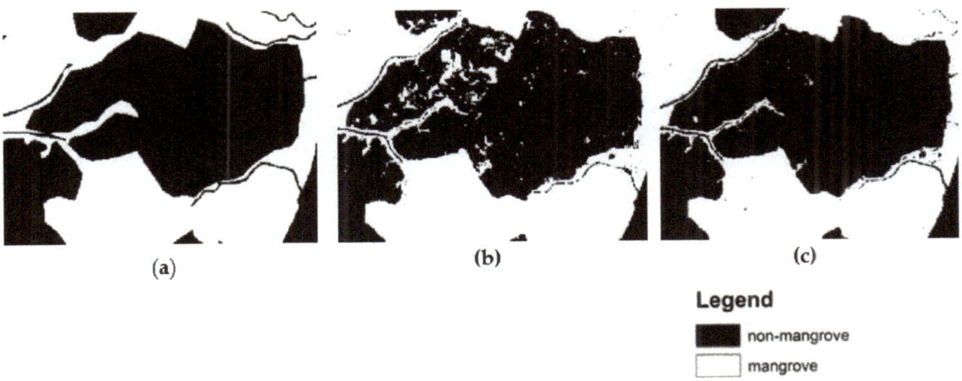

Figure 13. (**a**) Reference image of green paddy field area in 2015 dataset, (**b**) classified image of the original model and (**c**) classified image of T12 model.

Table 5. Accuracy, kappa, precision, recall, and F1 scores and statistical significance (Z-test) assessment of ANN model for mangrove classification.

Year	Accuracy	Kappa	Precision ([1] M)	Precision ([1] NM)	Recall (M)	Recall (NM)	F1 Score (M)	F1 Score (NM)
2020	95.98	0.92	0.95	0.97	0.96	0.96	0.96	0.96
2015	97.20	0.94	0.98	0.96	0.98	0.97	0.97	0.97

Z-value = 0.4753; *p*-value = 0.63122 (no significant difference in overall accuracies).

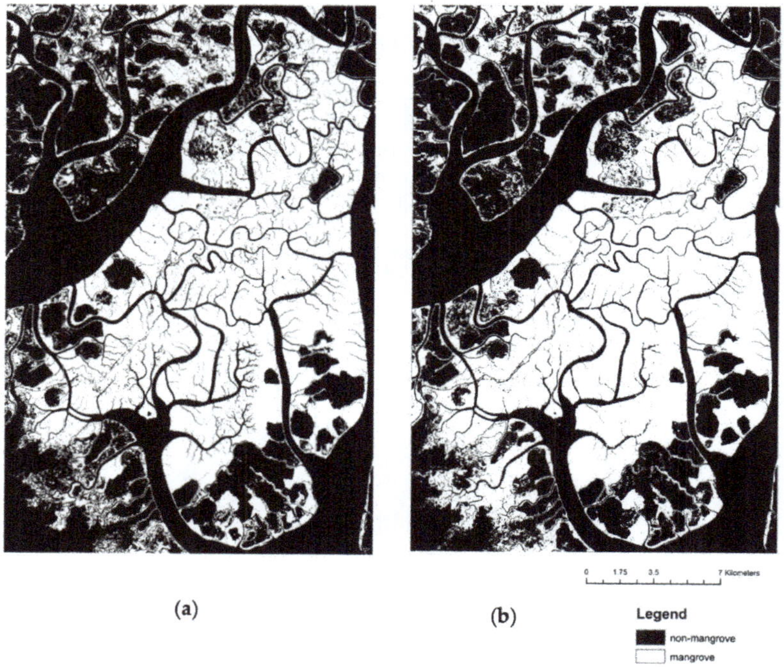

Figure 14. Mangrove distribution maps of Wunbaik Mangrove Forest in (**a**) 2020 and (**b**) 2015 predicted by the proposed ANN method.

Moreover, the statistical significance of the overall accuracies of the two classifications in 2020 and 2015 was assessed using Z-test [70], and the resultant Z-value and p-value were 0.4753 and 0.63122, respectively (Table 5). The difference in classification accuracies of the two periods was not statistically significant because the proposed method could achieve high accuracies in both classifications of 2020 and 2015 despite the slight improvement in the 2015 classification.

3.3. *Mangrove Changes and Drivers*

To provide updated information on the mangrove distribution in the WMF, we examined spatial changes in mangroves between 2015 and 2020. By differentiating between classified images in 2015 and 2020, the change detection results showed significant changes in mangrove extent in the WMF (Figure 15a). To focus on the major causes of mangrove loss and gain, only extensive change areas were examined and validated with Sentinel-2 true color imagery (10 m resolution) and very high resolution (1 m) Google Earth imagery.

Interpreting change detection results based on field information and mangrove losses in the study area between 2015 and 2020 was warranted because of the expansion of shrimp ponds (Figure 15b), whereas mangrove gain patches were detected at plantation sites (Figure 15c) and naturally regenerated areas (Figure 15d). The locations of artificial mangrove plantations established by the Forest Department were identified using high-resolution Google Earth imagery and GPS points collected during the field survey. Naturally recovered mangroves were found in active shrimp ponds, abandoned sites, and tidal flats. Between 2015 and 2020, the total mangrove area declined from 254.30–249.83 km^2 slightly, whereas the non-mangrove area increased from 284.30–288.77 km^2.

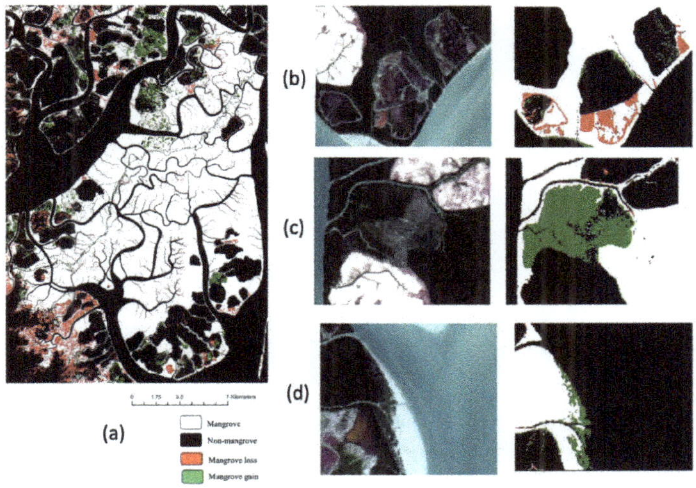

Figure 15. (**a**) Change detection result of Wunbaik Mangrove Forest (WMF), (**b**) Mangrove loss area due to shrimp pond expansion, (**c**) Mangrove gain area in artificial plantation, and (**d**) Mangrove gain area through natural regeneration.

3.4. Natural Recovery of Mangroves at Abandoned Sites

To delineate the naturally recovering mangroves in the study area, three abandoned shrimp ponds selected during the field survey were extracted from the results of change detection. The recovering mangrove area in each abandoned pond was then evaluated and compared to the extent of the abandoned sites in ArcGIS. During the abandonment period from 2015 to 2020, mangrove cover increased by 49.02% at Site 1, 55.93% at Site 2, and 50.00% at Site 3 (Table 6). The results showed that mangroves could naturally recover in approximately half of the abandoned sites without any artificial effort (Figure 16).

Table 6. Naturally recovering percentage of mangrove at different abandoned shrimp ponds.

Abandoned Sites	Site_Area (km^2)	Recovering Mangrove Area (km^2)	Recovering (%)
Site 1	1.02	0.5	49.02
Site 2	0.59	0.33	55.93
Site 3	0.14	0.07	50.00

The field survey and collaboration with Forest Department staff and shrimp pond owners confirmed that there was no artificial planting following the abandonment of these shrimp ponds. The recovering mangrove areas were validated using ground truth GPS points of 50 sample plots collected from the 3 recovery sites, and 12 different mangrove species were identified through the field inventory. *Avicennia officinalis* (Figure 17a) was the dominant species at Site 1, whereas *A. marina* (Figure 17b,c) dominated at sites 2 and 3.

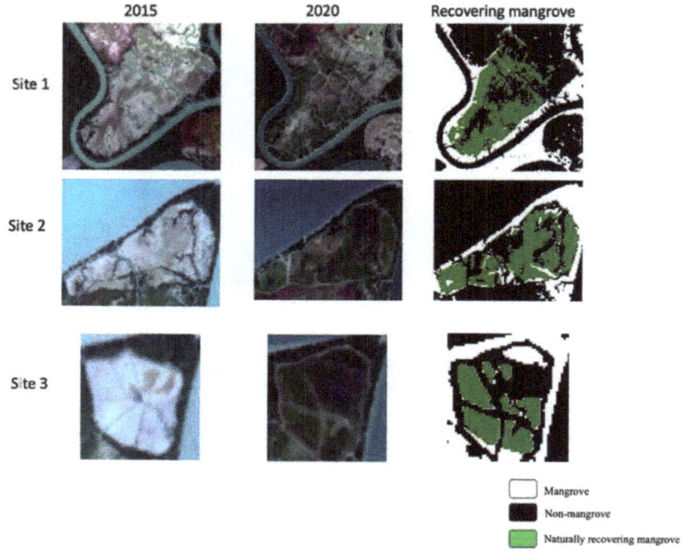

Figure 16. Naturally recovering mangrove at different abandoned sites between 2015 and 2020.

Figure 17. Ground checked photos of naturally recovering mangrove species at (**a**) Site 1 dominating *Avicennia officinalis*, (**b**) Site 2, and (**c**) Site 3 dominating *Avicennia marina*.

4. Discussion

The nature of mangrove intertidal ecosystems characterized by mixing with other vegetation in adjacent parts creates many challenges for the classification of mangrove distribution to produce accurate maps [14,41]. Selecting the most appropriate method in remote sensing analysis and an effective combination of input features is critical to overcoming such complex issues. Apart from the complex features of mangrove forests, different classification methods also have various drawbacks. In another analysis, Chun et al. compared the ANN method with the maximum likelihood classifier using Thailand Earth Observation System (THEOS) satellite data for mangrove mapping [44]. The study conducted by Toosi et al. on mangrove forest cover changes produced high overall accuracy (ranging 81–93%) by comparing machine learning algorithms (RF, RDA, and SVM) [17]. This study explored the ANN approach using Sentinel-2 satellite imagery for mangrove classification through various experiments with input features and hyper-parameter tuning.

The incorporation of topographic features in remote sensing classification generated a favorable outcome for mangrove mapping in other studies [71,72]. However, obtaining true topographic information remains challenging. This study integrated topographic information derived from the MERIT DEM to achieve high accuracy classification and to eliminate misclassified pixels from previous experiments. Despite these merits, the poor resolution of the MERIT DEM affected the quality of the output map with many noisy pixels.

The resulting accuracy of tree species classification was improved by applying the CHM derived from the differentiation of the DTM and DSM. The CHM obtained from multi-spectral and hyperspectral images and Light Detection and Ranging (LiDAR) data have been applied for mangrove species classification [73,74]. Cao et al. [28] showed that the CHM generated from a hyperspectral sensor mounted on a UAV was useful for mangrove species discrimination. However, owing to limited accessibility and the high investment involved with LiDAR, SAR data, and hyperspectral sensors, such as canopy height collection methods, are rarely deployed in remote sensing classification [75]. This study attempted to employ the CHM obtained from the freely and globally accessible SRTM DSM and MERIT DEM for mangrove classification. We found that mangrove distribution may be more accurately identified by integrating the DTM and CHM in remote sensing classification. Classification accuracy and image quality could improve future mangrove distribution mapping if the CHM was derived from higher resolution and more accurate DTM and DSM.

For the ANN model, a universal method to select the number of hidden layers and neurons for optimal model design is lacking [76,77]. Thus, a systematic experimental analysis presents a reliable path for solving a specific problem [78]. After conducting hyper-parameter tuning, the selected model with two hidden layers (544:322) outperformed the other models, obtaining the highest accuracy of 95.98%, whereas the basic model obtained an accuracy of 95.85% with the same dataset. The results of the analyses show that the setting of the deployed method is one of the pivotal factors in remote sensing classification, as it influences the selection of efficient input features. Although there were no dramatic improvements in accuracy, the behavior of the models was considered through the validation dataset to avoid overfitting problems for future predictions.

By developing advanced technologies in computer science, the remote sensing community has improved many classification methods by applying a state-of-the-art approach for time and cost-effective computation. Transfer learning is a cutting edge approach in neural networks and effective for future predictions through the use of a relatively small training dataset with reduced computational costs [79–81]. Guo [82] introduced and applied transfer learning to identify buildings in rural environments. This study confronted issues with the misclassification of green paddy fields by the trained model. Transfer learning was employed by delineating different layers of frozen data to overcome this issue. The original model was leveraged by re-training one hidden layer with a very small dataset. The resulting accuracy for the entire study area significantly improved from 94.50–97.20%. Through transfer learning, we built the model to be more applicable for mangrove classification in different seasons and regions. Despite satisfactory classification accuracy results, preparing a ground truth image to create a large amount of training data and a long-running time for ANN prediction limit the proposed approach. Moreover, our study has some uncertainties, as we could not apply a ground truth image to validate the performance of the pretrained ANN model in the 2015 classification. To address this necessity, we applied a reference image digitized using Google Earth imagery. Considering this limitation, future studies should explore the variable importance of highlighting the contributions of each variable in ANN classification for mangrove distribution.

Mangrove forests in Myanmar have declined, with an annual net reduction rate of 3.60–3.87% between 1996 and 2016, because of competition with and expansion of agricultural fields, aquaculture, palm oil, and rubber [83,84]. FAO [8] and Aye Saw [85] studied mangrove forest changes in WRMF and described shrimp ponds and paddy fields as the primary drivers of mangrove loss between 1990 and 2011, and 1990 to 2014, respectively. Similarly, this study observed that the mangrove area of WMF in 2015 declined slightly in 2020. The reason for mangrove loss was largely due to the expansion of shrimp ponds by local communities, whereas mangrove forests were able to expand by artificial plantation and natural recovery processes. Interestingly, the change detection results showed that more mangrove loss areas were discovered near local villages due to the expansion of shrimp farming than within the reserved forest (Figure 18).

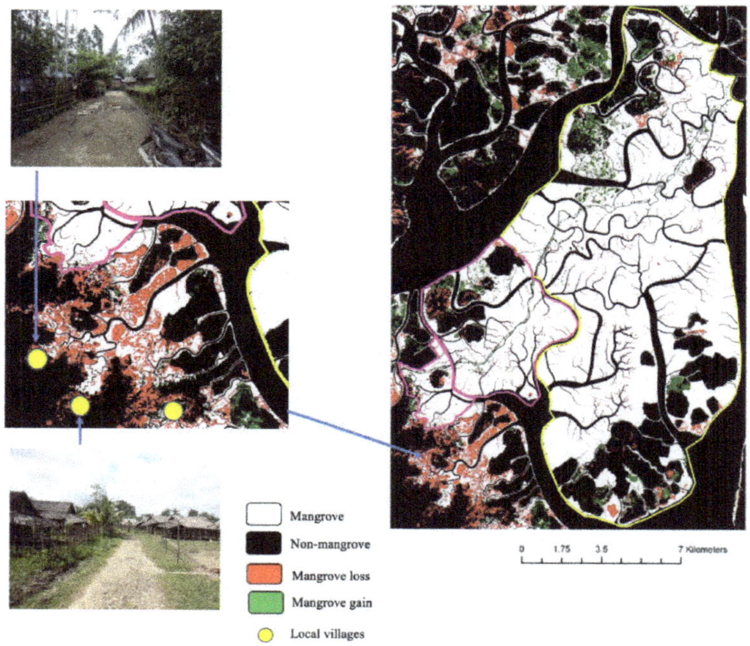

Figure 18. Shrimp pond expansion by local villages near the WRMF between 2015 and 2020.

One of the uncertainties is that the study could not distinguish shrub, planted, and natural recovering mangrove due to the medium resolution of the Sentinel-2 satellite images. A 10 m pixel range might include real shrub and weeds other than mangrove species, and high-resolution satellite or UAV images should be applied to validate mangrove classification and to change detection results. However, the only mangrove restoration projects in WMF are artificial plantations established by the Forest Department; other mangrove gain areas were assumed to appear as a result of natural recovery processes. Although recovering mangroves were also found in the active shrimp ponds of the WMF, the long-term survival of these mangroves is questionable, as they would likely be cleared if the owners wanted to repair their shrimp ponds or if the ponds were converted to paddy fields. However, these results highlight the natural recovery process in abandoned areas. The results show that mangroves are able to naturally recover in approximately 50% of the abandoned sites with a diverse range of species within a short period of abandonment from 2015 to 2020.

5. Conclusions

This study was the first attempt to employ the ANN classification using Sentinel-2 imagery for mangrove classification. We conducted two primary experiments relating to input features and hyper-parameter tuning. Based on these experiments, we proposed a promising ANN model applying Sentinel-2 satellite imagery and topographic and canopy height information for constraining mangrove distribution. Through optimization, we achieved the highest overall accuracy of 95.98%, a kappa coefficient of 0.92, mangrove and non-mangrove precisions of 0.95 and 0.98, respectively, recalls of 0.96, and F1 scores of 0.96. Moreover, we used the transfer learning approach to improve the ANN model performance and obtained a high overall accuracy of 97.20%, a kappa coefficient of 0.94, mangrove and non-mangrove precisions of 0.98 and 0.96, respectively, recalls of 0.98 and 0.97, and F1 scores of 0.96. The created model can be deployed for mangrove classification in different seasons and regions.

The outcomes of this study contribute to the production of up-to-date and reliable information regarding mangrove change in the WMF through post-classification change detection. The results highlight that the mangrove extent decreased in the WMF between 2015 and 2020, owing to the expansion of shrimp ponds. By integrating field observations with the change detection results, we highlight that the mangroves were able to naturally recover in approximately 50% of the abandoned sites within a short abandonment period. Further research on the natural recovery process of mangroves should be conducted, as it presents a valuable opportunity for mangrove rehabilitation after human disturbance.

Author Contributions: W.S.M. is responsible for this study. W.S.M. conducted a field survey and analyzed the preprocessing and main experiments. W.S.M. prepared the original draft manuscript. J.S. supervised the study, collaborated the field survey, provided a careful review and editing to the manuscript, and contributed to funding acquisition. All authors have read and agreed to the published version of the manuscript.

Funding: This research was partially funded by JSPS KAKENHI Grant Number JP20H02250.

Acknowledgments: The ANN computation was performed using the computer resources offered under the category of General Projects by the Research Institute for Information Technology, Kyushu University. We are thankful to Dai Yamzaki, Institute of Industrial Science, The University of Tokyo, for providing the MERIT DEM. The first author is deeply indebted to the Asian Development Bank for financial support during the study period in Japan. The authors would like to express our gratitude to the anonymous reviewers for careful reading of our manuscript and their insightful comments and suggestions.

Conflicts of Interest: The authors declare no conflict of interest.

References

1. Spalding, M. *World Atlas of Mangroves*; Routledge: Abingdon-on-Thames, UK, 2010.
2. Brander, L.M.; Wagtendonk, A.J.; Hussain, S.S.; McVittie, A.; Verburg, P.H.; De Groot, R.S.; Van Der Ploeg, S. Ecosystem service values for mangroves in Southeast Asia: A meta-analysis and value transfer application. *Ecosyst. Serv.* **2012**, *1*, 62–69. [CrossRef]
3. Alongi, D.M. Present state and future of the world's mangrove forests. *Environ. Conserv.* **2002**, *29*, 331–349. [CrossRef]
4. Richards, D.R.; Friess, D.A. Rates and drivers of mangrove deforestation in Southeast Asia, 2000–2012. *Proc. Natl. Acad. Sci. USA* **2016**, *113*, 344–349. [CrossRef]
5. Veettil, B.K.; Pereira, S.F.R.; Quang, N.X. Rapidly diminishing mangrove forests in Myanmar (Burma): A review. *Hydrobiologia* **2018**, *822*, 19–35. [CrossRef]
6. Lewis, R.R. Ecological engineering for successful management and restoration of mangrove forests. *Ecol. Eng.* **2005**, *24*, 403–418. [CrossRef]
7. Kamali, B.; Hashim, R. Mangrove restoration without planting. *Ecol. Eng.* **2011**, *37*, 387–391. [CrossRef]

8. Stanley, D.O.; Broadhead, J.; Aung Aung, M.; Burma. Forest Department; Food and Agriculture Organization of the United Nations. *The Atlas and Guidelines for Mangrove Management in Wunbaik Reserved Forest*; FAO-UN, Seed Division Compound, Insein Road, Yangon, Myanmar Publications: Yangon, Myanmar, 2011; Volume iii, 132p.
9. Giri, C. Observation and monitoring of mangrove forests using remote sensing: Opportunities and challenges. *Remote Sens.* **2016**, *8*, 783. [CrossRef]
10. Wang, L.; Jia, M.; Yin, D.; Tian, J. A review of remote sensing for mangrove forests: 1956–2018. *Remote Sens. Environ.* **2019**, *231*, 111223. [CrossRef]
11. Yoshino, K.; Pham, T.D.; Bui, D.T.; Friess, D.A.; Yokoya, N.; Bui, D.T.; Yoshino, K.; Friess, D.A. Remote sensing approaches for monitoring mangrove species, structure, and biomass: Opportunities and challenges. *Remote Sens.* **2019**, *11*, 1–24.
12. Lu, D.; Weng, Q. A survey of image classification methods and techniques for improving classification performance. *Int. J. Remote Sens.* **2007**, *28*, 823–870. [CrossRef]
13. Ahmad, A.; Quegan, S. Comparative analysis of supervised and unsupervised classification on multispectral data. *Appl. Math. Sci.* **2013**, *7*, 3681–3694. [CrossRef]
14. Green, E.P.; Clark, C.D.; Mumby, P.J.; Edwards, A.J.; Ellis, A.C. Remote sensing techniques for mangrove mapping. *Int. J. Remote Sens.* **1998**, *19*, 935–956. [CrossRef]
15. Rahman, M.; Ullah, R.; Lan, M.; Sri Sumantyo, J.T.; Kuze, H.; Tateishi, R. Comparison of Landsat image classification methods for detecting mangrove forests in Sundarbans. *Int. J. Remote Sens.* **2013**, *34*, 1041–1056. [CrossRef]
16. Nguyen, H.H.; Nghia, N.H.; Nguyen, H.T.T.; Le, A.T.; Tran, L.T.N.; Duong, L.V.K.; Bohm, S.; Furniss, M.J. Classification methods for mapping mangrove extents and drivers of change in Thanh Hoa province, Vietnam during 2005–2018. *For. Soc.* **2020**, *4*, 225–242. [CrossRef]
17. Toosi, N.B.; Soffianian, A.R.; Fakheran, S.; Pourmanafi, S.; Ginzler, C.; Waser, L.T. Comparing different classification algorithms for monitoring mangrove cover changes in southern Iran. *Glob. Ecol. Conserv.* **2019**, *19*. [CrossRef]
18. Kamal, M.; Jamaluddin, I.; Parela, A.; Farda, N.M. Comparison of Google Earth Engine (GEE)-based machine learning classifiers for mangrove mapping. In Proceedings of the 40th Asian Conference Remote Sensing, ACRS 2019, Daejeon, Korea, 14–18 October 2019; Volume 2000, pp. 1–8.
19. Campomanes, F.; Pada, A.V.; Silapan, J. Mangrove classification using support vector machines and random forest algorithm: A comparative study. In Proceedings of the GEOBIA 2016—Solutions & Synergies, Enschede, The Netherlands, 14–16 July 2016; pp. 2–6.
20. Altaei, M.S.M. Satellite Image Classification Using Artificial Neural Network. *Int. J. Res. Advent Technol.* **2019**, *7*, 459–462.
21. Toshniwal, M. Satellite image classification using neural networks. In Proceedings of the Third International Conference on Sciences of Electronic, Technology of Information and Telecommunications (SETIT 2005), Susa, Tunisia, 17–21 March 2005; pp. 1–5.
22. Ge, G.; Shi, Z.; Zhu, Y.; Yang, X.; Hao, Y. Land use/cover classification in an arid desert-oasis mosaic landscape of China using remote sensed imagery: Performance assessment of four machine learning algorithms. *Glob. Ecol. Conserv.* **2020**, *22*, e00971. [CrossRef]
23. Kawabata, D.; Bandibas, J. Landslide susceptibility mapping using geological data, a DEM from ASTER images and an Artificial Neural Network (ANN). *Geomorphology* **2009**, *113*, 97–109. [CrossRef]
24. Wang, L.; Sousa, W.P.; Gong, P.; Biging, G.S. Com. *Remote Sens. Environ.* **2004**, *91*, 432–440. [CrossRef]
25. Xia, J.; Yokoya, N.; Pham, T.D. Probabilistic mangrove species mapping with multiple-source remote-sensing datasets using label distribution learning in Xuan Thuy National Park, Vietnam. *Remote Sens.* **2020**, *12*, 3834. [CrossRef]
26. Poortinga, A.; Tenneson, K.; Shapiro, A.; Nquyen, Q.; Aung, K.S.; Chishtie, F.; Saah, D. Mapping plantations in Myanmar by fusing Landsat-8, Sentinel-2 and Sentinel-1 data along with systematic error quantification. *Remote Sens.* **2019**, *11*, 831. [CrossRef]
27. Li, Q.; Wong, F.K.K.; Fung, T. Classification of mangrove species using combined WordView-3 and LiDAR data in Mai Po Nature Reserve, Hong Kong. *Remote Sens.* **2019**, *11*, 2114. [CrossRef]
28. Cao, J.; Leng, W.; Liu, K.; Liu, L.; He, Z.; Zhu, Y. Object-Based mangrove species classification using unmanned aerial vehicle hyperspectral images and digital surface models. *Remote Sens.* **2018**, *10*, 89. [CrossRef]
29. European Space Agency ESA—About the Launch. Available online: https://www.esa.int/Applications/Observing_the_Earth/Copernicus/Sentinel-2/About_the_launch (accessed on 30 June 2020).
30. Astola, H.; Häme, T.; Sirro, L.; Molinier, M.; Kilpi, J. Comparison of Sentinel-2 and Landsat 8 imagery for forest variable prediction in boreal region. *Remote Sens. Environ.* **2019**, *223*, 257–273. [CrossRef]
31. García-Llamas, P.; Suárez-Seoane, S.; Fernández-Guisuraga, J.M.; Fernández-García, V.; Fernández-Manso, A.; Quintano, C.; Taboada, A.; Marcos, E.; Calvo, L. Evaluation and comparison of Landsat 8, Sentinel-2 and Deimos-1 remote sensing indices for assessing burn severity in Mediterranean fire-prone ecosystems. *Int. J. Appl. Earth Obs. Geoinf.* **2019**, *80*, 137–144. [CrossRef]
32. Lima, T.A.; Beuchle, R.; Langner, A.; Grecchi, R.C.; Griess, V.C.; Achard, F. Comparing Sentinel-2 MSI and Landsat 8 OLI Imagery for Monitoring Selective Logging in the Brazilian Amazon. *Remote Sens.* **2019**, *11*, 961. [CrossRef]
33. Connette, G.; Oswald, P.; Songer, M.; Leimgruber, P. Mapping distinct forest types improves overall forest identification based on multi-spectral landsat imagery for Myanmar'S Tanintharyi Region. *Remote Sens.* **2016**, *8*, 882. [CrossRef]
34. Ha, N.T.; Manley-Harris, M.; Pham, T.D.; Hawes, I. A comparative assessment of ensemble-based machine learning and maximum likelihood methods for mapping seagrass using sentinel-2 imagery in Tauranga Harbor, New Zealand. *Remote Sens.* **2020**, *12*, 355. [CrossRef]
35. Morgan, R.S.; El-Hady, M.A.; Rahim, I.S. Soil salinity mapping utilizing sentinel-2 and neural networks. *Indian J. Agric. Res.* **2018**, *52*, 524–529.

36. Mohite, J.; Twarakavi, N.; Pappula, S. Evaluating the Potential of Sentinel-2 for Low Severity Mites Infestation Detection in Grapes. In Proceedings of the 2018 IEEE International Geoscience and Remote Sensing Symposium, IGARSS 2018, Valencia, Spain, 22–27 July 2018; pp. 4651–4654.
37. Pereira-Pires, J.E.; Aubard, V.; Ribeiro, R.A.; Fonseca, J.M.; Silva, J.M.N.; Mora, A. Semi-automatic methodology for fire break maintenance operations detection with sentinel-2 imagery and artificial neural network. *Remote Sens.* **2020**, *12*, 909. [CrossRef]
38. Kristollari, V.; Karathanassi, V. Artificial neural networks for cloud masking of Sentinel-2 ocean images with noise and sunglint. *Int. J. Remote Sens.* **2020**, *41*, 4102–4135. [CrossRef]
39. Lee, Y.S.; Lee, S.; Jung, H.S. Mapping forest vertical structure in Gong-Ju, Korea using sentinel-2 satellite images and artificial neural networks. *Appl. Sci.* **2020**, *10*, 1666. [CrossRef]
40. Mondal, P.; Liu, X.; Fatoyinbo, T.E.; Lagomasino, D. Evaluating combinations of sentinel-2 data and machine-learning algorithms for mangrove mapping in West Africa. *Remote Sens.* **2019**, *11*, 2928. [CrossRef]
41. Bihamta Toosi, N.; Soffianian, A.R.; Fakheran, S.; Pourmanafi, S.; Ginzler, C.; Waser, L.T. Land Cover Classification in Mangrove Ecosystems Based on VHR Satellite Data and Machine Learning—An Upscaling Approach. *Remote Sens.* **2020**, *12*, 2684. [CrossRef]
42. Yu, X.; Shao, H.B.; Liu, X.H.; Zhao, D.Z. Applying neural network classification to obtain mangrove landscape characteristics for monitoring the travel environment quality on the Beihai Coast of Guangxi, P.R. China. *Clean Soil Air Water* **2010**, *38*, 289–295. [CrossRef]
43. Chun, B.B.; Jafri, M.Z.M.; San, L.H. Mangrove mapping in Penang Island by using artificial neural network technique. In Proceedings of the 2011 IEEE Conference on Open Systems ICOS 2011, Langkawi, Malaysia, 25–28 September 2011; pp. 251–255.
44. Chun, B.B.; Mat Jafri, M.Z.; San, L.H. Comparison of remote sensing approach for mangrove mapping over Penang Island. In Proceedings of the 4th International Conference on Computer and Communication Engineering 2012 (ICCCE 2012), Kuala Lumpur, Malaysia, 3–5 July 2012; pp. 258–262.
45. Hutchinson, C.F. Techniques for combining Landsat and ancillary data for digital classification improvement. *Photogramm. Eng. Remote Sens.* **1982**, *48*, 123–130.
46. Alganci, U.; Besol, B.; Sertel, E. Accuracy assessment of different digital surface models. *ISPRS Int. J. Geo-Inf.* **2018**, *7*, 114. [CrossRef]
47. Lee, S.K.; Fatoyinbo, T.E. TanDEM-X Pol-InSAR Inversion for Mangrove Canopy Height Estimation. *IEEE J. Sel. Top. Appl. Earth Obs. Remote Sens.* **2015**, *8*, 3608–3618. [CrossRef]
48. Wohlfart, C.; Winkler, K.; Wendleder, A.; Roth, A. TerraSAR-X and wetlands: A review. *Remote Sens.* **2018**, *10*, 916. [CrossRef]
49. Ghosh, S.M.; Behera, M.D.; Paramanik, S. Canopy height estimation using sentinel series images through machine learning models in a Mangrove Forest. *Remote Sens.* **2020**, *12*, 1519. [CrossRef]
50. Yamazaki, D.; Ikeshima, D.; Tawatari, R.; Yamaguchi, T.; O'Loughlin, F.; Neal, J.C.; Sampson, C.C.; Kanae, S.; Bates, P.D. A high-accuracy map of global terrain elevations. *Geophys. Res. Lett.* **2017**, *44*, 5844–5853. [CrossRef]
51. Saw, A.A.; Kanzaki, M. Local Livelihoods and Encroachment into a Mangrove Forest Reserve: A Case Study of the Wunbaik Reserved Mangrove Forest, Myanmar. *Procedia Environ. Sci.* **2015**, *28*, 483–492. [CrossRef]
52. Win, M. *The Mangrove Vegetation of Wubaik Reserved Forest*; FAO-UN, Seed Division Compound, Insein Road, Yangon, Myanmar Publications: Yangon, Myanmar, 2011; Volume iii, 191p.
53. Congedo, L. *Semi-Automatic Classification Plugin Documentation*; Lazio: Rome, Italy, 2016.
54. Hościło, A.; Lewandowska, A. Mapping Forest Type and Tree Species on a Regional Scale Using Multi-Temporal Sentinel-2 Data. *Remote Sens.* **2019**, *11*, 929. [CrossRef]
55. European Space Agency. *SENTINEL-2 User Handbook*; European Space Agency: Paris, France, 2015.
56. Teillet, P.M.; Staenz, K.; Williams, D.J. Effects of spectral, spatial, and radiometric characteristics on remote sensing vegetation indices of forested regions. *Remote Sens. Environ.* **1997**, *61*, 139–149. [CrossRef]
57. McFeeters, S.K. The use of the Normalized Difference Water Index (NDWI) in the delineation of open water features. *Int. J. Remote Sens.* **1996**, *17*, 1425–1432. [CrossRef]
58. Gupta, K.; Mukhopadhyay, A.; Giri, S.; Chanda, A.; Datta Majumdar, S.; Samanta, S.; Mitra, D.; Samal, R.N.; Pattnaik, A.K.; Hazra, S. An index for discrimination of mangroves from non-mangroves using LANDSAT 8 OLI imagery. *MethodsX* **2018**, *5*, 1129–1139. [CrossRef]
59. Yap, L.; Kandé, L.H.; Nouayou, R.; Kamguia, J.; Ngouh, N.A.; Makuate, M.B. Vertical accuracy evaluation of freely available latest high-resolution (30 m) global digital elevation models over Cameroon (Central Africa) with GPS/leveling ground control points. *Int. J. Digit. Earth* **2019**, *12*, 500–524. [CrossRef]
60. Lee, W.J.; Lee, C.W. Forest canopy height estimation using multiplatform remote sensing dataset. *J. Sens.* **2018**, *2018*, 1–9. [CrossRef]
61. Aggarwal, C.C. *Neural Networks and Deep Learning*; Gewerbestrasse 11: Cham, Switzerland, 2018; ISBN 9783319944623.
62. Grave, E.; Joulin, A.; Cissé, M.; Grangier, D.; Jégou, H. Efficient softmax approximation for GPUs. In Proceedings of the 34th International Conference on Machine Learning (ICML 2017), Sydney, Australia, 6–11 August 2017; Volume 3, pp. 2111–2119.
63. Uniqtech Understand the Softmax Function in Minutes—Data Science Bootcamp—Medium. Available online: https://medium.com/data-science-bootcamp/understand-the-softmax-function-in-minutes-f3a59641e86d (accessed on 15 June 2020).
64. Pan, S.J.; Yang, Q. A survey on transfer learning. *IEEE Trans. Knowl. Data Eng.* **2010**, *22*, 1345–1359. [CrossRef]
65. Rwanga, S.S.; Ndambuki, J.M. Accuracy Assessment of Land Use/Land Cover Classification Using Remote Sensing and GIS. *Int. J. Geosci.* **2017**, *8*, 611–622. [CrossRef]
66. Congalton, R.G. Assessing Landsat Classification Accuracy Using Discrete Multivariate Analysis Statistical Techniques. *Photogramm. Eng. Remote Sens.* **1983**, *49*, 1671–1678.
67. Cohen, J. A Coefficient of Aggrement for Normal Scales. *Curr. Contents* **1960**, *20*, 37–46.

68. Gandhi, S.; Jones, T.G. Identifying mangrove deforestation hotspots in South Asia, Southeast Asia and Asia-Pacific. *Remote Sens.* **2019**, *11*, 728. [CrossRef]
69. Afify, H.A. Evaluation of change detection techniques for monitoring land-cover changes: A case study in new Burg El-Arab area. *Alex. Eng. J.* **2011**, *50*, 187–195. [CrossRef]
70. Glen, S. Z Test: Definition & Two Proportion Z-Test—Statistics How To. Available online: https://www.statisticshowto.com/z-test/ (accessed on 19 December 2020).
71. Alsaaideh, B.; Al-Hanbali, A.; Tateishi, R.; Nguyen Thanh, H. The integration of spectral analyses of Landsat ETM+ with the DEM data for mapping mangrove forests. In Proceedings of the 2011 IEEE International Geoscience and Remote Sensing Symposium, Vancouver, BC, Canada, 24–29 July 2011; pp. 1914–1917.
72. Alsaaideh, B.; Al-Hanbali, A.; Tateishi, R.; Kobayashi, T.; Hoan, N.T. Mangrove Forests Mapping in the Southern Part of Japan Using Landsat ETM+ with DEM. *J. Geogr. Inf. Syst.* **2013**, *5*, 369–377. [CrossRef]
73. Chadwick, J. Integrated LiDAR and IKONOS multispectral imagery for mapping mangrove distribution and physical properties. *Int. J. Remote Sens.* **2011**, *32*, 6765–6781. [CrossRef]
74. Liu, X.; Bo, Y. Object-based crop species classification based on the combination of airborne hyperspectral images and LiDAR data. *Remote Sens.* **2015**, *7*, 922–950. [CrossRef]
75. Liu, M.; Cao, C.; Dang, Y.; Ni, X. Mapping forest canopy height in mountainous areas using ZiYuan-3 stereo images and Landsat data. *Forests* **2019**, *10*, 105. [CrossRef]
76. Stathakis, D. How many hidden layers and nodes? *Int. J. Remote Sens.* **2009**, *30*, 2133–2147. [CrossRef]
77. Mas, J.F.; Flores, J.J. The application of artificial neural networks to the analysis of remotely sensed data. *Int. J. Remote Sens.* **2008**, *29*, 617–663. [CrossRef]
78. Brownlee, J. How to Configure the Number of Layers and Nodes in a Neural Network. Available online: https://machinelearningmastery.com/how-to-configure-the-number-of-layers-and-nodes-in-a-neural-network/ (accessed on 1 July 2020).
79. Han, D.; Liu, Q.; Fan, W. A new image classification method using CNN transfer learning and web data augmentation. *Expert Syst. Appl.* **2018**, *95*, 43–56. [CrossRef]
80. Fchollet Transfer Learning & Fine-Tuning. Available online: https://keras.io/guides/transfer_learning/ (accessed on 19 June 2020).
81. Arrigoni, A. Transfer Learning in Tensorflow (VGG19 on CIFAR-10): Part 1. Available online: https://towardsdatascience.com/transfer-learning-in-tensorflow-9e4f7eae3bb4 (accessed on 19 June 2020).
82. Guo, Z. Identification of Buildings in Rural Environment Based on Convolutional Neural Networks. Master's Thesis, The University of Tokyo, Tokyo, Japan, 2017.
83. De Alban, J.D.T.; Jamaludin, J.; Wong, D.d.W.; Than, M.M.; Webb, E.L. Improved estimates of mangrove cover and change reveal catastrophic deforestation in Myanmar. *Environ. Res. Lett.* **2020**, *15*, 034034. [CrossRef]
84. Gaw, L.Y.F.; Linkie, M.; Friess, D.A. Mangrove forest dynamics in Tanintharyi, Myanmar from 1989–2014, and the role of future economic and political developments. *Singap. J. Trop. Geogr.* **2018**, *39*, 224–243. [CrossRef]
85. Saw, A.A. Deforestation and Local Livelihood Strategy: A Case of Encroachment into the Wunbaik Reserved Mangrove Forest, Myanmar. Ph.D. Thesis, Kyoto University, Kyoto, Japan, 2017.

remote sensing

Article

Remote Sensing to Study Mangrove Fragmentation and Its Impacts on Leaf Area Index and Gross Primary Productivity in the South of Peninsular Malaysia

Kasturi Devi Kanniah [1,2,*], Chuen Siang Kang [1], Sahadev Sharma [3] and A. Aldrie Amir [4]

1 Faculty of Built Environment and Surveying, Universiti Teknologi Malaysia, UTM Johor Bahru 81310, Johor, Malaysia; cskang@utm.my
2 Centre for Environmental Sustainability and Water Security (IPASA), Research Institute for Sustainable Environment (RISE), Universiti Teknologi Malaysia, UTM Johor Bahru 81310, Johor, Malaysia
3 Institute of Ocean and Earth Sciences (IOES), University of Malaya, Kuala Lumpur 50603, Wilayah Persekutuan, Malaysia; ssharma@um.edu.my
4 Institute for Environment and Development (LESTARI), Universiti Kebangsaan Malaysia, Bangi 43600, Selangor, Malaysia; aldrie@ukm.edu.my
* Correspondence: kasturi@utm.my

Citation: Kanniah, K.D.; Kang, C.S.; Sharma, S.; Amir, A.A. Remote Sensing to Study Mangrove Fragmentation and Its Impacts on Leaf Area Index and Gross Primary Productivity in the South of Peninsular Malaysia. *Remote Sens.* **2021**, *13*, 1427. https://doi.org/10.3390/rs13081427

Academic Editor: Chandra Giri

Received: 10 March 2021
Accepted: 2 April 2021
Published: 7 April 2021

Abstract: Mangrove is classified as an important ecosystem along the shorelines of tropical and subtropical landmasses, which are being degraded at an alarming rate despite numerous international treaties having been agreed. Iskandar Malaysia (IM) is a fast-growing economic region in southern Peninsular Malaysia, where three Ramsar Sites are located. Since the beginning of the 21st century (2000–2019), a total loss of 2907.29 ha of mangrove area has been estimated based on medium-high resolution remote sensing data. This corresponds to an annual loss rate of 1.12%, which is higher than the world mangrove depletion rate. The causes of mangrove loss were identified as land conversion to urban, plantations, and aquaculture activities, where large mangrove areas were shattered into many smaller patches. Fragmentation analysis over the mangrove area shows a reduction in the mean patch size (from 105 ha to 27 ha) and an increase in the number of mangrove patches (130 to 402), edge, and shape complexity, where smaller and isolated mangrove patches were found to be related to the rapid development of IM region. The Moderate Resolution Imaging Spectro-radiometer (MODIS) Leaf Area Index (LAI) and Gross Primary Productivity (GPP) products were used to inspect the impact of fragmentation on the mangrove ecosystem process. The mean LAI and GPP of mangrove areas that had not undergone any land cover changes over the years showed an increase from 3.03 to 3.55 (LAI) and 5.81 g C m^{-2} to 6.73 g C m^{-2} (GPP), highlighting the ability of the mangrove forest to assimilate CO_2 when it is not disturbed. Similarly, GPP also increased over the gained areas (from 1.88 g C m^{-2} to 2.78 g C m^{-2}). Meanwhile, areas that lost mangroves, but replaced them with oil palm, had decreased mean LAI from 2.99 to 2.62. In fragmented mangrove patches an increase in GPP was recorded, and this could be due to the smaller patches (<9 ha) and their edge effects where abundance of solar radiation along the edges of the patches may increase productivity. The impact on GPP due to fragmentation is found to rely on the type of land transformation and patch characteristics (size, edge, and shape complexity). The preservation of mangrove forests in a rapidly developing region such as IM is vital to ensure ecosystem, ecology, environment, and biodiversity conservation, in addition to providing economical revenue and supporting human activities.

Keywords: mangrove; coastal region; remote sensing; fragmentation; productivity; land cover change

1. Introduction

Changes in the extent of mangrove cover has resulted in fragmentation that can greatly affect its functions and services to the environment and human kind [1]. Fragmentation of mangrove areas refers to the sub-division of a continuous landscape into smaller units, which can be characterized based on patch size, edge, and shape modification. Fragmentation analysis and monitoring are important indicators of how mangrove patches influence ecosystem habitats, species diversity, carbon storage capacity, etc. [2]. Bryan-Brown, Connolly, Richards, Adame, Friess, and Brown [1] showed that Southeast Asia (SEA), which has the highest mangrove tree species diversity, is a hotspot for mangrove fragmentation. Among the SEA countries, Malaysia in particular has the highest reduction of mangrove mean patch size, ranked at first place in the nations of the world [1]. Fragmentation could impact habitat health and could cause species extinction, and the report of Ponnampalam et al. [3] showed that Sungai Pulai and Johor Straits in the southern part of Peninsular Malaysia have lost many of their dugong species as a consequence of mangrove loss and fragmentation. Mangroves that provide important economic, social, and ecosystem functions and services [4,5] warrant effective monitoring for appropriate management to avert further loss. The mangrove ecosystem is an essential component of nature-based solutions for climate change mitigation and adaptation [6–8]. While the prevention of deforestation and degradation of mangroves is the most effective strategy for climate change mitigation and adaptation, restoration can reverse the impacts of climate change after 25–30 years [9–11]. Mangroves can protect shorelines against storm surges, cyclones, and tsunami waves. [12,13]. Although recent studies revealed that mangrove loss has been reduced in last decade [14], further and continuous monitoring can help to accelerate the restoration and conservation of mangroves [15].

A review of the literature shows that, in the past decade, although studies on habitat loss is well reported, research on habitat fragmentation has been poorly documented [16]. The limited studies catalogued on habitat fragmentation have focused on nutrient cycling, dung removal, pollination, and seed dispersal [17]. Studies emphasizing the impacts of forest fragmentation on biomass, productivity, predation, parasitism, infection rate, and aquatic functioning account for only 3% of forest fragmentation studies [18]. This gap of information on the impact of fragmentation on mangrove productivity must be addressed and is one of the objectives of the current study. Gross primary production (GPP) of any ecosystem is often used as an indicator of the total amount of CO_2 assimilated by the ecosystem, particularly by the vegetation [19]. Thus, analysis of GPP and its relation to mangrove cover loss and fragmentation can provide useful information pertaining to the dynamics of the mangrove's function to assimilate CO_2. GPP is affected by forest vegetation structure and stand parameters such as the diameter, tree height, canopy area, basal area, leaf area index, and mean tree weight and density [20]. Leaf Area Index (LAI) refers to the total leaf area per unit surface, and this index has been used for assessing the rate of carbon assimilation, primary productivity, and respiration and growth of vegetation, including that of mangroves since the 1990s [21–25] because it is relatively easy to measure in the field and also to derive using remotely sensed data [26]. It can be hypothesized that, due to mangrove forest changes, the LAI of mangroves will be significantly affected, and this will be reflected in the amount of GPP accumulated by mangrove trees. LAI and GPP can therefore be considered as simplified indices to monitor the carbon storage of mangroves. Previous studies have mapped the carbon stock and biomass of mangroves using various remote sensing technology and unmanned aerial vehicle systems [25]. Mangrove biomass assessment and mapping can be done using LAI as an index [27,28].

There have been numerous studies mapping the spatial extent of mangroves and their spatio-temporal changes, and several studies on mangrove ecosystem functions and processes in the past decade [29–33]. However, research focusing on fragmentation and its impact on mangrove biodiversity, biophysical characteristics, and ecosystem processes and functions within the fragmented landscape are scarce. Mangrove fragmentation is assumed to distress ecological processes and ecosystem functioning; thus, establishing a relationship between mangrove cover change, fragmentation, and ecosystem processes will ultimately provide important baseline data for the formation of a stronger policy to protect the remaining mangroves and achieve Malaysia's Sustainable Development Goals, particularly goals 13, on climate action; 14, on life below water; and 15, on life on land.

Remote sensing and Geographic Information System (GIS) are cost- and time-effective technologies that can be used to support large scale mangrove mapping, monitoring its changes [34] and analyzing the landscape's fragmentation [1,2,29]. The launch of the Landsat mission promoted the creation of global and regional mangrove maps at 30 m spatial resolution. These maps can be further improved using freely available satellite images with better spatial resolution, such as that of Sentinel-2. Digital image classification algorithms have been widely developed for obtaining land use and land cover using remotely sensed images. Pixel-based classification involving the unsupervised or supervised method are very common and useful. In the last decade, object-oriented, neural network, decision tree, and machine learning based classification algorithms, with advantages of improved accuracy and reduced computational time, have become more popular. Reviews by Kuenzer et al. [35] and Pham, Xia, Ha, Bui, Le, and Takeuchi [25] concluded that most of the classification algorithms can give overall accuracy above 80%, while, with improvements such as training samples selection, satellite data quality etc., a higher classification accuracy (~90%) can be achieved. Better classification results can improve the mangrove extent mapping and the change detection analysis to accurately detect the gain and loss of mangroves.

This study aims to describe the mangrove loss and fragmentation in the Iskandar Malaysia region over the 2000–2019 period and analyze the loss and fragmentation effects on mangrove LAI and GPP values. More specifically, we intend to (1) document the loss and gain of mangroves over a period of two decades (2000–2019), (2) analyze mangrove fragmentation, and (3) relate and analyze the loss/fragmentation impacts on the LAI and GPP values in order to define the function of mangroves to assimilate CO_2 in relation to mangrove cover changes and fragmentation in the study area. It is hypothesized that a higher loss of mangrove canopy cover and fragmented areas will decrease the LAI and GPP of the ecosystem.

2. Materials and Methods

2.1. Study Area

Malaysia constitutes only 4% of global mangrove cover, but at least 70 true and associated species from 28 families have been identified in the country, and the majority of the mangrove trees are from the *Rhizophoraceae* family [36]. More than half of Malaysian mangroves are distributed in the state of Sabah (60%), 22% in Sarawak, and the remaining 18% are found in Peninsular Malaysia [36]. In Peninsular Malaysia, the majority of mangroves are located along the west coast, and the state of Johor alone holds 30% of the total mangroves (Figure 1). Three of the seven Ramsar Sites in Malaysia, namely Pulau Kukup (647 ha), Tanjung Piai (526 ha), and Sungai Pulai (9126 ha) are located in the Iskandar Malaysia region in Johor (Johor National Parks, https://www.johornationalparks.gov.my/v3/RAMSAR-site/, assessed on 12 December 2020).

Iskandar Malaysia (IM) is a fast-growing national special economic region with a total area of 221,700 ha, located in the southern part of Johor state, Peninsular Malaysia (Figure 1). Specific zones where rapid development has taken place in IM are the Port of Tanjung Pelepas, Tanjung Bin Power Plant, the Port of Pasir Gudang, and the rapid reclamation zone of Forest City, as shown in Figure 1. These activities are mainly caused by population growth that requires more land to be developed for housing establishments, transition to agricultural lands (particularly oil palm), and the building of infrastructure such as ports, dams, and waterfront cities may also lead to fragmentation and impact the carbon sink potential and primary productivity of the mangrove forests in IM. Thus, it is timely to investigate the impact of fragmentation on the primary productivity of the mangrove ecosystem. Although sustainable development policies are employed in the region, the decrease of mangrove forests was found to be at an alarming rate (33%) between 2005 and 2014 [32]. Nevertheless, the replantation and conservation efforts since 2014 may have also increased mangrove coverage; thus, it is important to monitor the changes and analyze their impact on mangrove services.

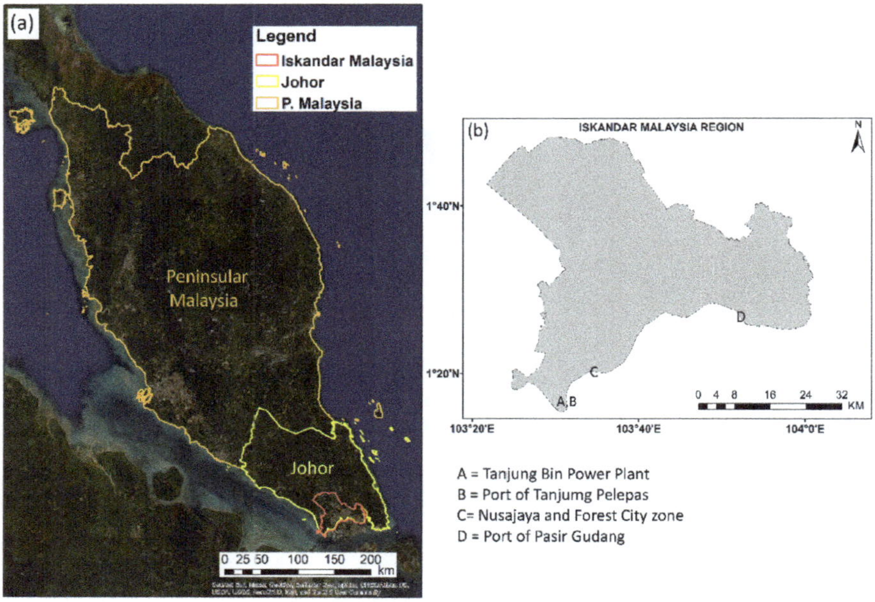

Figure 1. Map of Peninsular Malaysia (**a**) and the region of Iskandar Malaysia (**b**), with four rapidly developing zones (A, B, C, and D) in the region.

2.2. *Data and Methodology*

The methodology flow-chart (Figure 2) for this study and details on the data processing and the methods to achieve the aim of the study are described in the following sections. This study utilized medium-resolution remote sensing satellite imageries to monitor changes in mangrove cover and fragmentation, covering the years 2000 to 2019. Support Vector Machine (SVM) was selected as the classifier for the image classification process. A patch analysis plugin was implemented for fragmentation analysis based on ArcGIS software. The land cover, LAI, and GPP Products of Moderate Resolution Imaging Spectroradiometer (MODIS) Collection 006 were used to define the impact of loss and fragmentation on mangrove forests in IM. The MODIS Land Cover Product is based on a supervised decision-tree classification method using inputs from other MODIS products, including the land/water mask, nadir bidirectional reflectance distribution function (BRDF)-adjusted reflectance, Enhanced Vegetation Index (EVI), snow cover, land surface temperature, terrain elevation information, etc. [37]. The MODIS LAI product incorporated the MODIS Land Cover Product and MODIS Surface Reflectance Product into the radiative transfer model to derive LAI [38]. The MODIS GPP product is derived by relating the absorbed photosynthetically active radiation (APAR) calculated from LAI with the energy conversion efficiency (ε) [26]. More information of the MODIS products and the method to access these products are described below.

2.2.1. Remote Sensing Data and Pre-Processing

Landsat 7 Enhanced Thematic Mapper Plus (ETM+) Terrain Corrected (L1T), data dated 28 April 2000, was downloaded from the GloVis website (https://glovis.usgs.gov/ (accessed on 15 March 2021)). This data is registered on the Worldwide Reference System-2 (WRS-2), and one tile (Path 125, Row 59) is required to cover the whole Iskandar Malaysia zone. A sentinel-2 Level-1C multispectral image dated June 28, 2019, was chosen to present the latest conditions of the study area because no Landsat image with acceptable cloud cover was available. The image was downloaded from the Copernicus Open Access Hub (https://scihub.copernicus.eu/ (accessed on 15 March 2021)). For both Landsat 7 and

Sentinel-2, only images with less than 20% cloud cover were used. The Semi-Automatic Classification plug-in (SCP) provided in the QGIS software [39] was used to perform the preprocessing of Landsat 7 and the Sentinel-2 images.

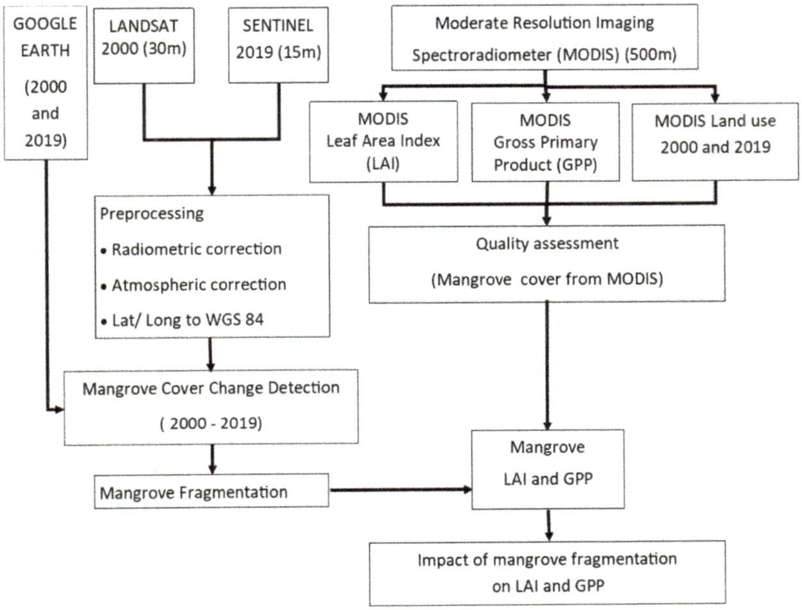

Figure 2. The work-flow chart for image processing, classification and assessment, fragmentation, LAI and GPP analysis.

Preprocessing, including radiometric correction and atmospheric correction based on the Dark Object Subtraction 1 (DOS 1) model and pan-sharpening model, was applied to the Landsat 7 image. The radiometric and atmospheric correction transferred the digital numbers (DN) to radiance then to reflectance. The pan-sharpening function, based on intensity computation, fused the 30 m Landsat multispectral bands (6 bands covering the visible (blue, green, red), near infrared, and shortwave infrared) with the panchromatic band to produce a 15 m spatial resolution reflectance image. The Sentinel-2 image underwent similar preprocessing steps (radiometric and atmospheric correction using DOS 1) based on the SCP. The visible bands (band 2—blue, 3—green, 4—red) and near infrared (band 8) of 10 m spatial resolution were stacked in this case. The output was a 4-band multispectral reflectance image at 10 m spatial resolution. A thresholding method based on the reflectance of the blue band (>0.25 for Landsat 7 and >0.21 for Sentinel-2) was applied to detect and mask out any cloud on the image. Note that, for both Landsat and Sentinel images, only a very low cloud cover was detected over the mangrove areas, which is the main focus of this study.

Three latest MODIS products (Collection 006) at 500 m spatial resolution were downloaded, namely the MOD15A2H (Leaf Area Index, LAI) [38], the MOD17A2H (Gross Primary Production, GPP) [26], and the MCD12Q1 (Land cover type, LC) [37]. Only one tile (H28.V08) is required to cover the whole IM region. Both LAI and GPP are 8-day composite products. Full year products of 2000 and 2019 were downloaded for LAI and GPP, which consisted of 39 tiles for year 2000 (MODIS started to provide data from 18 February 2000) and 46 tiles for year 2019. Meanwhile, the yearly composite (one product per year) LC product for years 2000 and 2019 were downloaded. All the MODIS products are available from the NASA Earthdata website (https://earthdata.nasa.gov/ (accessed on 15 March 2021)). The MODIS Conversion Toolkit (MCTK) was used to process the MODIS products, including re-projection and quality check (QC) assessment. In order to maintain the quality of the LAI and GPP products, pixels with a QC bit-converted value below 48 were kept and extracted, referring to the MODIS product user guides [26,38,40]. The products were re-projected into WGS 84 projection, and the QC considered the sensor, detectors, and cloud condition issues of the product. For the LC product, we selected the International Geosphere-Biosphere Program (IGBP) layer for this study. According to the IGBP layer, the permanent wetlands (PW) pixels (LC code 11) were identified. The corresponding LAI and GPP values located in the identified PW pixels were extracted with the strict consideration of the QC. The mean and standard deviation of the LAI and GPP at 8-day intervals were computed for further analysis.

2.2.2. Image Classification and Assessment

Support Vector Machine (SVM) classification has been one of the most applied classifiers in recent years. SVM uses the machine learning technique for multispectral and hyperspectral remotely sensed data classification. SVM relies on n-dimensional spectral space to create a hyperplane and identify and separate the classes based on user optimized parameters, including the selection of kernel type and function and the penalty parameter [41]. A total of 7 land use/land cover classes were detected and discriminated from the Landsat and Sentinel images based on the SVM, as follows: terrestrial forest, mangrove forest, oil palm plantation, rubber plantation, urban, water, and others (orchard, shrub, bush, aquaculture farms, etc.). The details of the parameters needed to run the SVM classifier are listed in Table 1. The classification was run based on a total of 140 training data (20 polygons per class with 50–60 pixels per polygon), where different sets of training data were selected during Landsat 7 and Sentinel 2 classification. Figure A1 in the Appendix A shows the training polygons for the Landsat and Sentinel classification. The accuracy assessment of the classified land covers/uses was based on the sample points selected from the Google Earth images as the "ground truth" samples. Google Earth is notably one of the most complete and useful datasets with acknowledged accuracy, and is also freely available for various types of research, including classification and its validation [42–44]. Classification assessment based on the error matrix was performed. The overall accuracy (OA) is the ratio between the correctly classified pixels to the total number of reference pixels. The producer accuracy (PA) is the accuracy of the map from the perspective of the map maker, computed by dividing the correctly classified samples of the class to the sum of the reference samples for the class. The user accuracy (UA) is the accuracy from the point of view of a map user, calculated by dividing the correct classification pixels to the total number of pixels mapped as the class. The kappa coefficient (KC) is a value computed based on the KHAT statistic to measure the agreement and accuracy of the classification. In addition, the Z-statistic has been computed to describe the significance of the classification results. The function of the Z-statistic is to determine if the agreement between the classification map significantly agrees with the reference data [45].

Table 1. Parameters for Support Vector Machine (SVM) classification.

Parameter	Type/Value
Kernel type	Radial Basis Function (RBF)
Gamma in kernel function	1.00
Penalty	100
Pyramid levels	0
Region of Interests (ROI)	140 (20 polygons for each class)

2.2.3. Mangrove Cover Change and Fragmentation

After the accuracy assessment of the 2000 and 2019 classification results, the classified mangrove areas were extracted from the classified map. Because the classification maps are in raster form, we converted the maps into vector polygons using the Raster to Polygon function in the ArcGIS software. Change detection was performed based on the polygons using ArcGIS software to detect the unchanged, gained, and lost mangrove areas in IM between 2000 and 2019 by superimposing the mangrove cover from 2000 onto the results of 2019.

The Patch Analyst extension in ArcGIS 10.5 was used as the tool for fragmentation analysis. Patch Analyst is based on the FRAGSTATS [46], which can be used for analysis and to provide attributes of the fragmented patches. Several indices were selected in this study to represent the fragmentation of the mangrove area in IM, including Class Area (CA), Total Landscape Area (TLA), Number of Patches (NP), Mean Patch Size (MPS), Total Edge (TE), Edge Density (ED), Mean Shape Index (MSI), Average Weighted MSI (AWMSI), Mean Perimeter Area Ratio (MPAR), and Mean Patch Fractal Dimension (MPFD). The group and details of each index is listed in Table 2. Because mangrove is the focus of the study, the fragmentation analysis was performed for mangrove areas only.

Table 2. List of metrics in the Patch Analyst extension used for fragmentation analysis [46].

Group	Fragmentation Index	Definition
Area	Class Area (CA)	Sum of areas of all patches of a given class
	Total Landscape Area (TLA)	Sum of areas of all patches in the landscape
Patch size	Number of Patches (NP)	Total number of patches in the class/landscape
	Mean Patch Size (MPS)	Average patch size
Edge	Total Edge (TE)	Total perimeter of patches
	Edge Density (ED)	Amount of edge relative to the landscape area
Shape	Mean Shape Index (MSI)	Average perimeter-to-area ratio
	Average Weighted MSI (AWMSI)	MSI divided by weighted patches area
	Mean Perimeter Area Ratio (MPAR)	Average of patch perimeter to patch area ratio
	Mean Patch Fractal Dimension (MPFD)	Average of log-transformed patch perimeter to log-transformed patch area ratio

2.2.4. Mangrove LAI and GPP

LAI and GPP are both useful for studying the coverage, health, and potential of carbon assimilation and storage [47]. Mangrove forests have been proven as the highest carbon stock medium; nevertheless, the role of mangrove forest in carbon assimilation and its processes from local to global scale is less understood. This is due to the difficulties of data collection in the mangrove forests. Plant leaves through the photosynthesis process convert the atmospheric CO_2 into nutrients and biomass to support plant growth. The use of LAI to estimate the ecosystem carbon uptake by linking it with GPP is common [48,49]. Remote sensing products were applied in this study to infer the leaf density and productivity of mangrove areas, aimed at detecting the impact of mangrove changes to the products (LAI and GPP) relevant to carbon stock.

To identify the impact of mangrove loss and fragmentation over years, the MODIS LAI and GPP of the permanent wetlands (PW) (based on MODIS IGBP land cover classification) were extracted. LAI can be used to describe the leaf density and therefore the vegetation condition/health/ photosynthesis rate and light use efficiency of the mangrove trees [33,34]. We chose LAI as the vegetation index because it is often widely applied to high leaf and tree density zones, including mangrove forests [50], and LAI has been proven to have a strong relationship with Normalized Difference Vegetation Index (NDVI) at different spatial resolutions [51,52]. Meanwhile, MODIS GPP is an indication of the amount of CO_2 assimilated by the ecosystem; thus, it can be used to identify carbon gain/loss due to any changes occurring in the ecosystem. Here, we used GPP rather than NPP because (1) GPP has more direct relations to the photosynthesis rate, leaves, and biomass growth, (2) NPP requires the input of respiration (data is very hard to acquire and not available in the study site), and (3) MODIS only distribute the NPP product at an annual timespan, while we aim to compare GPP and LAI at finer temporal scales (8-days). The values of LAI and GPP were extracted and their statistical values were computed to identify if any loss or gain had occurred in LAI or carbon intake by mangroves (to the selected permanent wetlands pixels).

3. Results

3.1. Classification Accuracy and Mangrove Cover Changes

Using the SVM classifier, seven land use/land cover classes, including forest, oil palm, mangrove, rubber, urban, water, and "others", were classified. The class "others" includes orchard, shrub, bush, farm, and aquaculture. Table 3 reports the accuracy of the classification from the two satellite images, including the producer accuracy (PA), user accuracy (UA), overall accuracy (OA) and the kappa coefficient (KC). The overall accuracy of both images is ~90%, with a KC of 0.85 and 0.88 for the 2000 and 2019 classifications, respectively. Both the Landsat 7 and Sentinel-2A classification results showed a high accuracy of around 90%, which may be because of their finer spatial resolution and multi spectral bands, and is also due to the machine learning technique of SVM. On the other hand, the pan-sharpened Landsat image also maintained high classification accuracy when more bands (more information) were considered during the classification process. Z-statistics were computed for both classification results, where both values were higher than 1.96, indicating that the classification is meaningful and significant and the classification results are better than random classification.

Table 3. Distribution of land use area and the producer accuracy (PA), user accuracy (UA), overall accuracy (OA), and kappa coefficient (KC) for respective class *.

Class	Landsat 7 (2000)		Sentinel-2A (2019)	
	PA (%)	UA (%)	PA (%)	UA (%)
Forest	89.25	92.45	78.22	87.93
Mangrove	97.52	90.64	87.85	99.70
Oil Palm	88.66	78.35	73.02	84.47
Rubber	43.17	78.15	96.40	53.49
Urban	64.56	93.12	95.44	93.88
Water Bodies	98.56	99.57	92.55	98.17
Others	46.32	19.19	91.42	62.67
OA	89.96%		90.51%	
KC	0.85		0.88	
Z-statistic	271.37		109.35	

* The definition of land cover types is given in Table A1.

Because the general land use land cover changes are similar to the previous findings in Kanniah, Sheikhi, Cracknell, Goh, Tan, Ho, and Rasli [32], we omitted the land use maps in this study and focused only on the mangrove cover in the following results and discussions. In this study, we further found an expansion of the urban area, especially at the eastern part of Sungai Pulai, where the new Nusajaya township is located, and in the central part of IM, where the Kulai–Senai township has expanded due to increased population. Rubber conversion to oil palm plantation and urban area is still found (in 2019 image) surrounding the Sultan Iskandar Reservoir in the eastern part of IM, showing similar results to [32] in their 2014 satellite data.

The use of the SVM classifier showed great capability by clearly distinguishing mangroves from forests and water bodies, which often can be easily misclassified. Both the user and producer accuracies are high for mangrove classification in the 2000 and 2019 images (Table 3). From the classification results, the loss of mangrove area is noticeable, particularly where it has been converted to urban area and oil palm plantations. Interestingly, the conversion of mangrove to aquaculture land use (classified as "others") was found at multiple locations, which showed the recent development and expansion of the economically valuable aquaculture activities in the IM region (Table 4 and Figure 4b). The performance of SVM classification in this study supports the accuracy of mangrove loss and gain detection (Table 4 and Figure 4), as well as clearly identifying fragmented mangrove patches (Table 6).

Table 4. The loss and gain of mangrove areas and their transition to other land use/cover types in IM between 2000 and 2019.

Gain ID.	Places/Location	Transition Type
1	Pulau Kukup	Eroded mangrove
2	Serkat Berkat Village	Oil palm
3	Tanjung Piai	Eroded mangrove
4	Perpat Pasir Village	Oil palm
5	Sungai Boh Village	Urban
6	Pulai River—Custom Administration Building	Urban
7	Forest City Golf Hotel/ Resort	Urban—resort
8	Sri Tulang Residential Project	Urban—residential
9	Bayu Senibong Residential zone	Bare land
10	Changkat Village	Aquaculture
11	Sri Aman Village	Aquaculture
12	Sri Aman Village	Aquaculture
13	Sri Aman Village	Aquaculture
14	Tanjung Langsat	Urban—port expansion
15	Tanjung Langsat	Urban—port expansion
16	Kampung Kong Kong	Aquaculture
17	Kampung Kong Kong	Aquaculture
18	Layang River	Eroded mangrove
19	Tiram River Fishing Village	Aquaculture
20	Tiram River Fishing Village	Aquaculture
21	Tiram River Fishing Village	Aquaculture
Gain ID.	**Places/Location**	**Transition Type**
1	Kong Kong River	Eroded mangrove
2	Kong Kong River	Eroded mangrove
3	Tanjung Langsat Village	Aquaculture
4	Melayu Village	Aquaculture

Based on the classification results, the total mangrove areas in 2000 and 2019 were 13,612.25 ha and 10,704.96 ha, respectively. Figure 3 shows the mangrove areas, their gain, and their loss, and unchanged mangrove areas between 2000 and 2019. The total unchanged mangrove areas over the 19-year period was 8483.44 ha, and the total loss was 5128.80 ha. Despite the loss, some new (2171.38 ha) mangrove areas were detected. This means that a net loss of 2907.29 ha (21.35%) has occurred in IM over the last 19 years. This is equal to a 1.12% annual mangrove loss in the IM region, which is higher than Malaysia's mean depletion rate of 0.41%, based on the estimation of Hamilton and Casey [53]. The southwestern part of the IM, where the Pulau Kukup, Tanjung Piai, and Sungai Pulai mangrove forests (Ramsar sites) are located (Figure 3c), faced significant loss over the study period. On the other hand, at the eastern side of IM, new patches of mangroves were noticed. Figure 4a shows the pinned locations of gain and loss of mangroves over the IM region. Table 4 lists the locations and types of transition from (loss) and to (gain) mangrove area, with Table 5 showing the conditions of some selected loss and gain areas, including when these transitions occurred. During the identification of the gain and loss locations, we also identified around 40 sites in IM that were used for aquaculture activities (Figure 4b).

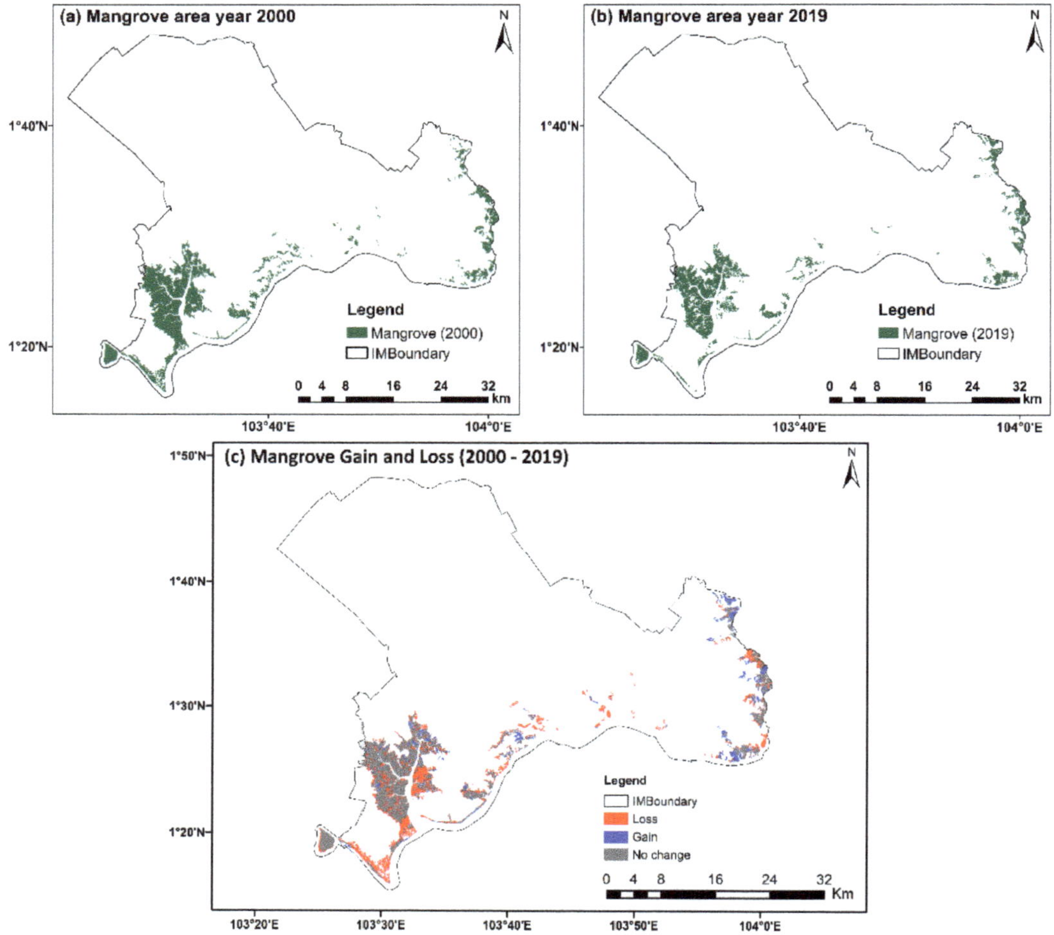

Figure 3. Mangrove area of (**a**) year 2000; (**b**) year 2019; and (**c**) the unchanged, gains, and losses of mangroves between 2000 and 2019.

3.2. Fragmentation Analysis

Fragmentation results in Table 6 show that the IM region underwent significant fragmentation over the two decades of the study period. Since we considered only one class (mangrove) in the fragmentation analysis, the area indices, including class area (CA) and total landscape area (TLA), are the same, which is 13,612 ha and 10,705 ha for years 2000 and 2019, respectively. There was a significant loss of mangrove area (21%) during the study period; thus, the analysis of the remaining fragmentation metrics will also be focused on the influence of the area loss on each metric.

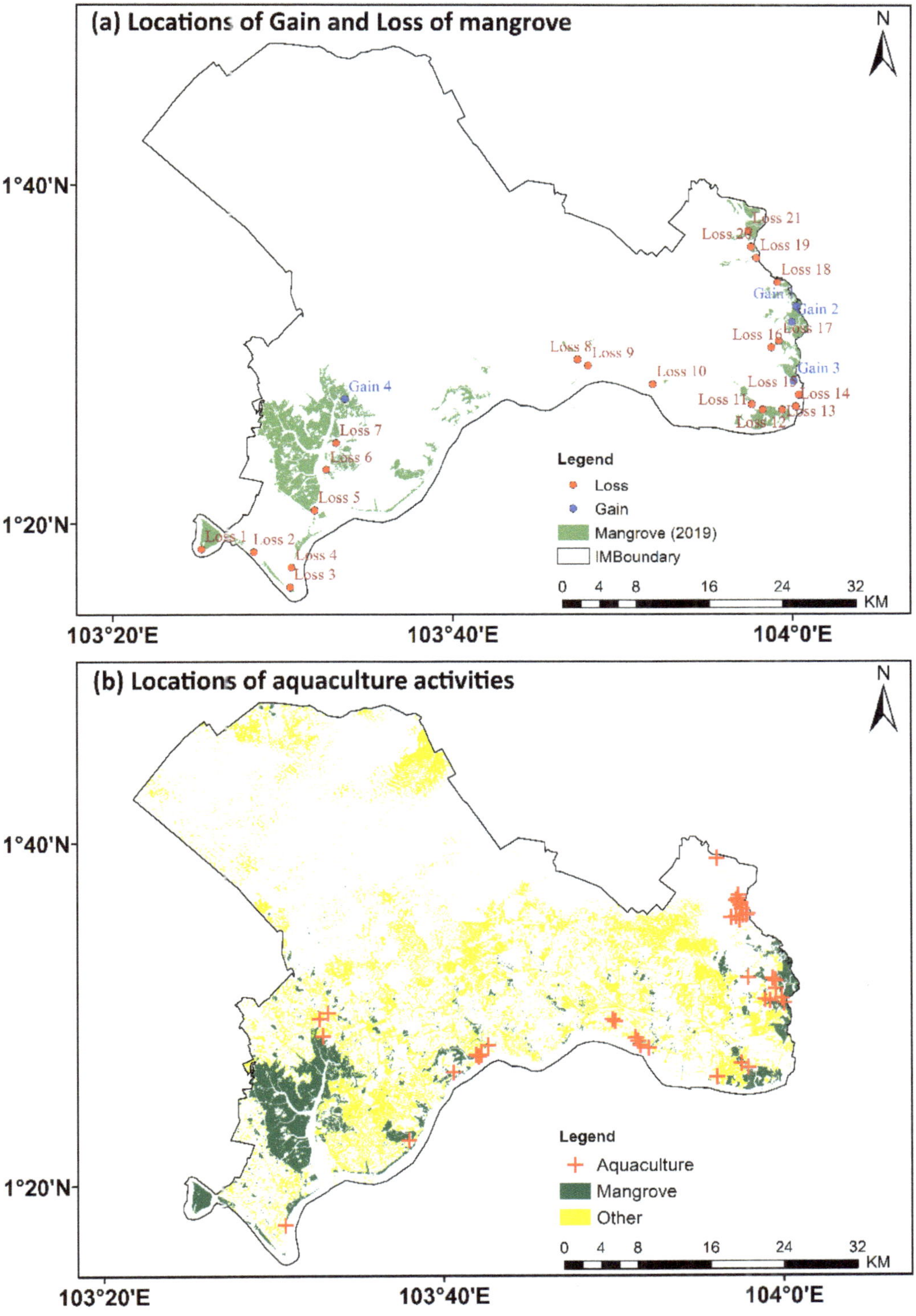

Figure 4. Identification of areas of (**a**) gain and loss and (**b**) aquaculture over Iskandar Malaysia (IM).

Table 5. Selected loss and gain of mangrove areas and their transition conditions. Different years are based on the availability of the Google Earth images. Pins shown on the images correspond to the location of gain and loss in Figure 4a.

Transition Type	ID.	Loss Area		
		Pulau Kukup		
			23 Aug 2006	2 Aug 2019
Eroded mangrove	1	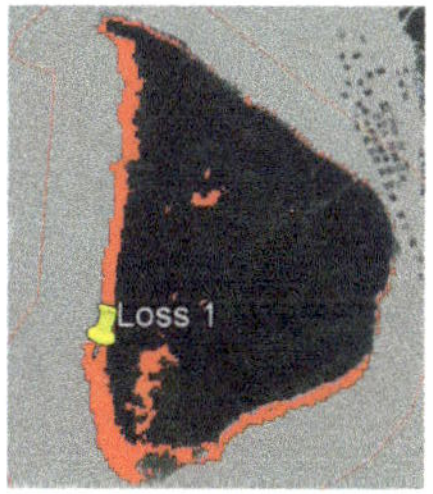	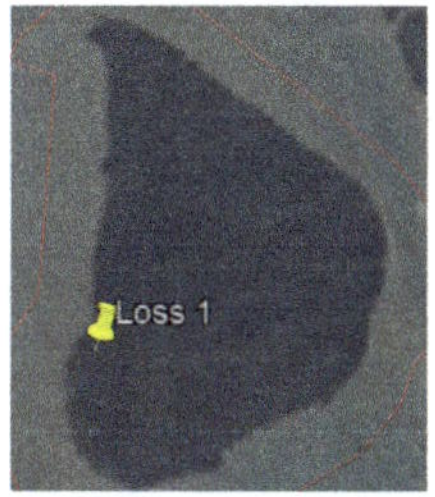	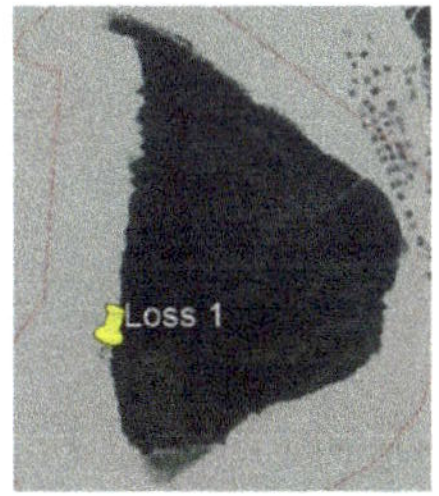
		Layang River		
			6 Apr 2003	30 June 2017
	18			

Table 5. *Cont.*

Transition Type	ID.	Loss Area		
		Sungai Boh Village		
			17 Jan 2008	10 Feb 2018
Urban	5	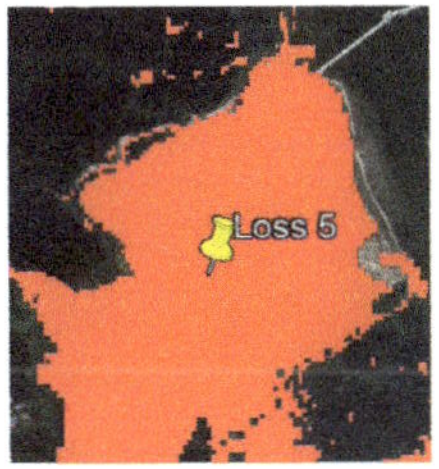	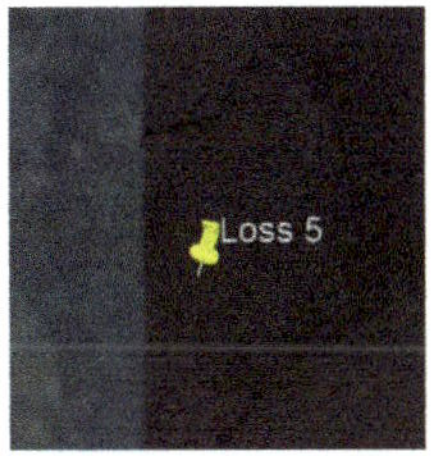	
		Pulai River—Custom Administration Building and Forest City Golf Hotel/Resort		
			July 2006	2 Aug 2019
	6 and 7	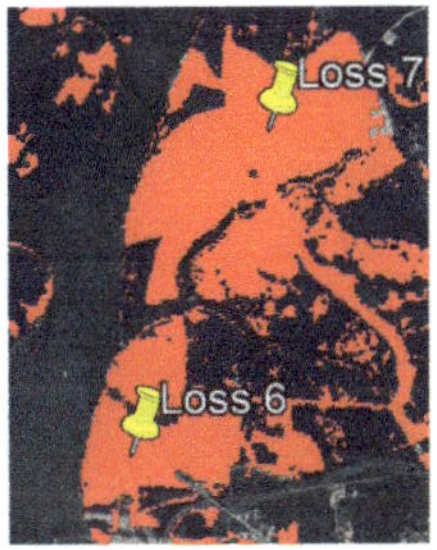	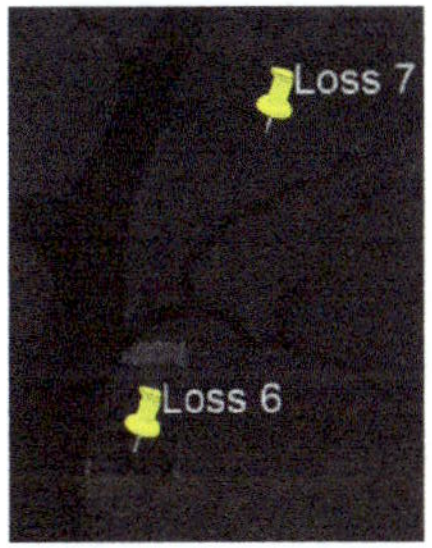	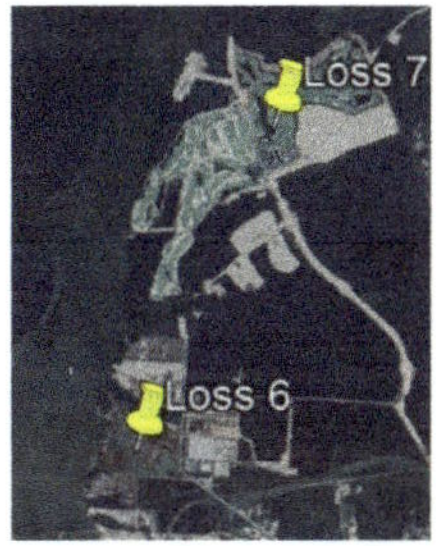

Table 5. *Cont.*

Transition Type	ID.	Loss Area		
		Sri Aman Village		
			6 Apr 2003	23 Jan 2012
	11 and 12	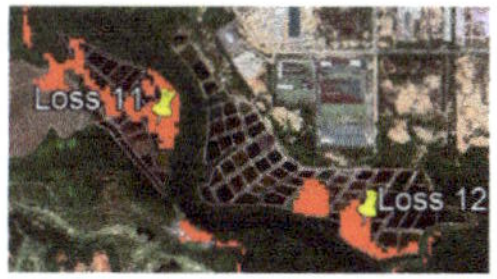		
		Kong Kong Village		
			6 Apr 2003	19 July 2020
Aquaculture	16 and 17	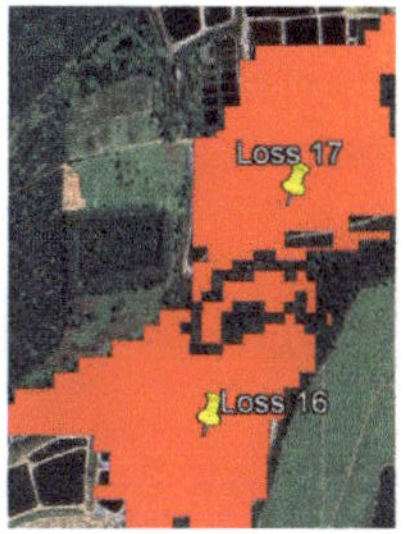	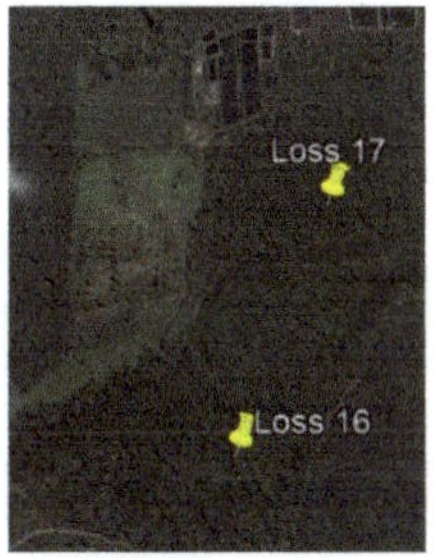	

Table 5. *Cont.*

Transition Type	ID.		Loss Area	
		Gain area		
			Kong Kong River	
			6 Apr 2003	30 June 2017
Eroded mangrove	1 and 2	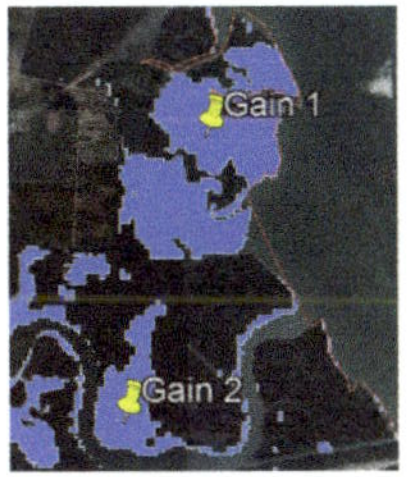	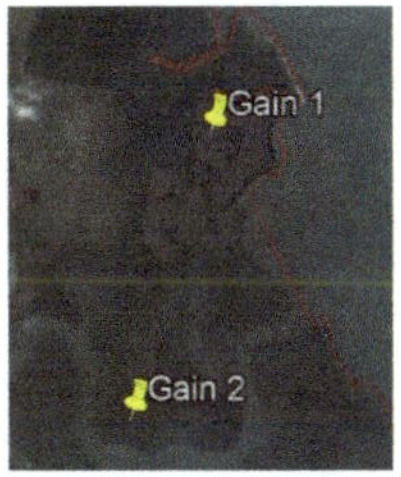	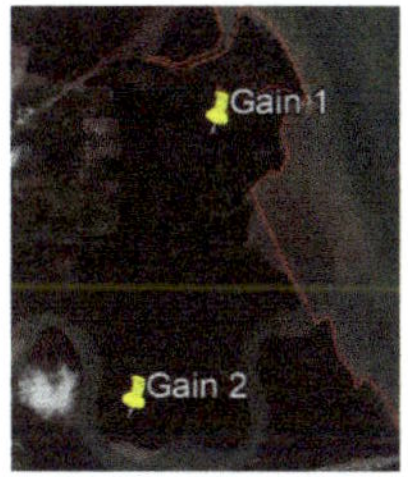
			Tanjung Langsat Village	
			6 April 2003	10 Nov 2019
Aquaculture	3			

Table 5. *Cont.*

Transition Type	ID.		Loss Area	
Gain area				
			Melayu Village	
	4		17 Jan 2008 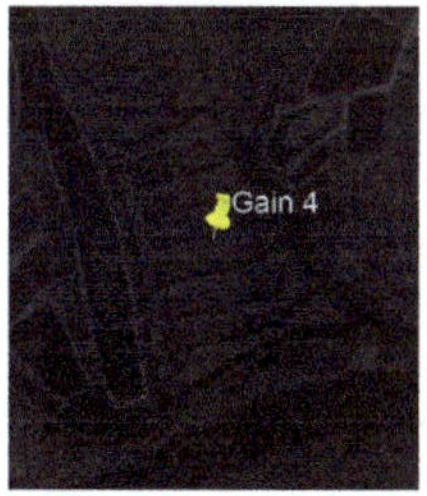	2 Aug 2019 

Table 6. Fragmentation values for the year 2000 and the year 2019.

Index	Unit	Year 2000	Year 2019
Class Area (CA)	ha	13,612	10,705
Total Landscape Area (TLA)	ha	13,612	10,705
Number of Patches (NP)	-	130	402
Mean Patch Size (MPS)	ha	104.71	26.63
Total Edge (TE)	m	1,200,480	1,364,970
Edge Density (ED)	m/ha	88.20	127.509
Mean Shape Index (MSI)	-	2.796	1.962
Average Weighted MSI (AWMSI)	-	8.802	8.761
Mean Perimeter Area Ratio (MPAR)	m/ha	245.95	1685.40
Mean Patch Fractal Dimension (MPFD)	-	1.366	1.444

The patch size index provides information about the Number of Patches (NP) and the Mean Patch Size (MPS). The patch size is one of the fragmentation parameters that has been studied most [17]. Increased NP from 130 to 402 and decreased MPS from 104.71 to 26.63 (reduction of −78.08 ha) between 2000 and 2019 indicates that the mangrove areas in the IM region had been broken down into smaller patches. The changes of minimum patch size (3.08 ha to 0.001 ha) and maximum patch size (6883.01 ha to 4067.96 ha) from years 2000 to 2019 showed a similar trend to the MPS values, indicating that the mangrove areas had been fragmented into smaller patches throughout the entire study area. In a previous study, Bryan–Brown, Connolly, Richards, Adame, Friess, and Brown [1] reported a smaller reduction of mean patch size of −7.2 ha for Malaysia between 2000 and 2012, which infers that the fragmentation in the IM region is significantly more serious than the country's average. The higher MPS could also be due to the more rapid development in IM compared to other cities located near to mangrove forests in other parts of Malaysia. Coincidentally, the largest mangrove patch in the year 2019 (4067.96 ha) was found located in the largest mangrove patch of the year 2000 (6883.01 ha). Analysis of this large mangrove patch showed that it had fragmented into 60 smaller patches (0.001 ha to 4067.96 ha, with a mean of 94.29 ha), with a loss of 1131.63 ha over the study period. This specific patch has been selected for detailed fragmentation analysis on its effect of LAI and GPP in the later part of this manuscript.

Similarly, the Edge indices (TE and ED) also increased from 2000 to 2019, primarily because of the separation of large patches into small patches. Figure A2 in the m shows the fragmentation of mangrove over a selected area within the study area. The increases in TE (from 1200.5 km to 1365 km) and ED (from 88.2 m/ha to 127.5 m/ha) demonstrate that the fragmentation is a result of man-made disturbances rather than natural effects. This can be explained by natural shaped patches normally having fewer edges than man-made urban areas and agriculture boundaries. The values of patch size and edge indices obtained in this study show typical mangrove forest fragmentation over a rapidly developing region, where the disturbance of a large mangrove area into smaller patches occurred due to the transformation of mangroves into urban, agriculture, and aquaculture lands. Generally, the decrease in any forest habitat area or MPS has been shown to create fragments of different sizes (large and small) and isolation among the fragmented patches. Hence, this can result in species extinction and biodiversity loss due to decreased colonization and population size [17,54].

The Shape metrics (MSI, AWMSI, and MPAR) explain the changes of the mangrove areas shape from regular to more complex patches. An MSI of 1 refers to a circular shape, and higher values indicate more irregular shape patterns. The MSI calculated in this study changed from 2.79 in 2000 to 1.96 in 2019, meaning that the complexity of the shape of the patches was reduced. This phenomenon indicates that the mangrove patches change from natural shaped patches (complex in shape) to man-made shape (regular in shape), due to anthropogenic activities [46,55]. The AWMSI refers to the area weighted MSI, which highly depends on the area of patches and the NP. Slightly lower AWMSI was found in 2019 compared to 2000 due to a decrease in the sum of patch area and a large increase of NP.

This indicates that the MSI was divided by a relatively larger value in 2019 than in 2000, meaning that there were more fragmented patches in the region in 2019. MPAR, which considers the mean perimeter-to-area ratio shows an increased value for 2019 compared to 2000. Smaller patches with decreasing areas and relatively longer perimeters are the main reason for the large increase in MPAR. Finally, the MPFD, which has values between 1 (simple shaped) to 2 (complex shaped) have increased between 2019 and 2000, showing that the fragmented patches of mangrove in 2019 were more irregular compared to larger mangrove patches in 2000.

3.3. LAI and GPP Analysis

LAI and GPP values for the MODIS IGBP Permanent Wetlands (PW) pixels were extracted. The classified mangrove and MODIS PW pixels of 2000 and 2019 are shown in Figure 5. There were 559 and 388 MODIS PW pixels (500 m) for years 2000 and 2019, respectively, and most of the MODIS PW pixels are located within the classified mangrove area (68% for the year 2000 and 65% for the year 2019). This means that MODIS PW pixels can be used to extract the LAI and GPP values and to analyze their changes over time due to mangrove area changes. The converted total area of MODIS PW is 13,975 ha for the year 2000 and 9700 ha for the year 2019, which was found to be close to the area based on the classification in this study (13,612.25 ha and 10,704.96 ha, in Section 3.1). Similar to the classification result (Figure 3a,b), most of the disappearance of PW pixels were noticed over the Sungai Pulai and Tanjung Piai area (Figure 5).

The mean and standard deviation of MODIS LAI and GPP at 8-day intervals for 2000 and 2019 are plotted and shown in Figures 6 and 7. Five cases are plotted and discussed namely: (1) mean LAI and mean GPP extracted over all mangrove areas—Figures 6a and 7a, (2) mean LAI and mean GPP extracted over areas where mangrove is unchanged—Figures 6b and 7b, (3) mean LAI and mean GPP extracted over areas where a mangrove gain is detected—Figures 6c and 7c, (4) mean LAI and mean GPP extracted over areas where a mangrove loss is detected—Figures 6d and 7d, and (5) mean LAI and mean GPP extracted over selected fragmented areas—Figures 6e and 7e. Table 7 summarizes the mean and standard deviation values of LAI and GPP for these cases. Note that there are some missing data in these graphs due to unsuccessful retrieval of the MODIS products on certain dates that could be due to cloud cover, failure of detectors, or failure of the main radiative transfer algorithm used to retrieve the LAI/GPP values. The analysis starts from day of year (DOY) 57 because the MODIS LAI and GPP products are only available after February 2000.

Table 7. Mean and standard deviation of LAI and GPP from MODIS products. Values in parentheses refer to the number of pixels involved in each extraction.

	LAI		GPP (g C m^{-2})	
	2000	2019	2000	2019
All PW	3.18 ± 1.56 (145)	3.46 ± 2.09 (84)	6.82 ± 15.26 (141)	5.49 ± 15.10 (77)
Unchanged	3.03 ± 1.94 (89)	3.55 ± 1.97 (67)	5.81 ± 14.05 (87)	6.73 ± 16.65 (65)
Gain	2.81 ± 1.79 (24)	3.51 ± 2.20 (16)	1.88 ± 6.75 (25)	2.78 ± 8.88 (16)
Loss	2.99 ± 2.02 (52)	2.62 ± 2.15 (37)	6.38 ± 14.34 (50)	6.88 ± 15.60 (37)
Fragmented	3.11 ± 1.71 (74)	3.41 ± 1.79 (42)	4.65 ± 12.76 (71)	6.37 ± 16.46 (42)

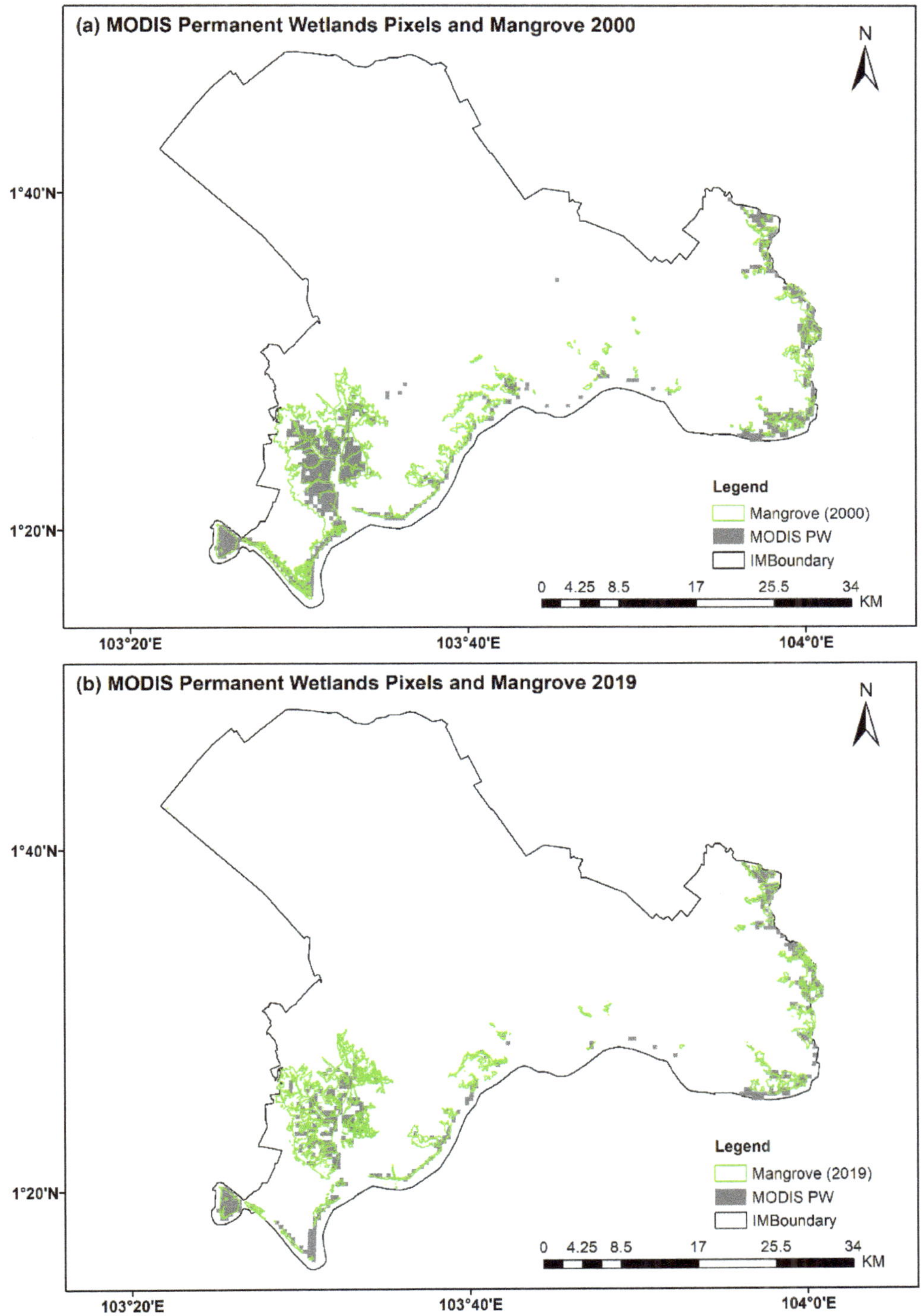

Figure 5. Mangrove area overlaid with the MODIS Permanent Wetlands pixels for year (**a**) 2000 and (**b**) year 2019.

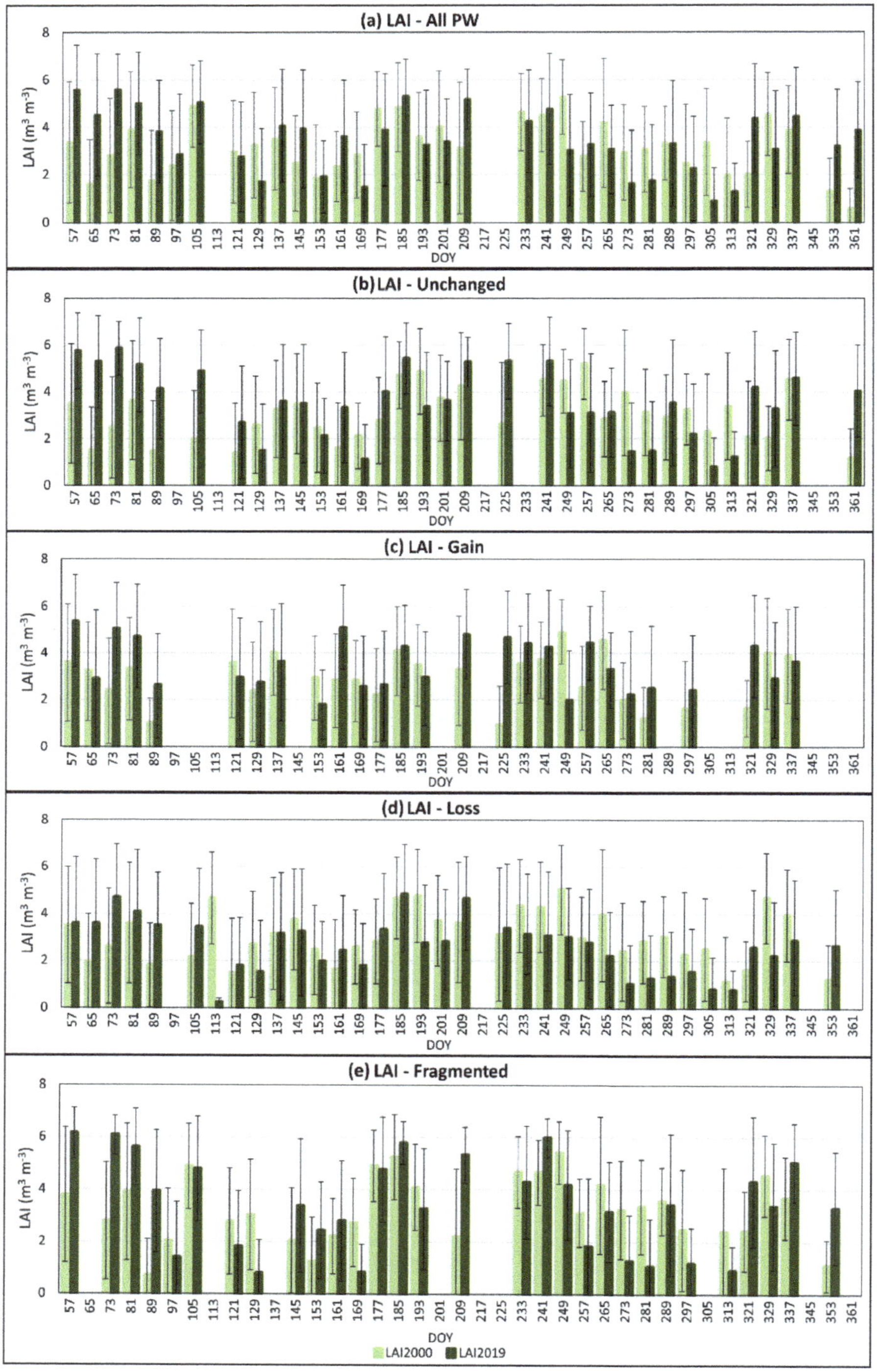

Figure 6. MODIS LAI values extracted based on (**a**) all mangrove area, (**b**) mangrove no change area, (**c**) mangrove gain area, (**d**) mangrove loss area, and (**e**) fragmented mangrove area.

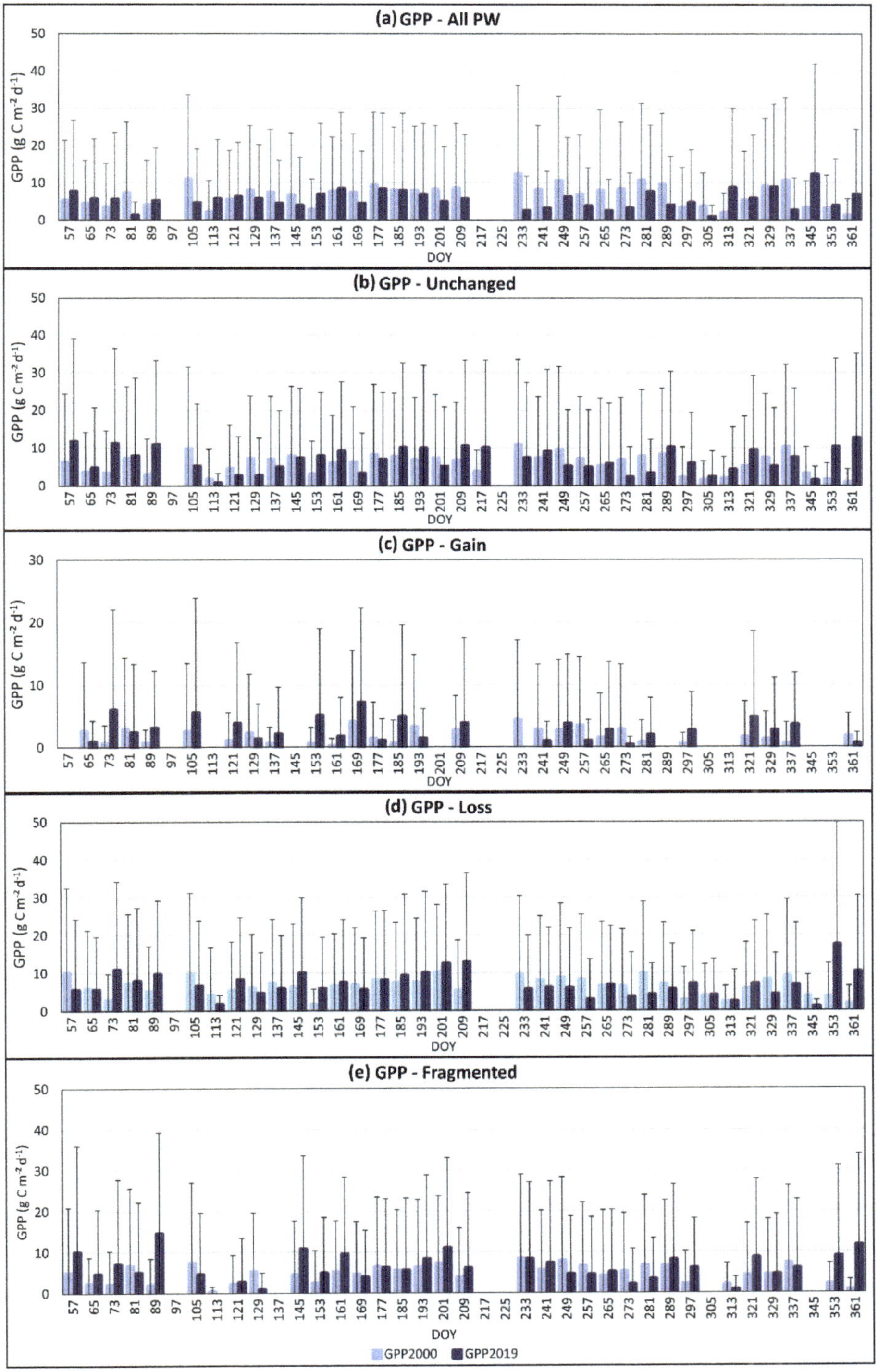

Figure 7. MODIS GPP values extracted based on (**a**) all mangrove area, (**b**) mangrove no change area, (**c**) mangrove gain area, (**d**) mangrove loss area, and (**e**) fragmented mangrove area.

Figure 6a shows the mean LAI extracted from the MODIS 8-day product overlaid with all of the PW pixels found in the classified mangrove area. The yearly mean LAI was 3.18 ± 1.56 (mean ± standard deviation) with minimum and maximum values of 0.10 and 6.90, respectively, for the year 2000; while it was 3.46 ± 2.09 (mean ± standard deviation) with minimum of 0.10 and maximum of 6.60 for the year 2019. Figure 6b shows that the values extracted from the PW pixels on the unchanged mangrove area also showed an increasing trend, increasing from 3.03 to 3.55 for the years 2000 and 2019, respectively. The increases of mean LAI for all PW and unchanged areas indicate that the mangrove leaf area has grown over the years. Previous analysis (Section 3.1 and Figure 3c) show that relatively less mangrove gain was noticed when compared to area of mangrove loss. Based on the values in Figure 6c,d, the mean LAI extracted from the mangrove gain increased from 2.81 to 3.51, while the mangrove loss area recorded mean LAI decrease from 2.99 to 2.62, respectively. The loss of native mangrove trees, which have dense leaves, to oil palm, which have a less dense canopy, may be the main reason for the reduced LAI values. The mean LAI values were found to correspond to the growth or loss of the mangrove in the study area. Note that the standard deviation values of the LAI are relatively high, which might be due to the changes in LAI values but are also dependent on species composition, forest structure, and vegetative phenology, which are all intrinsically related to temperature and rainfall, and this may need to be examined further.

GPP serves as one of the most important parameters in global carbon analysis, ecosystem assessment, and monitoring of environmental changes [56,57]. To date, studies relating to mangrove ecosystem processes (including Net Primary Production) have been few, and the majority of them depend on ground-based flux data (i.e., FLUXNET), which are restricted to certain locations [21,58,59]. Introducing remote sensing products such as MODIS GPP could be a solution that will help in improving and extending the study of mangrove carbon processes at a larger scale. Figure 7 shows the GPP values extracted in this study based on the five cases. For all PW pixels (Figure 7a), a reduction in mean GPP from 6.82 g C m^{-2} to 5.49 g C m^{-2} is noticed. This reduction could be due to the lower GPP contributions by the newly gained mangroves in the PW pixels, which may have lowered the productivity of the whole area. Figure 7b shows the GPP values of the area with no change. The mean GPP of these pixels increased from 5.81 g C m^{-2} to 6.73 g C m^{-2}. The increase of mean GPP over the no change area shows the great ability of mangrove forests to assimilate CO_2. Over the mangrove gain areas, the mean GPP shows an increasing trend from 1.88 g C m^{-2} to 2.78 g C m^{-2}. Most of the mangrove gained areas were found to be mangroves regrown from abandoned aquaculture ponds. Recent studies [60,61] confirmed that the regrowth of the *Rhizophoraceae* family on the abandoned ponds is feasible, thus supporting the increase of GPP over the gained areas. For the mangrove loss area, mean GPP increased only slightly from 6.38 g C m^{-2} to 6.68 g C m^{-2}. Based on the classification results, most of the mangrove losses are due to transformation of the mangrove areas to oil palm cultivation. The fast-growing oil palm cultivation surrounding the mangrove areas, which are mixed in the larger MODIS pixels (500m), could be the reason for increasing GPP despite the loss of mangrove.

The effect of fragmentation on CO_2 assimilation and biomass accumulation by ecosystems is dependent on the size and location of the fragmented patches [62]. To analyze the impact of mangrove fragmentation on the LAI and GPP, we selected the patch of mangrove area with the largest loss (6883.06 ha in 2000), with a loss 1131.68 ha (16.54%) of mangroves and which had fragmented into 60 smaller patches between 2019 and 2000. The mean LAI and GPP values extracted from the MODIS products (Figures 6e and 7e) show a slight increase from 3.11 to 3.41 and a large change from 4.65 g C m^{-2} to 6.37 g C m^{-2}, respectively. Of the fragmented patches, six are of a very large size—total of 5704.50 ha (82.88% of the total area) compared to the MPS (26.63—Table 6), one patch is 17.27 ha, and the remaining 53 patches are smaller than 9 ha. According to Laurance et al. [63], the interior of relatively large fragments tends to change less following fragmentation than it does at the forest margins. This may be due to the edge effects, such as altered microclimate, which do not penetrate into the core habitat of the forest interior. Peh, Lin, Luke, Foster, and Turner [18] concluded that productivity is stable in fragments of more than 9 ha. Therefore, despite serious fragmentation over this selected patch, the increases of LAI and GPP may be supported by the impact of fragmentation on productivity, and the survival rate of trees/plants may be higher in relatively smaller patches because the abundance of solar radiation along the edges of the patches may increase the productivity [63]. In smaller patches (<9 ha), the microclimate (temperature and rainfall) may also be changed, which may affect the productivity.

4. Discussion

4.1. Mangrove Classification and Loss

SVM classification has been widely explored on different types of remote sensing data and at global, regional, and local scales with high accuracy [64]. Mountrakis et al. [65] reviewed land cover classification using different techniques and concluded that SVM is superior compared to other classification techniques. Classification works related to mangrove area identification and mapping using SVM classifiers showed high accuracy [41,66,67]. SVM proved to be one of the best classifiers where high accuracy (~90%) was achieved based on the Landsat 7 ETM+ pan-sharpened image and the Sentinel-2A image. While several global mangrove maps are available at 30 m [34,53,68,69], the use of higher spatial resolution satellite images (15 m and 10 m) in this study aimed to increase the accuracy of mangrove extent mapping. The classification result showed clear delineation between mangrove, water bodies, and other land use and land cover, providing high quality mangrove cover in 2000 and 2019 for mangrove change identification. In a previous study, Kanniah, Sheikhi, Cracknell, Goh, Tan, Ho, and Rasli [32] performed land use classification of the IM region from 1989 to 2014 using multiple Landsat images. They showed that, during the 21st century, urban area significantly increased, converting from oil palm, rubber, mangrove, and forest.

The spatial pattern of mangrove changes in IM shows a significant loss in the southwestern part, where the three Ramsar sites are located (Figure 3c). The Ramsar sites have been protected since 2003; thus, the loss may be due to a high input of sediments generated via construction and land reclamation activities around the Sungai Pulai estuary [70], where the development of the Tanjung Bin Power Plant, the Forest City, and the expansion area of the Tanjung Pelepas Port (Figure 1) are located. Reports including the Marine Environment Protection Committee (MEPC) [71] stated that the increasing traffic of ships and vessels in this region generates wave currents that can destroy mangroves in coastal areas [72].

Within the mangrove region of IM, 40 sites with aquaculture activities were detected from the satellite images. Aquaculture conversion from mangrove forests is very popular, as can be seen in Indonesia, where there has been a 40% mangrove loss over the past three decades [73]. Figure 4b shows the locations of these aquaculture sites. Most of the aquaculture sites are located on the "others" land use, at the boundary of the mangrove area. Thus, we can conclude that aquaculture activities contributed greatly to the loss of mangrove areas during the study period. At the eastern side of IM (Figure 4b), more aquaculture sites were found compared to the western site (where the Ramsar sites are located). One of the reasons for this could be that the Ramsar sites are protected from these conversion activities while, on the eastern side, where the Johor River flows, the accessibility to water from the Johor River and the richness of nutrients at the site have encouraged the conversion of mangrove areas to aquaculture use.

4.2. Fragmentation

Based on all the fragmentation indices, the fragmentation of mangrove in the IM region indicates a serious impact on the ecosystem as the fragmented mangrove forests become ecologically and hydrologically isolated. A reduction of Class Area (CA) and Total Landscape Area (TLA) can decrease the movement zone of animals and discourage flora and fauna succession within or between fragments. An increased Number of Patches (NP) and decreased mean patch size (MPS) mean that increasingly isolated mangrove patches lead to a loss of connectivity between patches, thus prolonging the migration of animals to surrounding habitats (due to the decreased existing habitat size or increased species diversity), which is one of the factors that stimulates species extinction [74]. The fragmentation analysis of the edge and shape indices reveal that human-induced disturbance is significantly influencing the mangrove forests. The decreased MSI value showed a reduction of shape complexity, which could be due to the transformation of mangrove areas (irregular shaped) into regularly shaped/spaced patches of urban, agricultural, and aquaculture uses. The changes of edge and shape complexity could ease human access to the fragmented patches and exploit the remaining mangrove forests. This could accelerate the loss and fragmentation of mangrove areas in the future. Fragmentation analysis can support decision-making on the conservation of mangrove area. The impact of fragmentation to the mangrove ecosystem should be the focus of future study. Identification of the vulnerable mangrove patches and/or areas for immediate actions and further regrowth/preservation plans must be implemented to ensure the longevity of mangrove forests in the IM, as well as in Malaysia as a whole.

4.3. Mangrove LAI and GPP

The measurement of leaf area and biomass of mangrove at different scales (local, regional, and global) is very scarce. Traditional destructive methods require a high demand of manpower, cost, and time, and the allometric equations generated for biomass calculation are usually site specific [75]. The majority of the scientific data collection based on flux tower has only built on inland forests, and this kind of data is less available for wetland forests [76]. Therefore, we propose the initiative to relate MODIS LAI and GPP to show their capability of detecting leaf area depletion and GPP rate changes of a selected area with active mangrove loss and fragmentation.

The LAI and GPP values extracted from MODIS products showed reliable values when compared to previous studies. Clough et al. [77] reported the LAI of the main mangrove species (*Rhizophora apiculata*) in Malaysia to be in the range of 3.10–5.10, which is within the range reported in IM (mean LAI from All PW pixels, 3.18–3.46, respectively in 2000 and 2019). In another study, Shrestha et al. [78] showed MODIS GPP to have a good match with the modelled GPP of a mangrove area in India, which can be useful for carbon assimilation estimation. The mean GPP of mangroves in the IM region is 5.49 g C m^{-2} (based on 2019 data), and this value is close to the GPP value of the tropical forests in Asia and the Americas, which are in the range of 8.22 and 10.96 g C m^{-2}. Nevertheless, it should be noted that the sparse resolution of MODIS products (500 m) could have a mixing pixel issue, where the contribution of oil palm plantation can affect the accuracy of the LAI and GPP. Thus, data with higher spatial resolution can improve future analysis.

Mangrove forests serve multiple functions that are significant in terms of ecology and economy, and these functions will be lost or reduced due to mangrove fragmentation. The unique forms and rich biodiversity of mangrove forests can provide tourism and recreational activities, including wildlife-watching and fishing [79]. Mangrove forests can be transformed as habitats and nursery grounds for a variety of flora and fauna [57,80], while mangrove trees are also functional as a source of food, fuel, and building materials [81]. Mangrove forests are also effective at protecting shorelines, thus preventing soil erosion, shoreline depletion, and floods [82]. Recent studies have revealed the potential of mangrove ecosystems in sequestrating CO_2 from the atmosphere, thereby reducing the impact of global warming and climate change [21,83]. The complex structure of mangroves with trunk, branches, leaves, various types of roots, and soil enable the ecosystem to store a high volume of carbon (450 Mg ha^{-1}) [83]. Mangrove forests have an equal global averaged biomass (247 t ha^{-1}) to humid evergreen tropical forests [57,84]. Significant deforestation and degradation of mangrove forests also leads to the loss of a relatively large carbon pool [9,85].

5. Conclusions

Mangrove forest loss has become a significant environmental issue over the past few decades, mainly due to the disruption of their ecosystem functions and services. Although various mangrove conservation and preservation actions have been initiated worldwide, mangrove losses and fragmentation are still ongoing in order to fulfil the insatiable needs of human beings. In Iskandar Malaysia, mass development has been ongoing and has destroyed the mangrove cover and fragmented the landscape. Such changes have been shown in this study to affect the areal extent of mangroves, their leaf area, and thereby the potential of the ecosystem to assimilate and accumulate carbon dioxide from the atmosphere. In the last two decades, a net loss of 2907 ha of mangrove has been estimated in this study using freely available medium-high resolution remote sensing data. It is noteworthy that the annual loss rate of 1.12%, is higher than the world's mangrove loss rate, and the study identified urbanization, oil palm plantations, and aquaculture activities as the main reasons for the land conversions in this region. Land cover changes of mangroves involved a reduction in the mean patch size from 105 ha to 27 ha and an increase in the number of mangrove patches from 130 to 402. Other fragmentation indices, such as edge and shape complexity, also showed increased values. The impact of fragmentation on mangroves' primary productivity showed an increase of 37%, which could be related to the smaller patches (< 9 ha) that can allow more sunlight to penetrate the edges of the patches and hence increase the productivity. Thus, it can be concluded that the impact of fragmentation on mangrove productivity is also dependent on the fragmented patch characteristics. Despite most of the mangrove forests in IM being declared as Ramsar Sites, the land use change near these protected forests could aggravate other natural forms of destruction such as soil erosion and tidal changes, which could further affect small patches of mangroves. Nevertheless, the regrowth of mangrove trees has also been seen scattered over the IM region, which could be promoted in the immediate future. Accuracy in the monitoring of the LAI and GPP can assist in precision planting or rehabilitation of species and habitat. It can also be used as a powerful response tool for resource managers. This study introduced a novel approach, combining the mangrove classification results with fragmentation analysis and mangrove LAI and GPP data to reveal the impact of fragmentation on mangrove productivity in a region that hosts important mangrove (Ramsar) sites in Malaysia. Such results are expected to provide standard guidelines for a stronger policy formulation to protect the remaining mangroves in the region and in other mangrove areas. Further studies on how to prevent mangrove loss, how to reconnect the fragmented mangrove forests, and how to improve the carbon stock estimation to prove the value of mangrove to the ecosystem processes are equally important, now and in the future.

Author Contributions: Conceptualization, K.D.K.; formal analysis, K.D.K. and C.S.K.; data curation, K.D.K.; writing—original draft preparation, K.D.K. and C.S.K., writing—review and editing, K.D.K., C.S.K., S.S. and A.A.A.; supervision, K.D.K.; funding acquisition, K.D.K. All authors have read and agreed to the published version of the manuscript.

Funding: This research and the APC was funded by Universiti Teknologi Malaysia, grant numbers Q.J130000.2452.08G51, Q.J130000.3052.02M01, and Q.J130000.3052.02M11.

Institutional Review Board Statement: Not applicable.

Informed Consent Statement: Not applicable.

Data Availability Statement: The Landsat 7 ETM+ satellite image was downloaded from the GloVis website (https://glovis.usgs.gov/ (accessed on 15 March 2021)); the Sentinel-2A satellite image was downloaded from the Copernicus Open Access Hub (https://scihub.copernicus.eu/ (accessed on 15 March 2021)); and the MODIS LAI, GPP and Land cover type products were downloaded from the NASA Earthdata website (https://earthdata.nasa.gov/ (accessed on 15 March 2021)).

Acknowledgments: We acknowledge the Ministry of Education, Malaysia and Universiti Teknologi Malaysia (through research grants Q.J130000.2452.08G51, Q.J130000.3052.02M01, and Q.J130000.3052.02M11) for providing funding to conduct the study. We thank Nazarin Ezzaty and Fateen Nabilah for their help in downloading and processing the satellite data and the reviewers for their constructive comments and suggestions to improve the quality of this work.

Conflicts of Interest: The authors declare no conflict of interest.

Appendix A

Table A1. Characteristics of land cover types classified in the study area.

Land Cover Type	Characteristics
Forest	Primary and secondary forests area with high density of natural trees
Mangrove	Mangrove forests
Oil Palm	Oil palm trees at different ages
Rubber	Rubber trees
Urban	Urban area including residential, road networks, industrial and buildings
Water Bodies	Water surfaces including rivers and lakes
Others	Orchards, shrubs, bush, abandoned lands, aquaculture farms, etc.

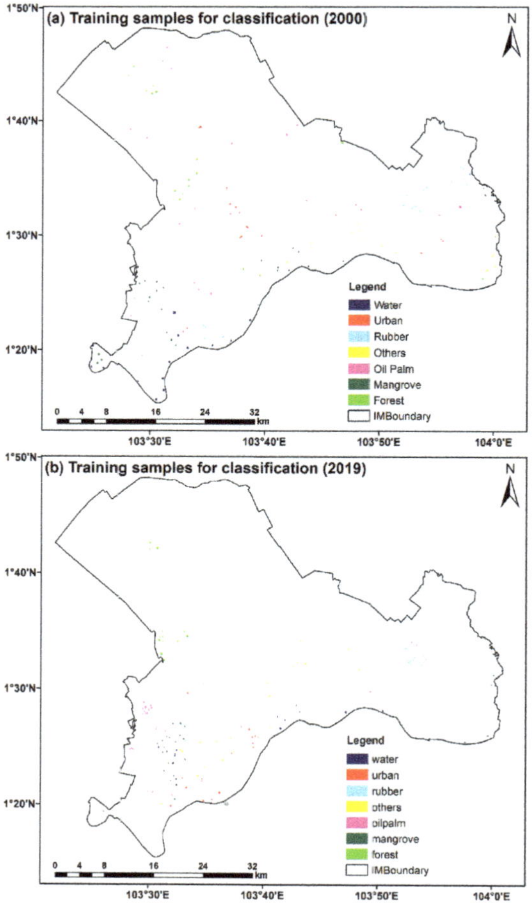

Figure A1. Training samples used for the SVM classification: (**a**) year 2000 and (**b**) year 2019.

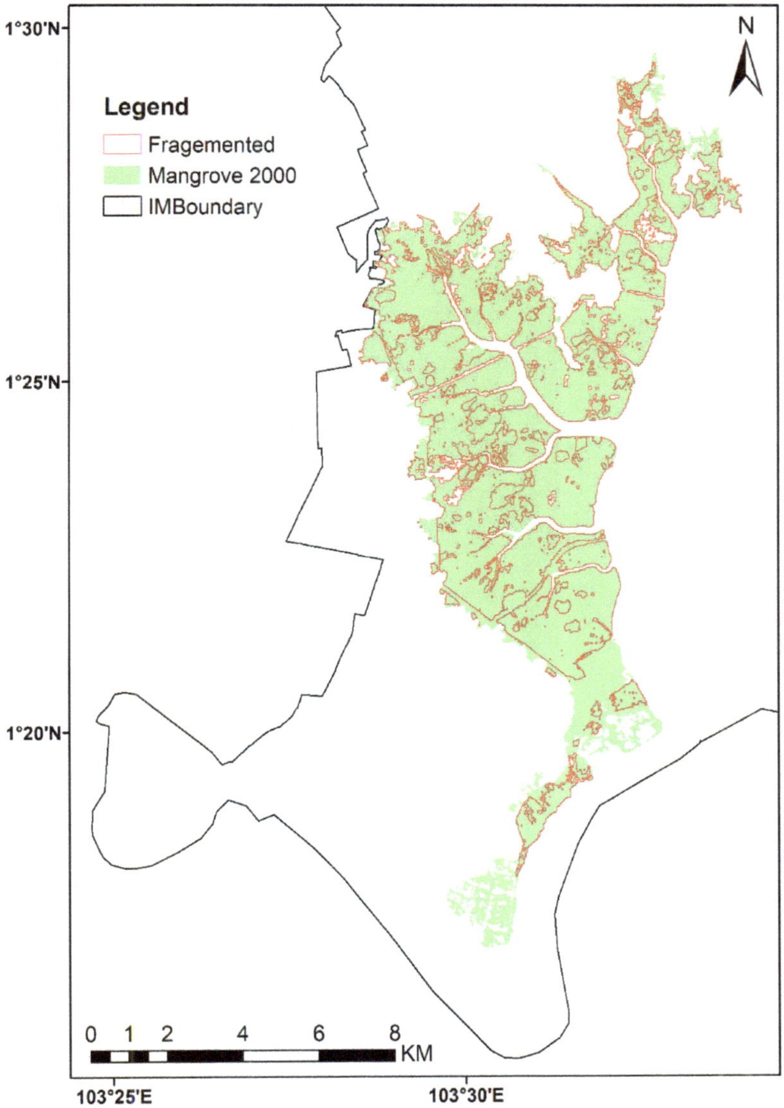

Figure A2. Selected area to show the fragmentation of the study area.

References

1. Bryan-Brown, D.N.; Connolly, R.M.; Richards, D.R.; Adame, F.; Friess, D.A.; Brown, C.J. Global trends in mangrove forest fragmentation. *Sci. Rep.* **2020**, *10*, 1–8. [CrossRef]
2. Herbeck, L.S.; Krumme, U.; Andersen, T.J.; Jennerjahn, T.C. Decadal trends in mangrove and pond aquaculture cover on Hainan (China) since 1966: Mangrove loss, fragmentation and associated biogeochemical changes. *Estuar. Coast. Shelf Sci.* **2020**, *233*, 106531. [CrossRef]
3. Ponnampalam, L.S.; Izmal, J.H.F.; Adulyanukosol, K.; Ooi, J.L.S.; Reynolds, J.E. Aligning conservation and research priorities for proactive species and habitat management: The case of dugongs Dugong dugon in Johor, Malaysia. *Oryx* **2014**, *49*, 743–749. [CrossRef]
4. Reyes-Arroyo, N.; Camacho-Valdez, V.; Saenz-Arroyo, A.; Infante-Mata, D. Socio-cultural analysis of ecosystem services provided by mangroves in La Encrucijada Biosphere Reserve, southeastern Mexico. *Local Environ.* **2021**, *26*, 86–109. [CrossRef]
5. Alemu, I.J.B.; Richards, D.R.; Gaw, L.Y.-F.; Masoudi, M.; Nathan, Y.; Friess, D.A. Identifying spatial patterns and interactions among multiple ecosystem services in an urban mangrove landscape. *Ecol. Indic.* **2021**, *121*, 107042. [CrossRef]

6. Seddon, N.; Daniels, E.; Davis, R.; Chausson, A.; Harris, R.; Hou-Jones, X.; Huq, S.; Kapos, V.; Mace, G.M.; Rizvi, A.R.; et al. Global recognition of the importance of nature-based solutions to the impacts of climate change. *Glob. Sustain.* **2020**, *3*, 1–12. [CrossRef]
7. Inácio, M.; Karnauskaitė, D.; Mikša, K.; Gomes, E.; Kalinauskas, M.; Pereira, P. *Nature-Based Solutions to Mitigate Coastal Floods and Associated Socioecological Impacts*; Metzler, J.B., Ed.; Springer: Berlin/Heidelberg, Germany, 2020; pp. 1–24.
8. Dahdouh-Guebas, F.; Ajonina, G.N.; Amir, A.A.; Andradi-Brown, D.A.; Aziz, I.; Balke, T.; Barbier, E.B.; Cannicci, S.; Cragg, S.M.; Cunha-Lignon, M.; et al. Public Perceptions of Mangrove Forests Matter for Their Conservation. *Front. Mar. Sci.* **2020**, *7*, 901. [CrossRef]
9. Sharma, S.; MacKenzie, R.A.; Tieng, T.; Soben, K.; Tulyasuwan, N.; Resanond, A.; Blate, G.; Litton, C.M. The impacts of degradation, deforestation and restoration on mangrove ecosystem carbon stocks across Cambodia. *Sci. Total. Environ.* **2020**, *706*, 135416. [CrossRef]
10. Sasmito, S.D.; Sillanpää, M.; Hayes, M.A.; Bachri, S.; Saragi-Sasmito, M.F.; Sidik, F.; Hanggara, B.B.; Mofu, W.Y.; Rumbiak, V.I.; Hendri; et al. Mangrove blue carbon stocks and dynamics are controlled by hydrogeomorphic settings and land-use change. *Glob. Chang. Biol.* **2020**, *26*, 3028–3039. [CrossRef]
11. Zeng, Y.; Friess, D.A.; Sarira, T.V.; Siman, K.; Koh, L.P. Global potential and limits of mangrove blue carbon for climate change mitigation. *Curr. Biol.* **2021**, in press. [CrossRef]
12. Sánchez-Núñez, D.A.; Pineda, J.E.M.; Osorio, A.F. From local-to global-scale control factors of wave attenuation in mangrove environments and the role of indirect mangrove wave attenuation. *Estuar. Coast. Shelf Sci.* **2020**, *245*, 106926. [CrossRef]
13. Kibler, K.M.; Kitsikoudis, V.; Donnelly, M.; Spiering, D.W.; Walters, L. Flow–Vegetation Interaction in a Living Shoreline Restoration and Potential Effect to Mangrove Recruitment. *Sustainability* **2019**, *11*, 3215. [CrossRef]
14. Friess, D.A.; Yando, E.S.; Abuchahla, G.M.; Adams, J.B.; Cannicci, S.; Canty, S.W.; Cavanaugh, K.C.; Connolly, R.M.; Cormier, N.; Dahdouh-Guebas, F.; et al. Mangroves give cause for conservation optimism, for now. *Curr. Biol.* **2020**, *30*, R153–R154. [CrossRef] [PubMed]
15. Worthington, T.A.; Andradi-Brown, D.A.; Bhargava, R.; Buelow, C.; Bunting, P.; Duncan, C.; Fatoyinbo, L.; Friess, D.A.; Goldberg, L.; Hilarides, L.; et al. Harnessing Big Data to Support the Conservation and Rehabilitation of Mangrove Forests Globally. *One Earth* **2020**, *2*, 429–443. [CrossRef]
16. Jacobson, A.P.; Riggio, J.; Tait, A.M.; Baillie, J.E.M. Global areas of low human impact ('Low Impact Areas') and fragmentation of the natural world. *Sci. Rep.* **2019**, *9*, 1–13. [CrossRef] [PubMed]
17. Rogan, J.E.; Lacher, T.E. Impacts of Habitat Loss and Fragmentation on Terrestrial Biodiversity. In *Reference Module in Earth Systems and Environmental Sciences*; Elsevier: Amsterdam, The Netherlands, 2018.
18. Peh, K.S.H.; Lin, Y.; Luke, S.H.; Foster, W.A.; Turner, E.C. Forest fragmentation and ecosystem function. *Glob. For. Fragm.* **2014**, 96–114. [CrossRef]
19. Chapin, F.S.; Woodwell, G.M.; Randerson, J.T.; Rastetter, E.B.; Lovett, G.M.; Baldocchi, D.D.; Clark, D.A.; Harmon, M.E.; Schimel, D.S.; Valentini, R.; et al. Reconciling Carbon-cycle Concepts, Terminology, and Methods. *Ecosystems* **2006**, *9*, 1041–1050. [CrossRef]
20. Braghiere, R.K.; Quaife, T.; Black, E.; He, L.; Chen, J.M. Underestimation of Global Photosynthesis in Earth System Models Due to Representation of Vegetation Structure. *Glob. Biogeochem. Cycles* **2019**, *33*, 1358–1369. [CrossRef]
21. Wang, L.; Jia, M.; Yin, D.; Tian, J. A review of remote sensing for mangrove forests: 1956–2018. *Remote Sens. Environ.* **2019**, *231*, 111223. [CrossRef]
22. Wicaksono, P.; Danoedoro, P.; Hartono; Nehren, U. Mangrove biomass carbon stock mapping of the Karimunjawa Islands using multispectral remote sensing. *Int. J. Remote Sens.* **2015**, *37*, 26–52. [CrossRef]
23. Hamdan, O.; Khairunnisa, M.; Ammar, A.; Hasmadi, I.M.; Aziz, H.K. Mangrove carbon stock assessment by optical satellite imagery. *J. Trop. For. Sci.* **2013**, *25*, 554–565.
24. Pham, T.D.; Yokoya, N.; Bui, D.T.; Yoshino, K.; Friess, D.A. Remote Sensing Approaches for Monitoring Mangrove Species, Structure, and Biomass: Opportunities and Challenges. *Remote Sens.* **2019**, *11*, 230. [CrossRef]
25. Pham, T.D.; Xia, J.; Ha, N.T.; Bui, D.T.; Le, N.N.; Tekeuchi, W. A Review of Remote Sensing Approaches for Monitoring Blue Carbon Ecosystems: Mangroves, Seagrassesand Salt Marshes during 2010–2018. *Sensors* **2019**, *19*, 1933. [CrossRef]
26. Running, S.; Mu, Q.; Zhao, M. MOD17A2H MODIS/Terra Gross Primary Productivity 8-Day L4 Global 500m SIN Grid VNASA EOSDIS Land Processes DAAC. 2015. Available online: https://doi.org/10.5067/MODIS/MOD17A2H.006 (accessed on 15 March 2021).
27. Patil, V.; Singh, A.; Naik, N.; Unnikrishnan, S. Estimation of Mangrove Carbon Stocks by Applying Remote Sensing and GIS Techniques. *Wetlands* **2015**, *35*, 695–707. [CrossRef]
28. Ishii, T.; Tateda, Y.; Ishil, T. Leaf area index and biomass estimation for mangrove plantation in Thailand. In Proceedings of the IGARSS 2004, 2004 IEEE International Geoscience and Remote Sensing Symposium, Anchorage, AK, USA, 20–24 September 2004; pp. 2323–2326.
29. Seto, K.C.; Fragkias, M. Mangrove conversion and aquaculture development in Vietnam: A remote sensing-based approach for evaluating the Ramsar Convention on Wetlands. *Glob. Environ. Chang.* **2007**, *17*, 486–500. [CrossRef]
30. Liu, D.; Li, S.; Fu, D.; Shen, C. Remote sensing analysis of mangrove distribution and dynamics in Zhanjiang from 1991 to 2011. *J. Oceanol. Limnol.* **2018**, *36*, 1597–1603. [CrossRef]

31. Shapiro, A.C.; Trettin, C.C.; Küchly, H.; Alavinapanah, S.; Bandeira, S. The Mangroves of the Zambezi Delta: Increase in Extent Observed via Satellite from 1994 to 2013. *Remote Sens.* **2015**, *7*, 16504–16518. [CrossRef]
32. Kanniah, K.D.; Sheikhi, A.; Cracknell, A.P.; Goh, H.C.; Tan, K.P.; Ho, C.S.; Rasli, F.N. Satellite Images for Monitoring Mangrove Cover Changes in a Fast Growing Economic Region in Southern Peninsular Malaysia. *Remote Sens.* **2015**, *7*, 14360–14385. [CrossRef]
33. De Alban, J.D.T.; Jamaludin, J.; De Wen, D.W.; Than, M.M.; Webb, E.L. Improved estimates of mangrove cover and change reveal catastrophic deforestation in Myanmar. *Environ. Res. Lett.* **2020**, *15*, 034034. [CrossRef]
34. Hansen, M.C.; Potapov, P.V.; Moore, R.; Hancher, M.; Turubanova, S.A.; Tyukavina, A.; Thau, D.; Stehman, S.V.; Goetz, S.J.; Loveland, T.R.; et al. High-Resolution Global Maps of 21st-Century Forest Cover Change. *Science* **2013**, *342*, 850–853. [CrossRef]
35. Kuenzer, C.; Bluemel, A.; Gebhardt, S.; Quoc, T.V.; Dech, S. Remote Sensing of Mangrove Ecosystems: A Review. *Remote Sens.* **2011**, *3*, 878–928. [CrossRef]
36. Omar, H.; Husin, T.M.; Parlan, I. *Status of Mangroves in Malaysia*; Forest Research Institute Malaysia: Selangor, Malaysia, 2020.
37. Friedl, M.; Sulla-Menashe, D. MCD12C1 MODIS/Terra+Aqua Land Cover Type Yearly L3 Global 0.05Deg CMG V006. NASA EOSDIS Land Processes DAAC. 2015. Available online: https://doi.org/10.5067/MODIS/MCD12C1.006 (accessed on 15 March 2021).
38. Myneni, R.; Knyazikhin, T.Y.; Park, T. MOD15A2H MODIS/Terra Leaf Area Index/FPAR 8-Day L4 Global 500m SIN Grid VNASA EOSDIS Land Processes DAAC. 2015. Available online: https://doi.org/10.5067/MODIS/MOD15A2H.006 (accessed on 15 March 2021).
39. Congedo, L. Semi-automatic classification plugin documentation. *Release* **2016**, *4*, 29.
40. Yuan, H.; Dai, Y.; Xiao, Z.; Ji, D.; Shangguan, W. Reprocessing the MODIS Leaf Area Index products for land surface and climate modelling. *Remote Sens. Environ.* **2011**, *115*, 1171–1187. [CrossRef]
41. Heumann, B.W. An Object-Based Classification of Mangroves Using a Hybrid Decision Tree—Support Vector Machine Approach. *Remote Sens.* **2011**, *3*, 2440–2460. [CrossRef]
42. Yu, L.; Gong, P. Google Earth as a virtual globe tool for Earth science applications at the global scale: Progress and perspectives. *Int. J. Remote Sens.* **2011**, *33*, 3966–3986. [CrossRef]
43. Dorais, A.; Cardille, J. Strategies for Incorporating High-Resolution Google Earth Databases to Guide and Validate Classifications: Understanding Deforestation in Borneo. *Remote Sens.* **2011**, *3*, 1157–1176. [CrossRef]
44. Kanniah, K.D. Quantifying green cover change for sustainable urban planning: A case of Kuala Lumpur, Malaysia. *Urban. For. Urban. Green.* **2017**, *27*, 287–304. [CrossRef]
45. Congalton, R.G.; Green, K. *Assessing the Accuracy of Remotely Sensed Data: Principles and Practices*; CRC Press: Boca Raton, FL, USA, 2019.
46. McGarigal, K. *FRAGSTATS: Spatial Pattern Analysis Program for Quantifying Landscape Structure*; US Forest Service Pacific Northwest Research Station: Portland, OR, USA, 1995; Volume 351.
47. Chen, J.M.; Ju, W.; Ciais, P.; Viovy, N.; Liu, R.; Liu, Y.; Lu, X. Vegetation structural change since 1981 significantly enhanced the terrestrial carbon sink. *Nat. Commun.* **2019**, *10*, 1–7. [CrossRef] [PubMed]
48. Li, Q.; Lu, X.; Wang, Y.; Huang, X.; Cox, P.M.; Luo, Y. Leaf area index identified as a major source of variability in modeled CO2 fertilization. *Biogeosciences* **2018**, *15*, 6909–6925. [CrossRef]
49. Vitale, L.; Di Tommasi, P.; D'Urso, G.; Magliulo, V. The response of ecosystem carbon fluxes to LAI and environmental drivers in a maize crop grown in two contrasting seasons. *Int. J. Biometeorol.* **2015**, *60*, 411–420. [CrossRef]
50. Amir, A.A.; Duke, N.C. Distinct characteristics of canopy gaps in the subtropical mangroves of Moreton Bay, Australia. *Estuar. Coast. Shelf Sci.* **2019**, *222*, 66–80. [CrossRef]
51. Kamal, M.; Phinn, S.; Johansen, K. Assessment of multi-resolution image data for mangrove leaf area index mapping. *Remote Sens. Environ.* **2016**, *176*, 242–254. [CrossRef]
52. Tian, Y.; Woodcock, C.E.; Wang, Y.; Privette, J.L.; Shabanov, N.V.; Zhou, L.; Zhang, Y.; Buermann, W.; Dong, J.; Veikkanen, B. Multiscale analysis and validation of the MODIS LAI productII. Sampling strategy. *Remote Sens. Environ.* **2002**, *83*, 431–441. [CrossRef]
53. Hamilton, S.E.; Casey, D. Creation of a high spatio-temporal resolution global database of continuous mangrove forest cover for the 21st century (CGMFC-21). *Glob. Ecol. Biogeogr.* **2016**, *25*, 729–738. [CrossRef]
54. Ewers, R.M.; Didham, R.K. The Effect of Fragment Shape and Species' Sensitivity to Habitat Edges on Animal Population Size. *Conserv. Biol.* **2007**, *21*, 926–936. [CrossRef] [PubMed]
55. Vaz, E. Managing urban coastal areas through landscape metrics: An assessment of Mumbai's mangrove system. *Ocean Coast. Manag.* **2014**, *98*, 27–37. [CrossRef]
56. Beer, C.; Reichstein, M.; Tomelleri, E.; Ciais, P.; Jung, M.; Carvalhais, N.; Rödenbeck, C.; Arain, M.A.; Baldocchi, D.; Bonan, G.B.; et al. Terrestrial gross carbon dioxide uptake: Global distribution and covariation with climate. *Science* **2010**, *329*, 834–838. [CrossRef] [PubMed]
57. Alongi, D.M. Carbon Cycling and Storage in Mangrove Forests. *Annu. Rev. Mar. Sci.* **2014**, *6*, 195–219. [CrossRef] [PubMed]
58. Gnanamoorthy, P.; Selvam, V.; Burman, P.K.D.; Chakraborty, S.; Karipot, A.; Nagarajan, R.; Ramasubramanian, R.; Song, Q.; Zhang, Y.; Grace, J. Seasonal variations of net ecosystem (CO_2) exchange in the Indian tropical mangrove forest of Pichavaram. *Estuar. Coast. Shelf Sci.* **2020**, *243*, 106828. [CrossRef]

59. Li, Q.; Lu, W.; Chen, H.; Luo, Y.; Lin, G. Differential Responses of Net Ecosystem Exchange of Carbon Dioxide to Light and Temperature between Spring and Neap Tides in Subtropical Mangrove Forests. *Sci. World J.* **2014**, *2014*, 1–11. [CrossRef]
60. Proisy, C.; Viennois, G.; Sidik, F.; Andayani, A.; Enright, J.A.; Guitet, S.; Gusmawati, N.; Lemonnier, H.; Muthusankar, G.; Olagoke, A.; et al. Monitoring mangrove forests after aquaculture abandonment using time series of very high spatial resolution satellite images: A case study from the Perancak estuary, Bali, Indonesia. *Mar. Pollut. Bull.* **2018**, *131*, 61–71. [CrossRef]
61. Bijsterveldt, C.E.J.V.; Wesenbeeck, B.V.; Wal, D.V.D.; Afiati, N.; Pribadi, R.; Brown, B.; Bouma, T. How to restore mangroves for greenbelt creation along eroding coasts with abandoned aquaculture ponds. *Estuar. Coast. Shelf Sci.* **2020**, *235*, 1–13. [CrossRef]
62. Collinge, S.K. Ecological consequences of habitat fragmentation: Implications for landscape architecture and planning. *Landsc. Urban. Plan.* **1996**, *36*, 59–77. [CrossRef]
63. Laurance, W.F.; Goosem, M.; Laurance, S.G. Impacts of roads and linear clearings on tropical forests. *Trends Ecol. Evol.* **2009**, *24*, 659–669. [CrossRef] [PubMed]
64. Sheykhmousa, M.; Mahdianpari, M.; Ghanbari, H.; Mohammadimanesh, F.; Ghamisi, P.; Homayouni, S. Support Vector Machine Versus Random Forest for Remote Sensing Image Classification: A Meta-Analysis and Systematic Review. *IEEE J. Sel. Top. Appl. Earth Obs. Remote Sens.* **2020**, *13*, 6308–6325. [CrossRef]
65. Mountrakis, G.; Im, J.; Ogole, C. Support vector machines in remote sensing: A review. *ISPRS J. Photogramm. Remote Sens.* **2011**, *66*, 247–259. [CrossRef]
66. Huang, X.; Zhang, L.; Wang, L. Evaluation of Morphological Texture Features for Mangrove Forest Mapping and Species Discrimination Using Multispectral IKONOS Imagery. *IEEE Geosci. Remote Sens. Lett.* **2009**, *6*, 393–397. [CrossRef]
67. Heumann, B.W. Satellite remote sensing of mangrove forests: Recent advances and future opportunities. *Prog. Phys. Geogr. Earth Environ.* **2011**, *35*, 87–108. [CrossRef]
68. Thomas, N.; Bunting, P.; Lucas, R.; Hardy, A.; Rosenqvist, A.; Fatoyinbo, T. Mapping Mangrove Extent and Change: A Globally Applicable Approach. *Remote Sens.* **2018**, *10*, 1466. [CrossRef]
69. Giri, C.; Ochieng, E.; Tieszen, L.L.; Zhu, Z.; Singh, A.; Loveland, T.; Masek, J.; Duke, N. Status and distribution of mangrove forests of the world using earth observation satellite data. *Glob. Ecol. Biogeogr.* **2011**, *20*, 154–159. [CrossRef]
70. Malaysia, S.A. *Impacts of Coastal Reclamation in Malaysia*; Sahabat Alam Malaysia: Penang, Malaysia, 2020.
71. MEPC. *Identification and Protection of Special Areas and Particularly Sensitive Sea Areas: Protection of Pulau Kukup (Kukup Island) and Tanjung Piai (Cape Piai)*; submitted by Malaysia, Singapore; Centre of International Law, National University of Singapore: Singapore, 28 April 2017.
72. McCreath, M. Burgeoning Practice of Southeast Asian States to Protect the Marine Environment from the Effects of International Shipping. *Asia-Pac. J. Ocean Law Policy* **2017**, *2*, 268–295. [CrossRef]
73. Murdiyarso, D.; Purbopuspito, J.; Kauffman, J.B.; Warren, M.W.; Sasmito, S.D.; Donato, D.C.; Manuri, S.; Krisnawati, H.; Taberima, S.; Kurnianto, S. The potential of Indonesian mangrove forests for global climate change mitigation. *Nat. Clim. Chang.* **2015**, *5*, 1089–1092. [CrossRef]
74. Haddad, N.M.; Brudvig, L.A.; Clobert, J.; Davies, K.F.; Gonzalez, A.; Holt, R.D.; Lovejoy, T.E.; Sexton, J.O.; Austin, M.P.; Collins, C.D.; et al. Habitat fragmentation and its lasting impact on Earth's ecosystems. *Sci. Adv.* **2015**, *1*, e1500052. [CrossRef] [PubMed]
75. Wang, D.; Wan, B.; Liu, J.; Su, Y.; Guo, Q.; Qiu, P.; Wu, X. Estimating aboveground biomass of the mangrove forests on northeast Hainan Island in China using an upscaling method from field plots, UAV-LiDAR data and Sentinel-2 imagery. *Int. J. Appl. Earth Obs. Geoinf.* **2020**, *85*, 101986. [CrossRef]
76. Rodda, S.R.; Thumaty, K.C.; Jha, C.S.; Dadhwal, V.K. Seasonal Variations of Carbon Dioxide, Water Vapor and Energy Fluxes in Tropical Indian Mangroves. *Forests* **2016**, *7*, 35. [CrossRef]
77. Clough, B.; Ong, J.; Gong, W. Estimating leaf area index and photosynthetic production in canopies of the mangrove Rhizophora apiculata. *Mar. Ecol. Prog. Ser.* **1997**, *159*, 285–292. [CrossRef]
78. Shrestha, S.; Miranda, I.; Kumar, A.; Pardo, M.L.E.; Dahal, S.; Rashid, T.; Remillard, C.; Mishra, D.R. Identifying and forecasting potential biophysical risk areas within a tropical mangrove ecosystem using multi-sensor data. *Int. J. Appl. Earth Obs. Geoinf.* **2019**, *74*, 281–294. [CrossRef]
79. Spalding, M.; Parrett, C.L. Global patterns in mangrove recreation and tourism. *Mar. Policy* **2019**, *110*, 103540. [CrossRef]
80. Glaser, M. Interrelations between mangrove ecosystem, local economy and social sustainability in Caeté Estuary, North Brazil. *Wetl. Ecol. Manag.* **2003**, *11*, 265–272. [CrossRef]
81. Giri, C.; Muhlhausen, J. Mangrove Forest Distributions and Dynamics in Madagascar (1975–2005). *Sensors* **2008**, *8*, 2104–2117. [CrossRef]
82. Das, S. Does mangrove plantation reduce coastal erosion? Assessment from the west coast of India. *Reg. Environ. Chang.* **2020**, *20*, 1–12. [CrossRef]
83. Ouyang, X.; Lee, S.Y. Improved estimates on global carbon stock and carbon pools in tidal wetlands. *Nat. Commun.* **2020**, *11*, 1–7. [CrossRef] [PubMed]
84. Kanniah, K.D.; Tan, K.P.; Cracknell, A.P.; Huete, A.R.; Idris, N.H.; Lau, A.M.S.; Rahman, M.Z.A.; Rasib, A.W.; Ahmad, A. Assessment of biophysical properties of Royal Belum tropical forest, Malaysia. *Singap. J. Trop. Geogr.* **2017**, *39*, 90–106. [CrossRef]
85. Richards, D.R.; Thompson, B.S.; Wijedasa, L. Quantifying net loss of global mangrove carbon stocks from 20 years of land cover change. *Nat. Commun.* **2020**, *11*, 1–7. [CrossRef] [PubMed]

remote sensing

MDPI

Article

Mangrove Ecosystem Mapping Using Sentinel-1 and Sentinel-2 Satellite Images and Random Forest Algorithm in Google Earth Engine

Arsalan Ghorbanian [1], Soheil Zaghian [1], Reza Mohammadi Asiyabi [2], Meisam Amani [3], Ali Mohammadzadeh [1] and Sadegh Jamali [4,*]

1 Department of Photogrammetry and Remote Sensing, Faculty of Geodesy and Geomatics Engineering, K. N. Toosi University of Technology, Tehran 19967-15433, Iran; a.ghorbanian@email.kntu.ac.ir (A.G.); soheil.zaghian@email.kntu.ac.ir (S.Z.); a_mohammadzadeh@kntu.ac.ir (A.M.)

2 Research Center for Spatial Information (CEOSpaceTech), University POLITEHNICA of Bucharest (UPB), Sector 1, 011061 Bucharest, Romania; reza.mohammadi@upb.ro

3 Wood Environment & Infrastructure Solutions, Ottawa, ON K2E 7L5, Canada; meisam.amani@woodplc.com

4 Department of Technology and Society, Faculty of Engineering, Lund University, P.O. Box 118, 221 00 Lund, Sweden

* Correspondence: sadegh.jamali@tft.lth.se

Abstract: Mangroves are among the most productive ecosystems in existence, with many ecological benefits. Therefore, generating accurate thematic maps from mangrove ecosystems is crucial for protecting, conserving, and reforestation planning for these valuable natural resources. In this paper, Sentinel-1 and Sentinel-2 satellite images were used in synergy to produce a detailed mangrove ecosystem map of the Hara protected area, Qeshm, Iran, at 10 m spatial resolution within the Google Earth Engine (GEE) cloud computing platform. In this regard, 86 Sentinel-1 and 41 Sentinel-2 data, acquired in 2019, were employed to generate seasonal optical and synthetic aperture radar (SAR) features. Afterward, seasonal features were inserted into a pixel-based random forest (RF) classifier, resulting in an accurate mangrove ecosystem map with average overall accuracy (OA) and Kappa coefficient (KC) of 93.23% and 0.92, respectively, wherein all classes (except aerial roots) achieved high producer and user accuracies of over 90%. Furthermore, comprehensive quantitative and qualitative assessments were performed to investigate the robustness of the proposed approach, and the accurate and stable results achieved through cross-validation and consistency checks confirmed its robustness and applicability. It was revealed that seasonal features and the integration of multi-source remote sensing data contributed towards obtaining a more reliable mangrove ecosystem map. The proposed approach relies on a straightforward yet effective workflow for mangrove ecosystem mapping, with a high rate of automation that can be easily implemented for frequent and precise mapping in other parts of the world. Overall, the proposed workflow can further improve the conservation and sustainable management of these valuable natural resources.

Keywords: mangrove ecosystem; random forest (RF); Google Earth Engine (GEE); Sentinel; synthetic aperture radar (SAR); optical; aerial roots

Citation: Ghorbanian, A.; Zaghian, S.; Asiyabi, R.M.; Amani, M.; Mohammadzadeh, A.; Jamali, S. Mangrove Ecosystem Mapping Using Sentinel-1 and Sentinel-2 Satellite Images and Random Forest Algorithm in Google Earth Engine. *Remote Sens.* **2021**, *13*, 2565. https://doi.org/10.3390/rs13132565

Academic Editor: Chandra Giri

Received: 19 May 2021
Accepted: 25 June 2021
Published: 30 June 2021

Publisher's Note: MDPI stays neutral with regard to jurisdictional claims in published maps and institutional affiliations.

1. Introduction

Mangroves are unique ecosystems that grow along tropical and sub-tropical coastlines. They provide many ecological benefits, including coastal protection, carbon sequestration, and waste and pollution assimilation [1–5]. Despite their significant environmental services, mangroves continue to disappear due to anthropogenic activities and climate change [6,7]. For instance, over the last five decades, approximately 20–30% of global mangroves have disappeared due to various phenomena, such as urban expansion, conversion to aquaculture, sea-level rise, and sediment alterations [8–13]. Therefore, accurate spatial and temporal mapping and monitoring of mangrove ecosystems are crucial for natural resource conservation and sustainable development goals [14].

In situ observations provide the most accurate information about mangroves. However, collecting in situ observations through field surveys is challenging, due to the limited accessibility of mangrove communities, as they are located in harsh and tidally inundated environments [15]. Therefore, remote sensing has been recognized as an efficient and cost-effective approach for mapping mangroves [16].

In this regard, multi-spectral [17–19], hyperspectral [20], light detection and ranging (LIDAR) [21], and synthetic aperture radar (SAR) [22] datasets have been utilized for mangrove studies. For instance, Manna and Raychaudhuri [23] examined the potential of Sentinel-2 images for mangrove mapping in Sunderban, India. Moreover, Zhu et al. [21] employed unmanned aerial vehicle (UAV) optical and LIDAR data to map mangrove-inundation patterns in Fujian, China. Furthermore, Kabiri [24] utilized RGB images, acquired by UAV, to classify the coastal ecosystem of Nayband Bay, Iran, into five classes of mangrove, shallow water, deep water, vegetation, and sand. The ortho-images and reference samples were fed to a maximum likelihood classifier to fulfill this task, producing a land cover map with an OA of 87.6%. Likewise, Toosi et al. [25] investigated the suitability of combining Sentinel-2 and WordView-2 images to produce a mangrove ecosystem map with eight classes. An upscaling approach in three stages was applied to produce a wall-to-wall land cover map based on a single-date Sentinel-2 image, resulting in the OA and Kappa coefficient (KC) of 65.5% and 0.63, respectively. The incorporation of single-date satellite imagery along with few reference samples were two limiting factors of their research, leading to moderate accuracy.

Most conducted studies on mangroves using remote sensing data have implemented traditional approaches on local computers, requiring manual acquiring, correcting, and processing of satellite images. Therefore, Cárdenas et al. [26] encouraged scholars to employ cloud-computing platforms, such as Google Earth Engine (GEE), making mangrove mapping more efficient [27,28]. This platform enables the automation of repetitive tasks (e.g., image acquisition, calibrating, and processing) and, thus, decreasing the dedicated time up to 60% [26].

GEE is a cloud platform enabling high-performance computing capabilities for geospatial data processing [29,30]. GEE hosts petabytes of satellite images, which have been efficiently employed in a variety of Earth science-related studies, such as land cover mapping [31–33], monitoring of volcanic thermal anomalies [34], monitoring of forest health [35,36], and wildfire damage assessment [37]. This cloud platform was also used in several mangrove studies, especially mangrove extent mapping [8,15,38–40]. For example, Mondal et al. [39] combined annual downscaled Sentinel-2 images and two machine learning algorithms of random forest (RF) and classification and regression trees (CART) within GEE for mangrove extent mapping along the coasts of Senegal and Gambia. The final mangrove extent maps of RF and CART had average OAs of about 93.44% and 92.18%, respectively. Furthermore, Sentinel-1 and Sentinel-2 data were employed in conjunction to produce a 10 m-resolution mangrove extent map of China [15]. In this regard, quantile synthesis features derived from both datasets were applied to an RF classifier using both pixel-based and object-based approaches. Finally, the classification results were improved by the constraint of tidal flats and visual manipulations. It was reported that the pixel-based approach obtained a higher OA, of approximately 95%, based on two-class (i.e., mangroves and non-mangroves) mapping.

Almost all of the mangrove-related studies within GEE, to the best of our knowledge, were dedicated to mangrove extent mapping [8,15,39–41] and, thus, the potential of this cloud platform for detailed mangrove ecosystem mapping was not fully explored. Additionally, other studies that were carried out to map a detailed mangrove ecosystem implemented traditional approaches on local computers using a few satellite observations [23,25]. Therefore, in this study, the GEE cloud computing platform was integrated with open-access satellite images to produce a detailed mangrove ecosystem map, advancing conservation and sustainable management of these valuable ecosystems through a higher automation level. The proposed approach uses an uncomplicated yet effective framework, allowing a consistent mangrove ecosystem mapping workflow for monitoring and assessment purposes. In this regard, the objectives of the present study are summarized as (1) incorporating multi-source images of Sentinel-1 and Sentinel-2 for accurate mangrove ecosystem mapping within GEE through an efficient and reproducible workflow, and (2) employing dense seasonal time-series observations to alleviate the tidal effect for more accurate mapping without any further tidal refinements. The robustness of the proposed method was then examined through cross-validation and consistency analyses. Furthermore, the contribution of multi-source remote sensing data and seasonal observations were also investigated.

2. Materials and Methods

2.1. Study Area

The study area covers the mangrove ecosystem in the Hara protected area of Qeshm Island, southern Iran (see Figure 1), which is under protection by different international conventions [42]. It is approximately centered at latitude and longitude of 26° 50′ N and 55° 44′ E, respectively, between the northwest estuaries of Qeshm Island and Hormozgan province. This area is the largest mangrove ecosystem in the Persian Gulf and Oman Sea coasts [43]. The study area includes grooved tidal channels, wherein tides are semi-diurnal. Furthermore, this region is affected by considerable tidal fluctuations, which necessitate taking the tidal effect into account for accurate mangrove mapping. There are generally two mangrove species, *Avicennia marina* and *Rhizophora mucronata*, in the Persian Gulf, the dominant of which is *Avicennia marina*, which grows in oxygen-poor sediments [44,45]. The local community uses this mangrove ecosystem for fishing, leaf-cutting, and regular boat journeys (i.e., tourism), which have negatively impacted the ecosystem, suggesting the necessity of frequent monitoring for conservation and natural resource management. Additionally, due to its proximity to the Strait of Hormuz, through which a large number of oil tankers pass, this area is also adversely impacted by oil leakage [46].

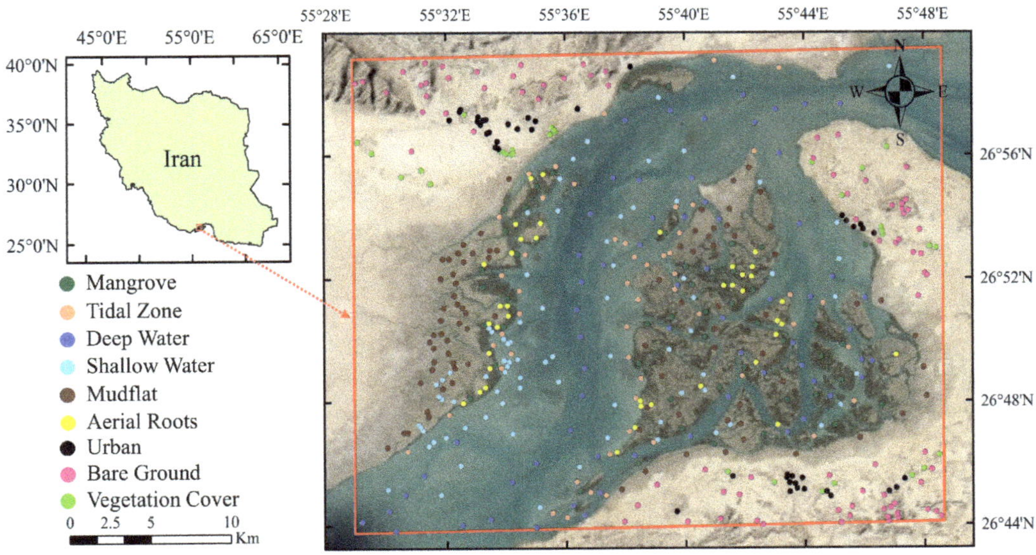

Figure 1. The geographical extent of the Hara protected area of Qeshm Island in southern Iran, along with the spatial distribution of reference samples over the study area (reference samples information is provided in Table 1).

Table 1. The number and area of training and test polygons for nine classes that were considered in this study.

ID	Class	Training Samples		Test Samples		Total	
		Polygon	Area (ha)	Polygon	Area (ha)	Polygon	Area (ha)
1	Mangrove	24	14.59	27	15.30	51	39.89
2	Tidal zone	30	17.04	22	13.38	52	30.42
3	Deep water	29	17.67	36	23.12	65	40.79
4	Shallow water	44	16.31	35	15.67	79	31.98
5	Mudflat	43	19.23	43	20.81	86	40.04
6	Aerial roots	20	10.01	20	9.05	40	19.06
7	Urban	18	7.65	24	9.82	42	17.47
8	Bare ground	40	17.61	41	18.41	81	36.20
9	Vegetation	17	5.11	16	4.82	33	9.93
	Total	265	125.22	264	130.38	529	529

2.2. Datasets

In this section, the datasets employed for mangrove ecosystem mapping are explained. First, a description of the collection and preparation of reference samples is provided, and then the Sentinel-1 and Sentinel-2 satellite datasets are described.

2.2.1. Reference Samples

In this study, precise visual interpretation depended on collecting reference samples from high-resolution satellite images available in ArcMap and Google Earth. Additionally, false-color composite satellite imagery and previous mangrove ecosystem maps were used. Homogenous sites were considered for reference sample collection to mitigate the challenge of mixed pixels by avoiding fragmented areas. In total, nine classes with adequate (i.e., in terms of land cover portion in the study area and possible complexity) reference samples and appropriate spatial distribution (i.e., distributed over the study area) were generated (see Figure 1). Reference samples were then randomly split into two groups of training (50%) and test (50%) samples. Random splitting leads to low bias in the performance of the final classification results [47]. However, the primary challenge of random sampling is the information leak between training and test samples [48]. In other words, random sampling at pixel unit causes the training and test datasets to include reference samples from the same polygons. This issue increases the spatial autocorrelation between training and test datasets, which affects the accuracy assessment results and decreases the generality of the classifier [49]. Therefore, to avoid this, the random splitting step was conducted at the polygon unit, which also spatially disjointed the training and test samples. It should be mentioned that the random splitting step was implemented ten times to enable applying a cross-validation procedure for performance evaluation, which can also prove the applicability and robustness of the proposed method for accurate and detailed mangrove ecosystem mapping [39]. Table 1 provides the number of training and test polygons (i.e., the average value in ten iterations) and their corresponding area. In total, 265 and 264 training and test polygons with an area of about 125.22 ha and 130.38 ha were generated, respectively.

2.2.2. Satellite Images

The time-series Sentinel-1 and Sentinel-2 satellite images were integrated to produce an accurate mangrove ecosystem map. Combining SAR and optical data allows the detection of different physical and spectral characteristics of land covers and, thus, their integration may achieve precise classification results [50–52]. Additionally, time-series data enables consideration of the water level fluctuations and tidal effects in the mangrove ecosystem, which can also increase the reliability of the classification results [53].

Sentinel-1 is a European SAR satellite, which acquires C-band data in dual-polarization in all-weather conditions with a 6-day temporal resolution. Level-1C ground range detected

(GRD) images with 10 m spatial resolution in both ascending and descending modes were employed [54,55]. In total, 86 Sentinel-1 scenes in the VV (vertical transmittance and receiving) and VH (vertical transmittance and horizontal receiving) polarizations, acquired in 2019 (i.e., from 1 January 2019 to 1 January 2020), were employed (see Table 2).

Sentinel-2 is also a European platform launched by the European Space Agency (ESA) and carries the MultiSpectral Instrument (MSI) sensor [54]. This sensor records the Earth's surface information in 13 spectral bands from visible to shortwave infrared (SWIR) regions, with different spatial resolutions ranging between 10 m and 60 m. In this study, only four bands of blue, green, red, and near infrared (NIR), which are captured with 10 m spatial resolution, were used. As it is acknowledged that satellite imagery with higher spatial resolutions improves mangrove ecosystem mapping, [25], here, only bands with 10 m spatial resolution were employed. This is mainly rooted in the fact that higher spatial resolution imagery improves the delineation of mangrove ecosystem classes, especially identifying mangrove patches with a small area or narrow shapes [56]. In total, 41 Sentinel-2 images, acquired in 2019 (i.e., from 1 January 2019 to 1 January 2020), were considered for mangrove ecosystem mapping (see Table 2).

Table 2. The number of Sentinel-1 and Sentinel-2 satellite images in each season.

Data	Season				Total	Date
	Spring	Summer	Autumn	Winter		
Sentinel-1	22	22	22	20	86	From 1 January 2019 to 1 January 2020
Sentinel-2	11	11	12	7	41	

3. Methodology

The schematic framework, providing an overview of the proposed approach, is presented in Figure 2. This section comprises three sub-sections, in which the satellite data preprocessing, classification workflow, and accuracy assessment procedure are explained in detail.

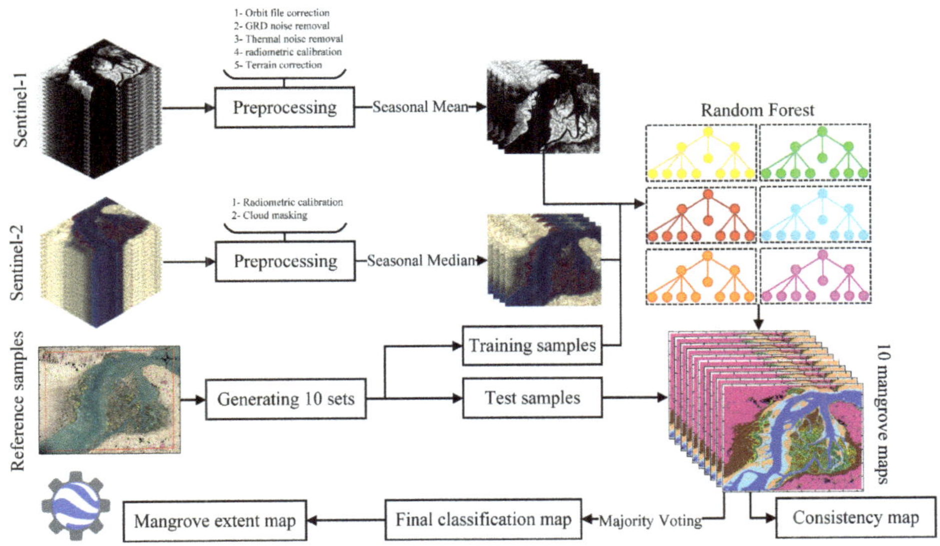

Figure 2. Framework of the proposed method for mangrove ecosystem mapping using the random forest (RF) algorithm and a combination of the Sentinel-1 and Sentinel-2 satellite images within the Google Earth Engine (GEE) platform.

3.1. Satellite Data Preprocessing

Sentinel-1 GRD data are available within GEE with the snippet of (Image Collection ID: *COPERNUCUS/S1_GRD*). They are generally ready-to-use data because several preprocessing steps are initially applied to them by the GEE developers. These data are already converted to the backscattering coefficient (σ°, dB) and are ortho-rectified. In particular, five preprocessing steps (1) applying orbit file correction, (2) GRD border noise removal, (3) thermal noise removal, (4) radiometric calibration, and (5) terrain correction were applied to each Sentinel-1 scene, while their detailed information is provided by the GEE developers (https://developers.google.com/earth-engine/guides/sentinel1 accessed on 29 June 2021). Afterward, all available Sentinel-1 scenes in 2019 were categorized based on the seasons of acquisition, followed by applying a *mean reducer* to generate seasonal time-series data. Downscaling time-series Sentinel-1 data by the *mean reducer* function produced seasonal datasets, which are less susceptible to image acquisition conditions, and reduced speckle noise [31].

Sentinel-2 top of atmosphere (TOA) reflectance data, which are available within GEE by the snippet (Image Collection ID: *COPERNICUS/S2*), were also used in this study. The TOA reflectance values were derived through radiometric calibration of raw data. Because of the importance of applying cloud masking, a filtering step was first implemented to remove Sentinel-2 scenes with a cloud cover percentage of higher than 5%. Subsequently, similar to Sentinel-1 data, a seasonal *median reducer* was applied to all remaining Sentinel-2 scenes, generating seasonal optical features for classification tasks. The *median reducer* function allows the production of cloud-free seasonal datasets, in which the noisy, very dark, and very bright pixels are also removed [31,57].

Ultimately, eight SAR features (4 VV + 4 VH) and 16 optical features (4 blue + 4 green + 4 red + 4 NIR bands) were used in synergy to produce mangrove ecosystem maps. It is well acknowledged that the quality of the classification results directly depends on the input features [58,59]. As such, the integration of multi-source (i.e., SAR + optical) data can increase the discriminative capability of the classifier [53]. Moreover, time-series satellite data can manifest the water level fluctuations in estuaries, such as mangrove ecosystems [53]. Consequently, seasonal datasets can mitigate the tidal effects in the study area and allow producing cloud-free mosaics.

3.2. Classification

Different classification algorithms have been employed for mangrove mapping using satellite images [23,39,60]. In this regard, choosing the most appropriate classifier, in addition to selecting discriminative features, is important, and directly affects the classification results. Among classifiers, random forest (RF) proved to be an efficient algorithm in mangrove mapping studies [15,44,61]. For instance, Toosi et al. [44] compared four frequently used non-parametric classifiers (i.e., RF, support vector machine (SVM) with linear and radial basis function kernels, and regularized discriminant analysis) for mangrove ecosystem mapping, and concluded that the RF classifier was superior.

RF is an authoritative non-parametric classifier, which employs the bootstrap aggregation technique to combine the classification results of various independent random decision trees and to predict the class label [62]. Each of these random decision trees is trained by a subset of training samples, called in-bag samples, and uses the remnant, called out-of-bag samples, for internal cross-validation. Later, their results are integrated to produce the final classification results. This enables the RF classifier to have a higher tolerance for noise, and also avoids overfitting possibilities [63,64]. Moreover, RF has proven its capability of handling high-dimensional data by resulting in promising maps [31].

In this study, a pixel-based RF classifier was implemented within GEE because of the high potential of the RF classifier for mangrove ecosystem mapping [15,44,61]. In this regard, the final image mosaics, comprising seasonal Sentinel-1 and Sentinel-2 data, were inserted into an RF classifier. Meanwhile, half of the reference samples were employed to train the RF classifier. The RF classifier has several tuning parameters that affect the training phase of the classification step and, thus, directly influence the classification results. Among these parameters, the number of trees and variables at each node are the most influential parameters [63], which were set at 100 and equal to the square root of the number of input features, respectively. It should be noted that these values were determined through several trial and error attempts based on computational efficacy, visual inspection of the classification results, and the average out-of-bag sampling error of the RF classifier in the training phase.

3.3. Accuracy Assessment

Any thematic map derived from remote sensing data should be subjected to a trustworthy accuracy assessment to ensure its quality and reliability [65]. Therefore, both visual interpretation and statistical approaches were conducted to evaluate the accuracy of the final mangrove ecosystem map. High-resolution satellite images available within Google Earth and ArcMap were employed for visual assessment of the classification results. In terms of statistical accuracy assessment, independent test samples were incorporated to generate the confusion matrix, and then other metrics, such as OA, KC, producer accuracy (PA), and user accuracy (UA), were derived. The PA represents how well a specific area can be mapped, whereas the UA is an indicator of how well the produced map represents what really exists in the study area [66]. Furthermore, the F-score, the harmonic average of recall and precision, was computed for each class, and then its macro-averaging was reported for the classification results [67]. As already mentioned in Section 2.2.1, the classification procedure was repeated ten times with different training and test sets to comprehensively assess the classification performance for mangrove ecosystem mapping through a cross-validation step.

4. Results

Figure 3 presents the resulting mangrove ecosystem map (i.e., only classes within the mangrove ecosystem) with 10 m spatial resolution using RF classifier and the integration of seasonal SAR and optical satellite data within GEE. It should be noted that this map was produced based on a majority voting step of ten classification results. The majority voting step allows the production of a more reliable and accurate mangrove ecosystem map [68]. This is rooted in the fact that combining decisions of several classification results can lead to better recognition results [69]. Visually, the thematic map had an acceptable accuracy, indicating the high potential of the proposed method for delineating different classes. Generally, mangrove areas were depicted precisely, and their surroundings were also classified as aerial roots, as these roots grow around mangroves. Furthermore, the middle parts of the tidal channels were correctly classified as deep water, while other parts located near the coastline areas (i.e., tidal zone) were successfully distinguished as shallow water. Additionally, along with their corresponding high-resolution satellite images, two zoomed-in areas are also provided in Figure 3 for better visual interpretation. For instance, Figure 3b,c illustrates a zoomed-in area of the mangrove ecosystem, in which the mangrove areas are successfully identified. Moreover, based on Figure 3d,e, the proposed workflow was highly capable of discriminating between different classes with acceptable precision. In particular, mangrove areas and their surroundings (i.e., aerial roots) were properly discriminated, and also shallow water, deep water, and tidal zone classes were delineated appropriately. Furthermore, other areas without specific covers, which are wet (due to the existence of narrow water channels), were also accurately classified as mudflat.

The consistency images of six classes within the mangrove ecosystem were also computed based on the classification results of ten iterations to determine the robustness of the proposed classification procedure and the consistency of each pixel. It should be noted that the consistency images were produced based on the number of label assignments of each pixel to different classes. In other words, the consistency images present how many times a pixel was assigned to a specific class. The consistency values ranged between one and ten, in which one means very low consistency, while ten means highly consistent. Figure 4 shows the consistency images of six classes within the mangrove ecosystem. It was observed that in almost all cases, the classifier was able to assign consistent labels to each pixel in ten iterations. In particular, mangrove pixels were consistently labeled as mangrove with a high consistency rate of over 85% (i.e., considering pixels with values of eight to ten). Additionally, the tidal zone, deep water, and mudflat classes achieved reasonable rates of stability of over 70%. In contrast, the classes of shallow water and aerial roots obtained lower stability rates. In general, most inconsistencies occurred at boundaries and fragmented locations, where mixed pixels exist, suggesting the necessity of satellite images with a higher spatial resolution for more accurate mapping. For instance, the aerial roots had the highest rate of inconsistency of about 13% (i.e., considering pixels with values of one). This is probably rooted in the fact that these roots grow around mangrove areas and in mudflat conditions, and thus, the classifier may encounter difficulties in distinguishing them. This was more serious in distinguishing aerial roots from mudflat. Furthermore, another source of inconsistency was found between the shallow water and deep water classes, which is in fact associated with their greater similarity. The inconsistency revealed the effect of different training sets and also implied that incorporating the majority voting step could increase the reliability of the final thematic map.

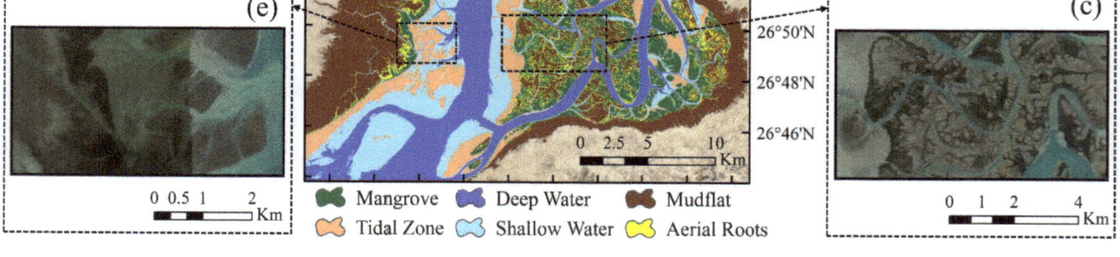

Figure 3. Mangrove ecosystem map of the study area produced using the random forest (RF) algorithm and a combination of the Sentinel-1 and Sentinel-2 satellite images within the Google Earth Engine (GEE) platform (**a**), two zoomed areas of the produced map (**b**,**d**), and their corresponding high-resolution satellite images for better visualization (**c**,**e**).

Figure 5 presents the average values of PA and UA along with their standard deviation computed for ten iterations of producing mangrove ecosystem maps. The proposed classification framework obtained high average OA, KC, and F-scores of 93.23% (±1.1), 0.92 (±0.012), and 0.92 (±0.011), respectively. Furthermore, almost all classes achieved high PAs and UAs of over 90%, demonstrating the high potential and applicability of the proposed method for detailed mangrove ecosystem mapping. It is evident that the mangrove classes obtained significant average PA and UA of about 94.4% and 94.5%, indicating the applicability of the implemented approach for accurate mangrove delineation. In fact, the results proved that the proposed approach can not only be used for detailed mangrove ecosystem mapping but also can be effectively employed for accurate mangrove extent mapping. The highest and lowest accuracies were related to deep water and aerial roots, respectively. The aerial roots acquired moderate average PA and UA of approximately 77.6% and 78.2%, respectively, illustrating the challenging task of delineating aerial roots with high accuracy, which is also in accordance with [25]. Furthermore, Figure 5 demonstrates that the proposed approach obtained stable PAs and UAs in all ten iterations, since their standard deviation values ranged between 1.6% and 3.8% for all classes, except for the aerial roots, for which the standard deviation of PAs and UAs were 7.4% and 5.9%, respectively.

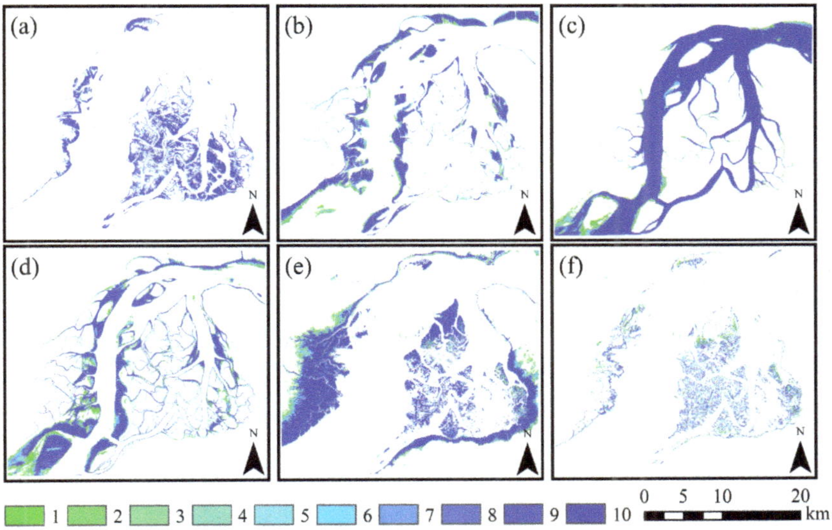

Figure 4. Consistency images of six classes (**a**) mangrove, (**b**) tidal zone, (**c**) deep water, (**d**) shallow water, (**e**) mudflat, and (**f**) aerial roots in the mangrove ecosystem. The consistency values ranged between one and ten, indicating very low and very high consistency, respectively.

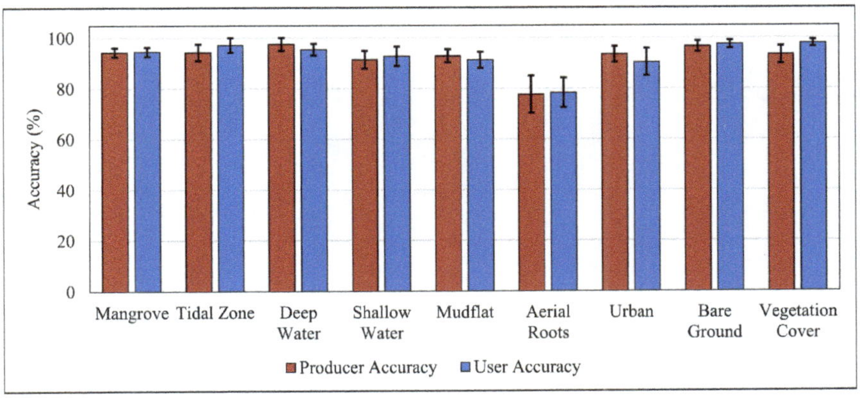

Figure 5. Average producer accuracy (PA) and user accuracy (UA) of different classes along with their standard deviation (black error bars) computed from ten iterations.

The confusion matrices of the classifications were also investigated, to provide an in-depth statistical assessment in each iteration (see Figure 6). Generally, the proposed approach obtained acceptable accuracy since less confusion happened between classes, and confusion matrices were almost diagonal. However, there are several spots where some confusions are observable in Figure 6, such as the confusion between deep water and shallow water, which occurred in all iterations. Additionally, relatively high confusions were also found between aerial roots and other classes of mangrove and mudflat, which were the reasons for the moderate accuracy of aerial roots.

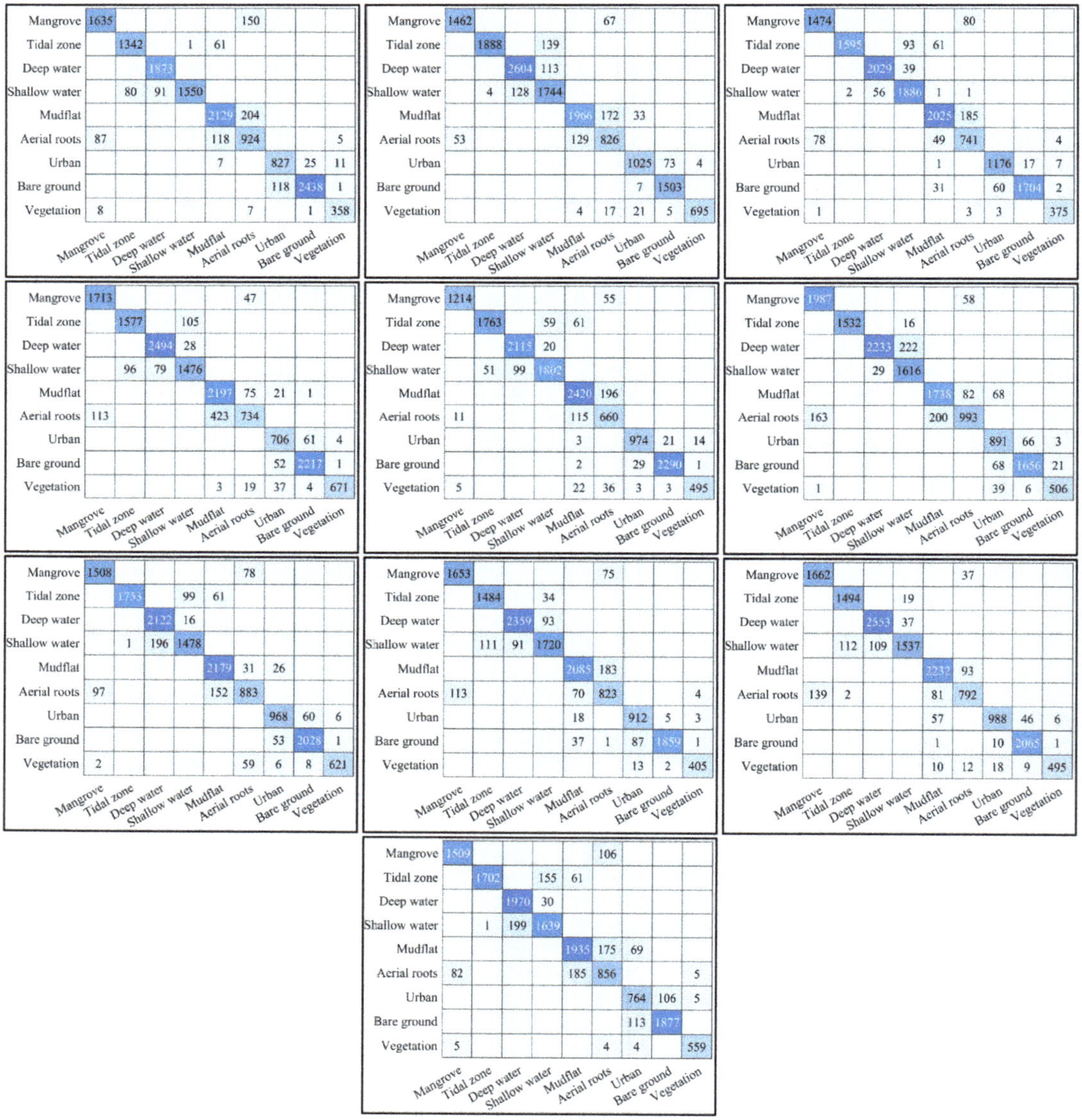

Figure 6. Confusion matrices of mangrove ecosystem maps for ten iterations.

An average confusion matrix (i.e., by averaging and rounding of the corresponding elements in 10 confusion matrices) is also provided in Table 3 to provide a comprehensive overview of all ten confusion matrices at a glance. Table 3 shows that the proposed approach obtained high average OA and KC values of 93.23% and 0.92, respectively, indicating the high potential of this approach for detailed mangrove ecosystem mapping. Moreover, the average omission error (OE) and commission error (CE) values were respectively 7.56% and 7.18%. The dominant confusion was associated with the aerial roots class, causing the highest OE and CE values of 23.06% and 21.92%, respectively, which are also in accordance with [25]. In particular, the two highest confusions occurred interchangeably between aerial roots/mudflat and aerial roots/mangrove. This was because these roots are grown in mudflat areas, exactly around mangroves. Therefore, it was challenging to separate aerial roots from two other classes due to the spectral/backscattering similarities and also the existence of mixed pixels at 10 m spatial resolution. Additionally, the second highest confusion was also observed between deep water and shallow water, which was actually because of their spectral/backscattering similarity.

Table 3. Average confusion matrix of mangrove ecosystem mapping using the random forest (RF) algorithm and a combination of the Sentinel-1 and Sentinel-2 satellite images processed within the Google Earth Engine (GEE) platform.

	Mangrove	Tidal Zone	Deep Water	Shallow Water	Mudflat	Aerial Roots	Urban	Bare Ground	Vegetation
Mangrove	1582	0	0	0	0	75	0	0	0
Tidal Zone	0	1613	0	72	31	0	0	0	0
Deep Water	0	0	2235	60	0	0	0	0	0
Shallow Water	0	46	108	1645	0	0	0	0	0
Mudflat	0	0	0	0	2091	140	22	0	0
Aerial Roots	94	0	0	0	152	823	0	0	2
Urban	0	0	0	0	9	0	923	48	6
Bare Ground	0	0	0	0	7	0	60	1964	3
Vegetation	2	0	0	0	4	16	14	4	518
PA (%)	95.47	93.99	97.38	91.44	92.81	76.94	94.47	96.56	92.83
UA (%)	94.27	97.22	95.39	92.57	91.51	78.08	90.57	97.42	98.29
OE (%)	4.53	6.01	2.62	8.56	7.19	23.06	5.53	3.44	7.17
CE (%)	5.73	2.78	4.61	7.43	8.49	21.92	9.43	2.58	1.71
Overall Accuracy (OA) = 93.23%					Kappa Coefficient (KC) = 0.92				

5. Discussion

5.1. General Findings

Mangrove ecosystems are routinely inundated and are located in inter-tidal zones [21]. Therefore, it is difficult to carry out field surveys to collect reference samples with global positioning system (GPS) devices [15]. The availability of highly accurate reference samples through field surveys is critical for reliable mapping; however, precise visual interpretation of high-resolution satellite images can compensate for this limitation. The proposed method was able to achieve high average OA and KC values, suggesting the suitability of applying this method when conducting fieldwork is difficult. It should be noted that generating reference samples even from high-resolution satellite imagery through precise visual interpretations can cause uncertainties and does not obviate the importance of in situ reference sample collection for more reliable mapping.

Despite the importance of mapping the mangrove extents, which have been conducted more frequently using either traditional or cloud computing-based approaches, fewer studies were devoted to producing detailed mangrove ecosystem maps. However, mapping relevant classes within a mangrove ecosystem (i.e., aerial roots and mudflat) is required to enhance condition assessment and also to suggest appropriate solutions for protection and conservation [25]. In particular, mapping the distribution and condition of aerial roots are highly required for mangrove status monitoring since these roots provide different morphological and physiological adaptions to adverse environmental conditions [70]. For instance, these roots facilitate gas exchange to adapt mangroves to live in oxygen-poor locations [71], protect mangroves from flood effects and hurricanes [70], and mitigate shoreline erosion to prevent mangrove loss [72]. Furthermore, these roots support mangrove growth and accretion by stabilizing mudflat conditions [71,73]. Therefore, obtaining precise information about aerial roots helps to conduct more profound monitoring of mangroves and also to determine the locations for efficient reforestation programs.

In this study, the GEE cloud computing platform was employed for mangrove ecosystem mapping. GEE provides an unprecedented opportunity to employ dense time-series data from free-of-charge satellite images, and also it contains many built-in machine learning and image processing algorithms for satellite data manipulations [29,30]. Furthermore, the existence of ready-to-use data, along with other characteristics, contributes towards developing efficient methods with higher rates of automation and, thus, decreases the dedicated time for data acquisition, calibrating, and preprocessing [26]. Despite these advantages, GEE also has several limitations, such as restrictions on using numerous features and training samples (i.e., encountering limit exceed), which were also addressed in previous studies [32].

5.2. Comparison with the Latest Global Mangrove Maps

The obtained results indicated the high applicability of the proposed method for mangrove ecosystem mapping and also mangrove extent delineation. Therefore, the mangrove extent derived from the proposed method was compared with the recent available global mangrove extent map [74]. This global mangrove map was produced at 25 m (0.8 arc-second) spatial resolution using multi-source remote sensing data of the Advanced Land Observation Satellite (ALOS) Phased-Array L-band SAR (PALSAR) and optical data of Landsat-5 and Landsat-7 through an Extremely Randomized Trees (ERT) classifier [74]. Figure 7 presents the mangrove extent resultants of two products, along with the corresponding high-resolution satellite images for better visualization. Based on Figure 7, it is obvious that the proposed approach outperformed the global mangrove product. Clearly, based on Figure 7a–c, the proposed method was resistant to misclassification, probably of mudflat and aerial roots (i.e., generally unvegetated areas between mangrove patches). Likewise, Figure 7d–f also illustrates that the proposed method was successful at delineating narrow mangrove patches which occurred along the tidal zone areas. Despite the subtle differences in the classification algorithms and the considered classes in each workflow, the main discrepancies between the two maps were almost associated with the spatial resolution of remote sensing satellite data, which suggests the use of Sentinel-1 and Sentinel-2 data. Therefore, the synergistic use of Sentinel-1 and Sentinel-2 can definitely enhance the recent global mangrove extent mapping [53]. Although the utility of 10 m spatial resolution indeed enhanced the mangrove extent mapping, as the results suggest, it is still necessary to use satellite images with a higher spatial resolution to permit the production of more precise detailed mangrove ecosystem maps. This is rooted in the fact that higher spatial resolutions are necessary to discriminate between different classes that are located near each other, and to reduce the mixed pixel effect, such as separating aerial roots from two other classes of mangrove and mudflat. However, the current high cost of acquiring commercial high-resolution images is the main limiting factor.

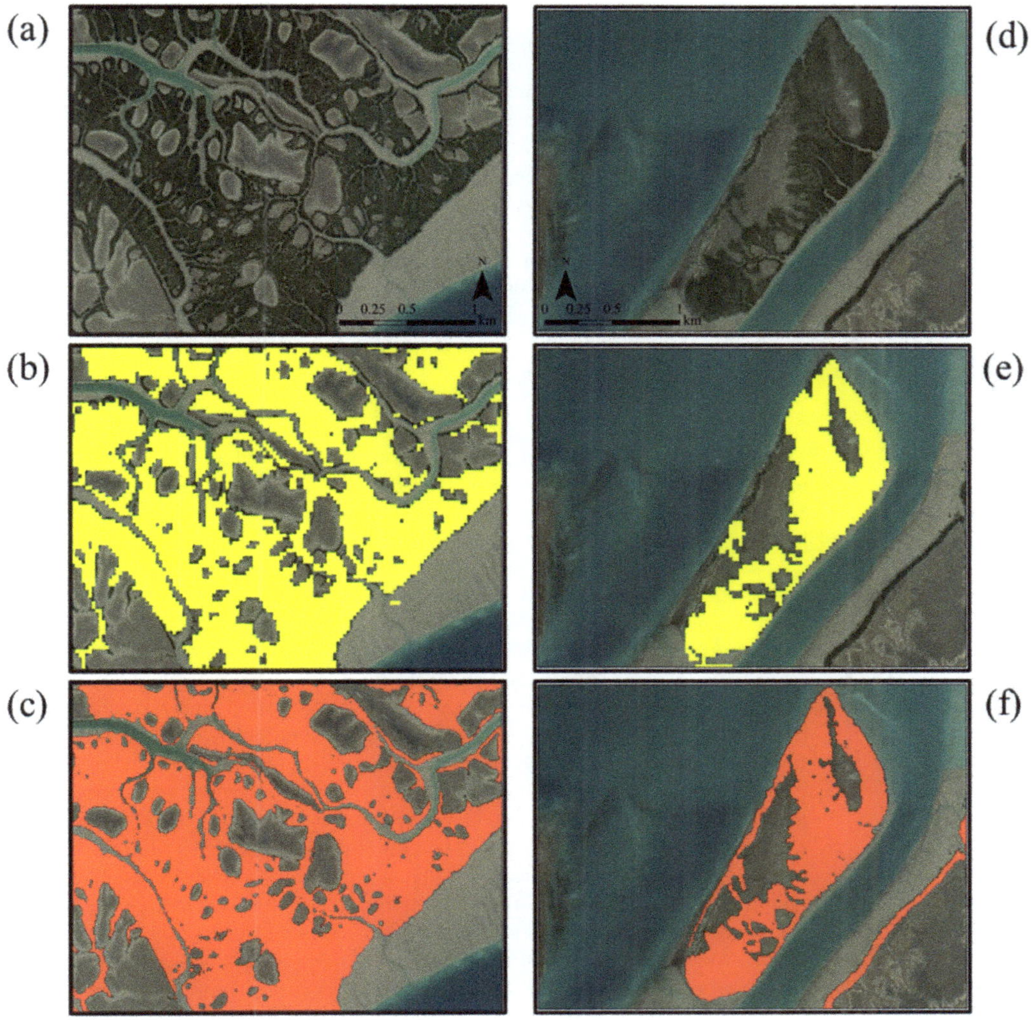

Figure 7. High-resolution satellite images of two mangrove areas (**a**,**d**), mangrove extents from the recent global mangrove map (**b**,**e**) [74], and mangrove extents derived from the proposed method (**c**,**f**).

5.3. Contribution of Multi-Source Remote Sensing Data

As mentioned earlier, the Sentinel-1 SAR and Sentinel-2 optical data were combined to produce an accurate mangrove ecosystem map. It has been well acknowledged that a combination of optical and SAR remote sensing data could enhance the classification results of land cover mapping [75]. However, few studies have combined optical and SAR data for detailed mangrove ecosystem mapping, and they were mainly conducted based on optical data [25,76]. Therefore, here, the contribution of multi-source remote sensing data for detailed mangrove ecosystem mapping was evaluated based on two metrics of OA and KC. In this regard, the explained workflow (Section 3) was repeated with two other scenarios of using only optical or SAR data. The obtained results for the multi-source data processing were comprehensively discussed in the previous section, and the average OA and KC were 93.23% and 0.92, respectively. The average OA and KC achieved for the scenario, in which only optical data were incorporated, were 90.67% and 0.89, respectively.

Although the accuracies based on the optical data were still satisfactory, the decline in the accuracies indicates the contribution of multi-source remote sensing data for more accurate mangrove ecosystem mapping. In particular, the use of only optical data resulted in over 1.5% (i.e., averaged over ten iterations) loss of class accuracies of each tidal zone, deep water, shallow water, and mudflat classes. Additionally, the results using only SAR data revealed a weak performance (i.e., OA = 62.98% and KC = 0.58); however, the SAR data was successful at delineating mangrove areas with a high accuracy of over 90% in almost all iterations. When using only SAR data, the lower accuracy was mainly associated with the low capability of separating different classes (i.e., except mangrove) within the mangrove ecosystem, with an average loss of 31.94% in class accuracies. The obtained results suggest that the inclusion of SAR data in the classification task, along with optical data, leads to producing a precise and accurate mangrove ecosystem map. This is mainly due to the fact that these two data sources can provide complementary information about the spectral and physical properties of different classes [77,78].

5.4. Comparison with Annual Downscaling

The seasonal downscaling approach was implemented for mangrove ecosystem mapping. In fact, the dense time-series SAR and optical data, acquired in 2019, were downscaled to seasonal features for further analysis. This approach takes the benefits of multi-temporal image analysis, as well as reducing the computational complexity of considering all available images without downscaling. Furthermore, since the mangrove ecosystems are routinely located in the inter-tidal zone and are inundated [21], the classifier may be influenced by the inundation intensity [79], biasing the mangrove extent and area. One solution is to use a single date image, in which the lowest tidal condition occurred; however, finding an image with such criteria may be challenging for several reasons, such as the presence of cloud and low temporal resolution of satellites. Therefore, it is recommended to apply time-series data to acquire more reliable classification results. In this study, seasonal features were used; however, other researchers applied annual downscaling for mangrove mapping [39]. Consequently, the usefulness of seasonal downscaling is compared with the annual downscaling approach. In this regard, the mentioned workflow was also applied to annual downscaled SAR and optical data, and the obtained results were compared with the seasonal approach. As is clear from Figure 8, incorporating seasonal downscaling achieved higher OAs and KCs in all iterations and, thus, it suggests that the seasonal downscaling can provide more discriminative information compared to annual downscaling [80]. This may be rooted in the fact that annual downscaling would override the tidal fluctuations, while the seasonal downscaling includes seasonal variations of water level (i.e., dry and wet seasons), reflecting tidal fluctuations [78].

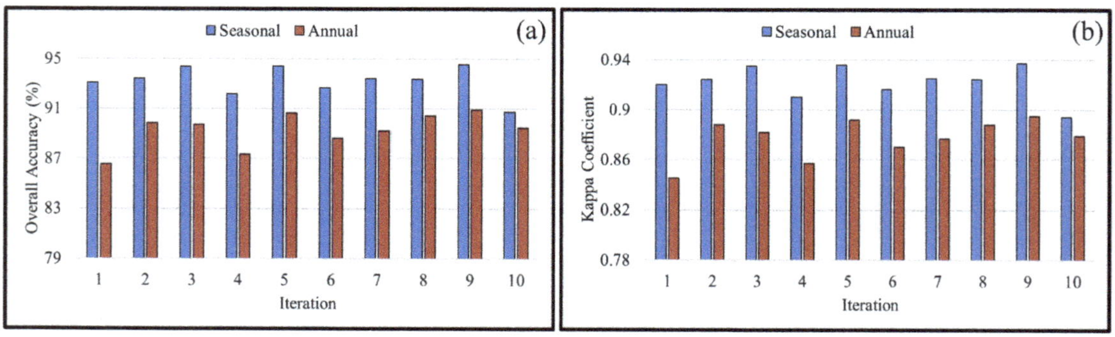

Figure 8. Overall accuracy (OA; (**a**)) and Kappa coefficient (KC; (**b**)) of annual and seasonal downscaling of SAR and optical data for mangrove ecosystem mapping.

6. Conclusions

Producing detailed mangrove ecosystem maps is essential for natural resource monitoring and sustainable development goals tracking. In this paper, a straightforward yet robust workflow was proposed to produce mangrove ecosystem maps with high accuracies. For this purpose, Sentinel-1 and Sentinel-2 data were inserted into an RF classifier within the GEE cloud computing platform. The final classification results were comprehensively evaluated through different analyses, and further discussions were provided. The classification results obtained high average OA and KC of 93.23% and 0.92, respectively. Moreover, all the classes obtained high accuracies, except the aerial roots, which achieved moderate accuracies in all cases, suggesting the consideration of other possible solutions (i.e., higher resolution images) for more accurate delineation of this class. Altogether, taking both qualitative (i.e., visual interpretation) and quantitative (i.e., statistical accuracy assessment) evaluation criteria into account, the proposed method confirmed its applicability in producing a detailed mangrove ecosystem map. Furthermore, the comparison results proved the contribution of multi-source remote sensing data (i.e., SAR + optical), as well as the effectiveness of seasonal downscaling. The proposed workflow was implemented based on an open-source platform and free-of-charge satellite data, making it appealing and applicable in almost all countries.

Author Contributions: Conceptualization, A.G. and S.Z.; methodology, A.G. and S.Z.; software, A.G.; validation, A.G., S.Z. and R.M.A.; formal analysis, A.G.; investigation, A.G.; data curation, A.G. and S.Z.; writing—original draft preparation, A.G. and R.M.A.; writing—review and editing, A.G., S.Z., R.M.A., M.A., S.J., A.M.; visualization, A.G.; supervision, A.G., A.M. and S.J.; project administration, A.G.; funding acquisition, M.A. and S.J. All authors have read and agreed to the published version of the manuscript.

Funding: S.J. was partially funded by Knut and Alice Wallenberg grant 99007-RFh2018-2016.

Data Availability Statement: The produced mangrove ecosystem map using Sentinel-1, Sentinel-2, and Random Forest algorithm and the derived mangrove extent map are available upon request from the first author.

Acknowledgments: The authors would like to thank the European Space Agency (ESA) to freely provide Sentinel-1 and Sentinel-2 satellite data. Additionally, the authors thank the anonymous three reviewers for their comments.

Conflicts of Interest: The authors declare no conflict of interest.

References

1. Chai, M.; Li, R.; Shi, C.; Shen, X.; Li, R.; Zan, Q. Contamination of polybrominated diphenyl ethers (PBDEs) in urban mangroves of Southern China. *Sci. Total Environ.* **2019**, *646*, 390–399. [CrossRef]
2. Veettil, B.K.; Pereira, S.F.R.; Quang, N.X. Rapidly diminishing mangrove forests in Myanmar (Burma): A review. *Hydrobiologia* **2018**, *822*, 19–35. [CrossRef]
3. Doughty, C.L.; Langley, J.A.; Walker, W.S.; Feller, I.C.; Schaub, R.; Chapman, S.K. Mangrove range expansion rapidly increases coastal wetland carbon storage. *Estuar. Coasts* **2016**, *39*, 385–396. [CrossRef]
4. Debrot, A.O.; Veldhuizen, A.; Van Den Burg, S.W.K.; Klapwijk, C.J.; Islam, M.; Alam, M.; Ahsan, M.; Ahmed, M.U.; Hasan, S.R.; Fadilah, R.; et al. Non-Timber Forest Product Livelihood-Focused Interventions in Support of Mangrove Restoration: A Call to Action. *Forests* **2020**, *11*, 1224. [CrossRef]
5. Giri, C.; Ochieng, E.; Tieszen, L.L.; Zhu, Z.; Singh, A.; Loveland, T.; Masek, J.; Duke, N. Status and distribution of mangrove forests of the world using earth observation satellite data. *Glob. Ecol. Biogeogr.* **2011**, *20*, 154–159. [CrossRef]
6. Bhargava, R.; Sarkar, D.; Friess, D.A. A cloud computing-based approach to mapping mangrove erosion and progradation: Case studies from the Sundarbans and French Guiana. *Estuar. Coast. Shelf Sci.* **2020**. [CrossRef]
7. Mafi-Gholami, D.; Jaafari, A.; Zenner, E.K.; Kamari, A.N.; Bui, D.T. Spatial modeling of exposure of mangrove ecosystems to multiple environmental hazards. *Sci. Total Environ.* **2020**, *740*. [CrossRef] [PubMed]
8. Chen, B.; Xiao, X.; Li, X.; Pan, L.; Doughty, R.; Ma, J.; Dong, J.; Qin, Y.; Zhao, B.; Wu, Z.; et al. A mangrove forest map of China in 2015: Analysis of time series Landsat 7/8 and Sentinel-1A imagery in Google Earth Engine cloud computing platform. *ISPRS J. Photogramm. Remote Sens.* **2017**, *131*, 104–120. [CrossRef]
9. Romañach, S.S.; DeAngelis, D.L.; Koh, H.L.; Li, Y.; Teh, S.Y.; Barizan, R.S.R.; Zhai, L. Conservation and restoration of mangroves: Global status, perspectives, and prognosis. *Ocean. Coast. Manag.* **2018**, *154*, 72–82. [CrossRef]

10. Primavera, J.H. Mangroves, fishponds, and the quest for sustainability. *Science* **2005**, *310*, 57–59. [CrossRef]
11. Duncan, C.; Owen, H.J.F.; Thompson, J.R.; Koldewey, H.J.; Primavera, J.H.; Pettorelli, N. Satellite remote sensing to monitor mangrove forest resilience and resistance to sea level rise. *Methods Ecol. Evol.* **2018**, *9*, 1837–1852. [CrossRef]
12. Lovelock, C.E.; Cahoon, D.R.; Friess, D.A.; Guntenspergen, G.R.; Krauss, K.W.; Reef, R.; Rogers, K.; Saunders, M.L.; Sidik, F.; Swales, A.; et al. The vulnerability of Indo-Pacific mangrove forests to sea-level rise. *Nature* **2015**, *526*, 559–563. [CrossRef] [PubMed]
13. Mafi-Gholami, D.; Zenner, E.K.; Jaafari, A. Mangrove regional feedback to sea level rise and drought intensity at the end of the 21st century. *Ecol. Indic.* **2020**, *110*. [CrossRef]
14. Jaramillo, F.; Desormeaux, A.; Hedlund, J.; Jawitz, J.W.; Clerici, N.; Piemontese, L.; Rodríguez-Rodriguez, J.A.; Anaya, J.A.; Blanco-Libreros, J.F.; Borja, S.; et al. Priorities and interactions of sustainable development goals (SDGs) with focus on wetlands. *Water* **2019**, *11*, 619. [CrossRef]
15. Zhao, C.; Qin, C.-Z. 10-m-resolution mangrove maps of China derived from multi-source and multi-temporal satellite observations. *ISPRS J. Photogramm. Remote Sens.* **2020**, *169*, 389–405. [CrossRef]
16. Kuenzer, C.; Bluemel, A.; Gebhardt, S.; Quoc, T.V.; Dech, S. Remote sensing of mangrove ecosystems: A review. *Remote Sens.* **2011**, *3*, 878–928. [CrossRef]
17. Wicaksono, P.; Danoedoro, P.; Hartono; Nehren, U. Mangrove biomass carbon stock mapping of the Karimunjawa Islands using multispectral remote sensing. *Int. J. Remote Sens.* **2016**, *37*, 26–52. [CrossRef]
18. Giri, C.; Long, J. Is the geographic range of mangrove forests in the conterminous United States really expanding? *Sensors* **2016**, *16*, 2010. [CrossRef] [PubMed]
19. Hauser, L.T.; An Binh, N.; Viet Hoa, P.; Hong Quan, N.; Timmermans, J. Gap-Free Monitoring of Annual Mangrove Forest Dynamics in Ca Mau Province, Vietnamese Mekong Delta, Using the Landsat-7-8 Archives and Post-Classification Temporal Optimization. *Remote Sens.* **2020**, *12*, 3729. [CrossRef]
20. Cao, J.; Leng, W.; Liu, K.; Liu, L.; He, Z.; Zhu, Y. Object-based mangrove species classification using unmanned aerial vehicle hyperspectral images and digital surface models. *Remote Sens.* **2018**, *10*, 89. [CrossRef]
21. Zhu, X.; Hou, Y.; Weng, Q.; Chen, L. Integrating UAV optical imagery and LiDAR data for assessing the spatial relationship between mangrove and inundation across a subtropical estuarine wetland. *ISPRS J. Photogramm. Remote Sens.* **2019**, *149*, 146–156. [CrossRef]
22. Abdel-Hamid, A.; Dubovyk, O.; El-Magd, A.; Menz, G. Mapping mangroves extents on the Red Sea coastline in Egypt using polarimetric SAR and high resolution optical remote sensing data. *Sustainability* **2018**, *10*, 646. [CrossRef]
23. Manna, S.; Raychaudhuri, B. Mapping distribution of Sundarban mangroves using Sentinel-2 data and new spectral metric for detecting their health condition. *Geocarto Int.* **2020**, *35*, 434–452. [CrossRef]
24. Kabiri, K. Mapping coastal ecosystems and features using a low-cost standard drone: Case study, Nayband Bay, Persian gulf, Iran. *J. Coast. Conserv.* **2020**, *24*, 1–8. [CrossRef]
25. Bihamta Toosi, N.; Soffianian, A.R.; Fakheran, S.; Pourmanafi, S.; Ginzler, C.; Waser, L.T. Land Cover Classification in Mangrove Ecosystems Based on VHR Satellite Data and Machine Learning—An Upscaling Approach. *Remote Sens.* **2020**, *12*, 2684. [CrossRef]
26. Cárdenas, N.Y.; Joyce, K.E.; Maier, S.W. Monitoring mangrove forests: Are we taking full advantage of technology? *Int. J. Appl. Earth Obs. Geoinf.* **2017**, *63*, 1–14. [CrossRef]
27. Giri, C.; Pengra, B.; Long, J.; Loveland, T.R. Next generation of global land cover characterization, mapping, and monitoring. *Int. J. Appl. Earth Obs. Geoinf.* **2013**, *25*, 30–37. [CrossRef]
28. Giri, C. Observation and monitoring of mangrove forests using remote sensing: Opportunities and challenges. *Remote Sens.* **2016**, *8*, 783. [CrossRef]
29. Gorelick, N.; Hancher, M.; Dixon, M.; Ilyushchenko, S.; Thau, D.; Moore, R. Google Earth Engine: Planetary-scale geospatial analysis for everyone. *Remote Sens. Environ.* **2017**, *202*, 18–27. [CrossRef]
30. Amani, M.; Ghorbanian, A.; Ahmadi, S.A.; Kakooei, M.; Moghimi, A.; Mirmazloumi, S.M.; Moghaddam, S.H.A.; Mahdavi, S.; Ghahremanloo, M.; Parsian, S.; et al. Google Earth Engine Cloud Computing Platform for Remote Sensing Big Data Applications: A Comprehensive Review. *IEEE J. Sel. Top. Appl. Earth Obs. Remote Sens.* **2020**, *13*, 5326–5350. [CrossRef]
31. Ghorbanian, A.; Kakooei, M.; Amani, M.; Mahdavi, S.; Mohammadzadeh, A.; Hasanlou, M. Improved land cover map of Iran using Sentinel imagery within Google Earth Engine and a novel automatic workflow for land cover classification using migrated training samples. *ISPRS J. Photogramm. Remote Sens.* **2020**, *167*, 276–288. [CrossRef]
32. Amani, M.; Mahdavi, S.; Afshar, M.; Brisco, B.; Huang, W.; Mohammad Javad Mirzadeh, S.; White, L.; Banks, S.; Montgomery, J.; Hopkinson, C. Canadian wetland inventory using google earth engine: The first map and preliminary results. *Remote Sens.* **2019**, *11*, 842. [CrossRef]
33. Zhang, M.; Huang, H.; Li, Z.; Hackman, K.O.; Liu, C.; Andriamiarisoa, R.L.; Ny Aina Nomenjanahary Raherivelo, T.; Li, Y.; Gong, P. Automatic High-Resolution Land Cover Production in Madagascar Using Sentinel-2 Time Series, Tile-Based Image Classification and Google Earth Engine. *Remote Sens.* **2020**, *12*, 3663. [CrossRef]
34. Genzano, N.; Pergola, N.; Marchese, F. A Google Earth Engine tool to investigate, map and monitor volcanic thermal anomalies at global scale by means of mid-high spatial resolution satellite data. *Remote Sens.* **2020**, *12*, 3232. [CrossRef]
35. Cao, B.; Domke, G.M.; Russell, M.B.; Walters, B.F. Spatial modeling of litter and soil carbon stocks on forest land in the conterminous United States. *Sci. Total Environ.* **2019**, *654*, 94–106. [CrossRef] [PubMed]

36. Pérez-Romero, J.; Navarro-Cerrillo, R.M.; Palacios-Rodriguez, G.; Acosta, C.; Mesas-Carrascosa, F.J. Improvement of remote sensing-based assessment of defoliation of Pinus spp. caused by Thaumetopoea Pityocampa Denis and Schiffermüller and related environmental drivers in Southeastern Spain. *Remote Sens.* **2019**, *11*, 1736. [CrossRef]
37. Seydi, S.T.; Akhoondzadeh, M.; Amani, M.; Mahdavi, S. Wildfire Damage Assessment over Australia Using Sentinel-2 Imagery and MODIS Land Cover Product within the Google Earth Engine Cloud Platform. *Remote Sens.* **2021**, *13*, 220. [CrossRef]
38. Besselly, S.M.; van der Wegen, M.; Grueters, U.; Reyns, J.; Dijkstra, J.; Roelvink, D. Eleven Years of Mangrove–Mudflat Dynamics on the Mud Volcano-Induced Prograding Delta in East Java, Indonesia: Integrating UAV and Satellite Imagery. *Remote Sens.* **2021**, *13*, 1084. [CrossRef]
39. Mondal, P.; Liu, X.; Fatoyinbo, T.E.; Lagomasino, D. Evaluating Combinations of Sentinel-2 Data and Machine-Learning Algorithms for Mangrove Mapping in West Africa. *Remote Sens.* **2019**, *11*, 2928. [CrossRef]
40. Baloloy, A.B.; Blanco, A.C.; Ana, R.R.C.S.; Nadaoka, K. Development and application of a new mangrove vegetation index (MVI) for rapid and accurate mangrove mapping. *ISPRS J. Photogramm. Remote Sens.* **2020**, *166*, 95–117. [CrossRef]
41. Yancho, J.M.M.; Jones, T.G.; Gandhi, S.R.; Ferster, C.; Lin, A.; Glass, L. The Google Earth Engine Mangrove Mapping Methodology (GEEMMM). *Remote Sens.* **2020**, *12*, 3758. [CrossRef]
42. Milani, A.S. Mangrove forests of the Persian Gulf and the Gulf of Oman. In *Threats to Mangrove Forests*; Springer: Berlin/Heidelberg, Germany, 2018; pp. 53–75.
43. Milani, S.A.; Beglu, J.M. Satellite based assessment of the area and changes in the Mangrove ecosystem of the QESHM island, Iran. *J. Environ. Res. Dev.* **2012**, *7*, 1052–1060.
44. Toosi, N.B.; Soffianian, A.R.; Fakheran, S.; Pourmanafi, S.; Ginzler, C.; Waser, L.T. Comparing different classification algorithms for monitoring mangrove cover changes in southern Iran. *Glob. Ecol. Conserv.* **2019**, *19*, e00662. [CrossRef]
45. Hajializadeh, P.; Safaie, M.; Naderloo, R.; Shojaei, M.G.; Gammal, J.; Villnäs, A.; Norkko, A. Species Composition and Functional Traits of Macrofauna in Different Mangrove Habitats in the Persian Gulf. *Front. Mar. Sci.* **2020**, *7*, 809. [CrossRef]
46. Dadashi, M.; Ghaffari, S.; Bakhtiari, A.R.; Tauler, R. Multivariate curve resolution of organic pollution patterns in mangrove forest sediment from Qeshm Island and Khamir Port—Persian Gulf, Iran. *Environ. Sci. Pollut. Res.* **2018**, *25*, 723–735. [CrossRef]
47. Lohr, S.L. *Sampling: Design and Analysis*; CRC Press: Bocaton, FL, USA, 2019.
48. Mahdianpari, M.; Jafarzadeh, H.; Granger, J.E.; Mohammadimanesh, F.; Brisco, B.; Salehi, B.; Homayouni, S.; Weng, Q. A large-scale change monitoring of wetlands using time series Landsat imagery on Google Earth Engine: A case study in Newfoundland. *GIScience Remote Sens.* **2020**, *57*, 1102–1124. [CrossRef]
49. Geiß, C.; Pelizari, P.A.; Schrade, H.; Brenning, A.; Taubenböck, H. On the effect of spatially non-disjoint training and test samples on estimated model generalization capabilities in supervised classification with spatial features. *IEEE Geosci. Remote Sens. Lett.* **2017**, *14*, 2008–2012.
50. Dong, D.; Wang, C.; Yan, J.; He, Q.; Zeng, J.; Wei, Z. Combing Sentinel-1 and Sentinel-2 image time series for invasive Spartina alterniflora mapping on Google Earth Engine: A case study in Zhangjiang Estuary. *J. Appl. Remote Sens.* **2020**, *14*, 44504. [CrossRef]
51. Mahdavi, S.; Salehi, B.; Amani, M.; Granger, J.; Brisco, B.; Huang, W. A dynamic classification scheme for mapping spectrally similar classes: Application to wetland classification. *Int. J. Appl. Earth Obs. Geoinf.* **2019**, *83*. [CrossRef]
52. Xia, J.; Yokoya, N.; Pham, T.D. Probabilistic mangrove species mapping with multiple-source remote-sensing datasets using label distribution learning in Xuan Thuy National Park, Vietnam. *Remote Sens.* **2020**, *12*, 3834. [CrossRef]
53. Wang, L.; Jia, M.; Yin, D.; Tian, J. A review of remote sensing for mangrove forests: 1956–2018. *Remote Sens. Environ.* **2019**, *231*. [CrossRef]
54. Malenovsky, Z.; Rott, H.; Cihlar, J.; Schaepman, M.E.; Garcia-Santos, G.; Fernandes, R.; Berger, M. Sentinels for science: Potential of Sentinel-1,-2, and-3 missions for scientific observations of ocean, cryosphere, and land. *Remote Sens. Environ.* **2012**, *120*, 91–101. [CrossRef]
55. Torres, R.; Snoeij, P.; Geudtner, D.; Bibby, D.; Davidson, M.; Attema, E.; Potin, P.; Rommen, B.; Floury, N.; Brown, M.; et al. GMES Sentinel-1 mission. *Remote Sens. Environ.* **2012**, *120*, 9–24. [CrossRef]
56. Liu, X.; Fatoyinbo, T.E.; Thomas, N.M.; Guan, W.W.; Zhan, Y.; Mondal, P.; Lagomasino, D.; Simard, M.; Trettin, C.C.; Deo, R.; et al. Large-scale High-resolution Coastal Mangrove Forests Mapping across West Africa with Machine Learning Ensemble and Satellite Big Data. *Front. Earth Sci.* **2021**, *8*, 677. [CrossRef]
57. Naboureh, A.; Li, A.; Bian, J.; Lei, G.; Amani, M. A Hybrid Data Balancing Method for Classification of Imbalanced Training Data within Google Earth Engine: Case Studies from Mountainous Regions. *Remote Sens.* **2020**, *12*, 3301. [CrossRef]
58. Ghorbanian, A.; Mohammadzadeh, A. An unsupervised feature extraction method based on band correlation clustering for hyperspectral image classification using limited training samples. *Remote Sens. Lett.* **2018**. [CrossRef]
59. Ghorbanian, A.; Maghsoudi, Y.; Mohammadzadeh, A. Clustering-Based Band Selection Using Structural Similarity Index and Entropy for Hyperspectral Image Classification Clustering-Based Band Selection Using Structural Similarity Index and Entropy for Hyperspectral Image Classification. *Trait. Signal.* **2020**, *37*, 785–791. [CrossRef]
60. Quang, N.H.; Quinn, C.H.; Stringer, L.C.; Carrie, R.; Hackney, C.R.; Van Hue, L.T.; Van Tan, D.; Nga, P.T.T. Multi-Decadal Changes in Mangrove Extent, Age and Species in the Red River Estuaries of Viet Nam. *Remote Sens.* **2020**, *12*, 2289. [CrossRef]
61. Nababa, I.I.; Symeonakis, E.; Koukoulas, S.; Higginbottom, T.P.; Cavan, G.; Marsden, S. Land Cover Dynamics and Mangrove Degradation in the Niger Delta Region. *Remote Sens.* **2020**, *12*, 3619. [CrossRef]
62. Breiman, L. Random forests. *Mach. Learn.* **2001**, *45*, 5–32. [CrossRef]

63. Belgiu, M.; Dragut, L. Random forest in remote sensing: A review of applications and future directions. *ISPRS J. Photogramm. Remote Sens.* **2016**, *114*, 24–31. [CrossRef]
64. Morell-Monzó, S.; Sebastiá-Frasquet, M.-T.; Estornell, J. Land Use Classification of VHR Images for Mapping Small-Sized Abandoned Citrus Plots by Using Spectral and Textural Information. *Remote Sens.* **2021**, *13*, 681. [CrossRef]
65. Stehman, S. V Sampling designs for accuracy assessment of land cover. *Int. J. Remote Sens.* **2009**, *30*, 5243–5272. [CrossRef]
66. Story, M.; Congalton, R.G. Accuracy assessment: A user's perspective. *Photogramm. Eng. Remote Sens.* **1986**, *52*, 397–399.
67. Berger, A.; Guda, S. Threshold optimization for F measure of macro-averaged precision and recall. *Pattern Recognit.* **2020**, *102*. [CrossRef]
68. Ruta, D.; Gabrys, B. Classifier selection for majority voting. *Inf. Fusion* **2005**, *6*, 63–81. [CrossRef]
69. Amani, M.; Salehi, B.; Mahdavi, S.; Brisco, B.; Shehata, M. A Multiple Classifier System to improve mapping complex land covers: A case study of wetland classification using SAR data in Newfoundland, Canada. *Int. J. Remote Sens.* **2018**, *39*, 7370–7383. [CrossRef]
70. Srikanth, S.; Lum, S.K.Y.; Chen, Z. Mangrove root: Adaptations and ecological importance. *Trees* **2016**, *30*, 451–465. [CrossRef]
71. Al-Khayat, J.A.; Alatalo, J.M. Relationship Between Tree Size, Sediment Mud Content, Oxygen Levels, and Pneumatophore Abundance in the Mangrove Tree Species Avicennia Marina (Forssk.) Vierh. *J. Mar. Sci. Eng.* **2021**, *9*, 100. [CrossRef]
72. Duke, N.C.; Kovacs, J.M.; Griffiths, A.D.; Preece, L.; Hill, D.J.E.; Van Oosterzee, P.; Mackenzie, J.; Morning, H.S.; Burrows, D. Large-scale dieback of mangroves in Australia's Gulf of Carpentaria: A severe ecosystem response, coincidental with an unusually extreme weather event. *Mar. Freshw. Res.* **2017**, *68*, 1816–1829. [CrossRef]
73. Okello, J.A.; Kairo, J.G.; Dahdouh-Guebas, F.; Beeckman, H.; Koedam, N. Mangrove trees survive partial sediment burial by developing new roots and adapting their root, branch and stem anatomy. *Trees* **2020**, *34*, 37–49. [CrossRef]
74. Bunting, P.; Rosenqvist, A.; Lucas, R.M.; Rebelo, L.-M.; Hilarides, L.; Thomas, N.; Hardy, A.; Itoh, T.; Shimada, M.; Finlayson, C.M. The global mangrove watch—A new 2010 global baseline of mangrove extent. *Remote Sens.* **2018**, *10*, 1669. [CrossRef]
75. Amani, M.; Salehi, B.; Mahdavi, S.; Granger, J.; Brisco, B. Wetland classification in Newfoundland and Labrador using multi-source SAR and optical data integration. *GIScience Remote Sens.* **2017**, *54*, 779–796. [CrossRef]
76. Wang, D.; Wan, B.; Qiu, P.; Su, Y.; Guo, Q.; Wang, R.; Sun, F.; Wu, X. Evaluating the performance of sentinel-2, landsat 8 and pléiades-1 in mapping mangrove extent and species. *Remote Sens.* **2018**, *10*, 1468. [CrossRef]
77. Chauhan, S.; Srivastava, H.S. Comparative evaluation of the sensitivity of multi-polarized SAR and optical data for various land cover classes. *Int. J. Adv. Remote Sens. GIS Geogr.* **2016**, *4*, 1–14.
78. Zhao, C.; Qin, C.-Z.; Teng, J. Mapping large-area tidal flats without the dependence on tidal elevations: A case study of Southern China. *ISPRS J. Photogramm. Remote Sens.* **2020**, *159*, 256–270. [CrossRef]
79. Wang, X.; Xiao, X.; Zou, Z.; Hou, L.; Qin, Y.; Dong, J.; Doughty, R.B.; Chen, B.; Zhang, X.; Chen, Y.; et al. Mapping coastal wetlands of China using time series Landsat images in 2018 and Google Earth Engine. *ISPRS J. Photogramm. Remote Sens.* **2020**, *163*, 312–326. [CrossRef]
80. Wang, L.; Silván-Cárdenas, J.L.; Sousa, W.P. Neural network classification of mangrove species from multi-seasonal Ikonos imagery. *Photogramm. Eng. Remote Sens.* **2008**, *74*, 921–927. [CrossRef]

remote sensing

MDPI

Article

Global Sensitivity Analysis for Canopy Reflectance and Vegetation Indices of Mangroves

Chunyue Niu *, Stuart Phinn and Chris Roelfsema

Remote Sensing Research Centre, School of Earth and Environmental Sciences, The University of Queensland, Brisbane, QLD 4072, Australia; s.phinn@uq.edu.au (S.P.); c.roelfsema@uq.edu.au (C.R.)
* Correspondence: c.niu@uqconnect.edu.au

Abstract: Remote sensing has been applied to map the extent and biophysical properties of mangroves. However, the impact of several critical factors, such as the fractional cover and leaf-to-total area ratio of mangroves, on their canopy reflectance have rarely been reported. In this study, a systematic global sensitivity analysis was performed for mangroves based on a one-dimensional canopy reflectance model. Different scenarios such as sparse or dense canopies were set up to evaluate the impact of various biophysical and environmental factors, together with their ranges and probability distributions, on simulated canopy reflectance spectra and selected Sentinel-2A vegetation indices of mangroves. A variance-based method and a density-based method were adopted to compare the computed sensitivity indices. Our results showed that the fractional cover and leaf-to-total area ratio of mangrove crowns were among the most influential factors for all examined scenarios. As for other factors, plant area index and water depth were influential for sparse canopies while leaf biochemical properties and inclination angles were more influential for dense canopies. Therefore, these influential factors may need attention when mapping the biophysical properties of mangroves such as leaf area index. Moreover, a tailored sensitivity analysis is recommended for a specific mapping application as the computed sensitivity indices may be different if a specific input configuration and sensitivity analysis method are adopted.

Keywords: global sensitivity analysis; PAWN; canopy reflectance model; vegetation index (VI); mangroves

Citation: Niu, C.; Phinn, S.; Roelfsema, C. Global Sensitivity Analysis for Canopy Reflectance and Vegetation Indices of Mangroves. *Remote Sens.* **2021**, *13*, 2617. https://doi.org/10.3390/rs13132617

Academic Editors: Lars T. Waser and Chandra Giri

Received: 13 May 2021
Accepted: 20 June 2021
Published: 3 July 2021

1. Introduction

The extent and biophysical properties of mangroves have been mapped from multi-spectral remote sensing (RS) images [1–4]. However, several factors such as the woody material and tidal height could affect the canopy reflectance [5,6], which was often ignored. The scientific community has been aware of these, but quantitative assessments are still needed to identify essential factors that have a major impact on the spectral responses of mangroves.

Sensitivity analysis (SA) is the study of how to attribute the variation or uncertainty in the output of a model to variations in the model inputs [7,8]. It can be employed to rank the input factors based on their contributions to the variations of the model output, to identify factors that have an insignificant impact on output variations and to find out regions in the input space that produce particular output values [7,8]. SA can be divided into global SA (GSA) or local SA according to whether input values are allowed to vary across the entire input space or around a reference point in the input space [8].

SA has been broadly employed in RS to recognise the input factors that had a major impact on vegetation canopy reflectance using various canopy reflectance models (CRMs). The PROSPECT model [9], the SAIL family including SAILH, 4SAIL2 and soil–leaf–canopy (SLC) [10–12] and the PROSAIL (PROSPECT + SAIL) model [13] were among the CRMs that had been extensively explored. Variance-based SA (VBSA) methods have attracted a great deal of attention. Its fundamental concept is to decompose the direct contribution of a factor to output variance and its overall contribution when interaction effects with other factors are considered [8,14,15]. Corresponding metrics are usually termed as first- and total-order sensitivity indices. Liu et al. [16] applied the extended Fourier amplitude sensitivity test (EFAST) [17] to calculate the first order indices and interaction effects for analysing the sensitivities of vegetation indices (VIs) to the leaf area index (LAI) and other interference factors, such as leaf chlorophyll content (C_{ab}) and average leaf inclination angle (θ_l), and evaluating the uncertainties caused by these factors. This study found that leaf parameters such as C_{ab} and leaf dry matter content (C_m) had an increasing impact on the uncertainty of estimated LAI towards higher LAI and θ_l was the most critical factor for LAI estimation if an ellipsoidal distribution [18] was used. Xiao et al. [19] also adopted EFAST and model simulations to investigate the sensitivities of reflectance and VIs to biophysical and biochemical parameters at the leaf, canopy and regional levels. The importance of leaf parameters reduced at the canopy level, especially for a LAI value of 0–3, where LAI dominated except for the absorption bands of C_{ab} and leaf water content (C_w). Although the contributions of LAI and soil were still evident for a sparse canopy, the fractional cover of vegetation (fCv) dominated at the regional level. The importance of LAI and soil dropped significantly for LAI > 3 at both the canopy and regional levels. Mousivand et al. [20] proposed an improved design and sampling for SA to identify influential and non-influential factors on canopy reflectance based on the SLC model [12], in which soil moisture (SM) and fCv were explicitly considered. fCv, LAI, θ_l and SM were recognised as the most influential factors for a LAI range of 0–6.

The studies mentioned above mainly focused on terrestrial vegetation, while only limited SA studies were carried out for aquatic vegetation. Villa et al. [21] introduced two new VIs specifically for aquatic vegetation: the normalised difference aquatic vegetation index (NDAVI) and the water adjusted vegetation index (WASI) and performed VBSA to evaluate their sensitivities to LAI, C_{ab} and θ_l. Both radiative transfer simulations and linear spectral mixture simulations based on real endmembers were applied. The new VIs were found to have a higher sensitivity to LAI and θ_l and better capacity in separating aquatic and terrestrial vegetation compared with previous VIs for terrestrial vegetation such as NDVI. Unfortunately, the fCv in 4SAIL2 was fixed to 1, and thus its impact was not discussed. Zhou et al. [22] conducted EFAST for emergent and submerged vegetation based on a CRM model for aquatic vegetation [23]. Four scenarios were adopted according to the combinations of shallow or deep water and sparse or dense canopies. Canopy reflectance spectra and four VIs, including NDAVI and WAVI for Sentinel-2A (S2A) images, were simulated. The most influential factors were found to be different for two vegetation types, with leaf and canopy parameters being dominant for emergent vegetation while water components accounted for most variability in canopy reflectance of submerged vegetation. For emergent vegetation, the total order sensitivity indices of all four VIs to LAI were high for a sparse canopy but dropped notably for a dense canopy. NDAVI outperformed the other three VIs for a dense emergent canopy in terms of their sensitivities to LAI.

Although extensively applied, VBSA was based on an implicit assumption that variance was sufficient to represent output variability [24]. This assumption was not appropriate if the output was highly skewed or had multiple modals [25]. Moreover, the probability distribution of an input factor also mattered, and the total order indices could not be computed if there were correlated inputs [24]. However, input probability distributions were usually assumed to be uniform, e.g., [20,22]; the potential correlations between input factors and the output distributions were rarely examined in many VBSA studies.

These limitations of VBSA could be avoided by adopting moment-independent sensitivity indices that relied on the entire output distribution rather than a particular moment [26]. These were also referred to as density-based SA (DBSA) as they employed the probability density function (PDF) or cumulative distribution function (CDF) of the output to characterise its uncertainty [8,27]. The basic idea was to evaluate the sensitivity by calculating the divergence between unconditional output PDF and conditional output PDF, where "unconditional" represented varying all input factors while "conditional" meant that the corresponding input factor was fixed [8]. Despite its advantages, DBSA was not widely applied, partly due to the challenges in obtaining conditional PDFs. Thus, using CDF instead of PDF could reduce the computational costs and make it relatively easy to implement [25].

In summary, current studies on SA of canopy reflectance of terrestrial vegetation and their conclusions might not be applied to mangroves directly, partly because the adopted CRMs ignored essential factors for mangroves such as fCv, water body and woody material. Additionally, a few recently proposed VIs for aquatic vegetation, e.g., [21], would possibly be effective for the RS of mangroves, but quantitative assessments were required. In addition, the impact of potential correlations between input factors, e.g., plant area index (PAI) and fCv [28], input probability distributions and the adopted SA methods on the produced sensitivity indices also needed more evaluations. Therefore, the objective of this study is to perform a systematic GSA for mangroves to quantitatively assess the impact of various biophysical and environmental factors on their canopy reflectance and selected VIs, and to identify the most influential factors under different scenarios, e.g., sparse or dense canopies. The findings may help to improve our understanding of the spectral characteristics of mangroves and recognise the factors that may be mapped from multispectral satellite images.

2. Methodology

2.1. Overview

In this study, 16 factors were selected based on a priori knowledge, similar studies such as [19,20,22] and trial tests. As only images captured by nadir or near-nadir viewing satellites, e.g., Sentinel-2, would be applied at this stage, the observation zenith angle was set as 0°, and the solar zenith angle was fixed as 30° to avoid distraction. Then various mangrove scenarios such as sparse or dense canopies were set up according to one or more properties such as the crown PAI and fCv. This corresponded to different mangrove formations such as open or closed forests. Samples of the 16 input factors were generated using VBSA or DBSA and following uniform or normal probability distributions. The canopy reflectance spectra from 400 to 2500 nm with a 1 nm interval were then simulated using a CRM of mangrove [28]. Additionally, values of corresponding S2A bands (Bands 2–8, 8a, 11 and 12) were simulated by convolving the simulated reflectance spectra with the spectral response functions of S2A. Then twelve VIs were calculated. Finally, both VBSA and DBSA were performed and the computed sensitivity indices were examined and compared. The GSA flowchart is shown in Figure 1.

2.2. Canopy Reflectance Model of Mangroves

The adopted CRM of mangroves was a one-dimensional multiple-layer radiative transfer model [28]. It considered essential biophysical and biochemical properties of mangroves such as C_{ab}, PAI, fCv, L2T ratio, inclination distributions of leaves and woody material and environmental factors such as the water body. It could be regarded as an extension of a previous CRM for aquatic vegetation [23], which was a revision of the PROSAIL model. Mangroves were divided into three layers as in [28]: the understory (L1), the stem (L2) and the crown (L3) layers (Figure 2). The layer index would be added to the variable names in the following to distinguish them from each other, e.g., PAI(3) for the PAI of the crown layer. The height of the understory was assumed to be the same as water depth (H_w), and water was not allowed to immerse the crown as that could change canopy reflectance significantly [29]. Stems were assumed to be covered by the crowns and hence only had a minor contribution to canopy reflectance. Thus, the stem layer was assigned a low PAI value of 0.5 and a low fCv value of 0.15, according to [28]. Similarly, most factors with regard to the optically active components in the water body and other environmental factors such as bottom reflectance were also fixed as their impact on canopy reflectance was minor when covered by the canopy. For example, the contribution of suspended particles in the background water on canopy reflectance was around 10% or lower and mainly in the visible region [22,28]. Therefore, more attention was paid to the crown, the understory and H_W, and other factors were beyond the scope of this study. H_W corresponded to the changing tidal height. PAI(1) and fCv(1) were employed to explore if there were any potential contributions from the understory to the observed canopy reflectance, especially for a sparse canopy.

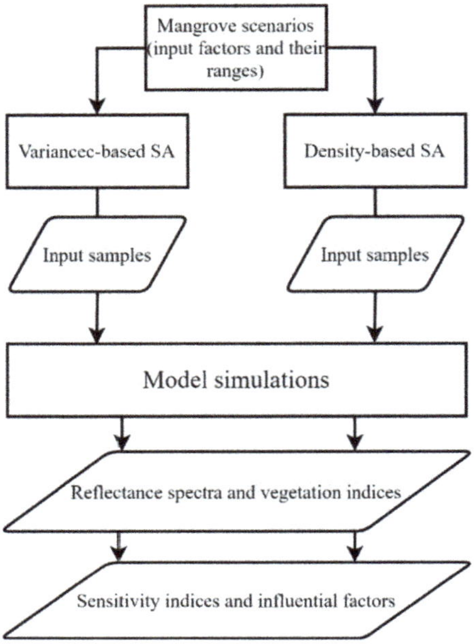

Figure 1. The flowchart of global sensitivity analyses for mangroves.

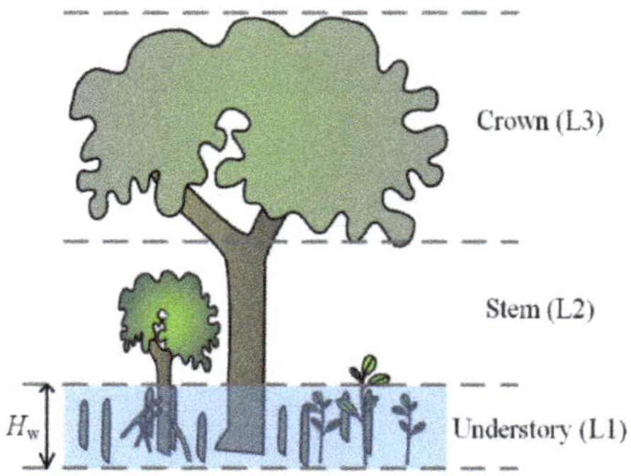

Figure 2. Illustrations of the understory (L1), stem (L2) and crown (L3) layers for mangroves. H_W is the water depth and is assumed to be the same as the height of the understory for simplicity (The symbols are courtesy of the Integration and Application Network, University of Maryland Center for Environmental Science (ian.umces.edu/symbols/, accessed on 14 June 2018)).

2.3. Mangrove Scenarios

If not specified, the input factors were independent, as they were treated in other SA studies [19,20,22]. All samples discussed in Section 2.3.1 were generated using the Latin hypercube sampling (LHS) [30] following a uniform or normal probability distribution. Although independent input samples were widely adopted in SA studies, there might be an issue with some biophysical properties. For example, it was possible to obtain a high PAI(3) value together with a low fCv(3) value or vice versa if they were assumed to be independent (Figure 3a). However, this was not reasonable in nature and the field data in [28] revealed a positive correlation with a coefficient of determination of 0.73 between the PAI and fCv of mangroves. Therefore, another input dataset with correlated PAI(3) and fCv(3) was also designed in Section 2.3.2.

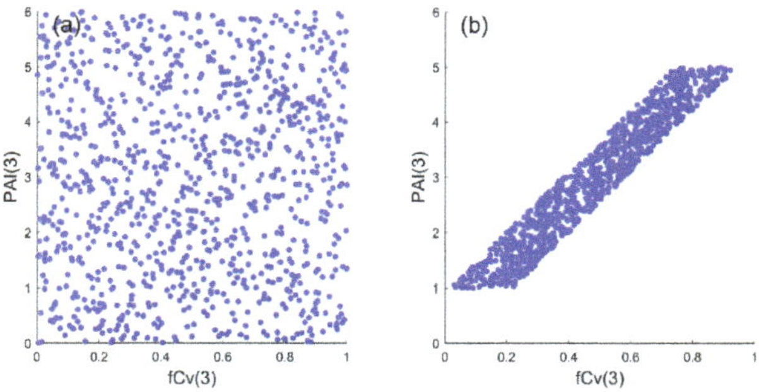

Figure 3. Illustrations of generated samples (N = 1000) if the crown plant area index (PAI(3)) and fractional cover (fCv(3)) are assumed to be independent (using Latin hypercube sampling) (**a**) or linearly correlated (**b**).

2.3.1. Factors, Their Ranges and Distributions

GSA was performed for three different mangrove scenarios defined by the ranges of PAI(3) and fCv(3): general canopy (GEN), sparse canopy (SPS) and dense canopy (DEN) (Table 1), as previous studies demonstrated that different ranges of inputs could result in different sensitivity values [19,22]. PAI(3) ranges were set as 0–6 for GEN, 0–3 for SPS and 3–6 for DEN according to the literature [19,31]. Correspondingly, fCv(3) ranges were 0–1, 0–0.7 and 0.5–1. The overlap between the fCv(3) ranges for SPS and DEN was to reflect the variability in mangrove forests. The SPS and DEN scenarios corresponded to open and closed mangrove forests, while the GEN scenario included both situations and implied a significant variation of mangrove formations. Ranges of other factors such as leaf parameters were not changed across these scenarios.

Table 1. Considered input factors and configurations for mangrove scenarios.

Factors	Unit	Definitions	General Uniform		Sparse Uniform		Sparse Normal		Dense Uniform		Dense Normal	
			Min	Max	Min	Max	Mean	STD	Min	Max	Mean	STD
Leaf												
N	-	Leaf structural properties	1	4	1	4	3	0.3	1	4	3	0.3
C_{ab}	$\mu g{\cdot}cm^{-2}$	Leaf chlorophyll content	0	100	0	100	35	3.5	0	100	35	3.5
C_w	$\mu g{\cdot}cm^{-2}$	Leaf water content	0	0.2	0	0.2	0.07	0.007	0	0.2	0.07	0.007
C_m	$g{\cdot}cm^{-2}$	Leaf dry matter content	0	0.05	0	0.05	0.01	0.001	0	0.05	0.01	0.001
Canopy												
PAI(1)	-	Plant area index (PAI) of Layer 1 (L1, understory)	0	3	0	3	1	0.1	0	3	1	0.1
PAI(3)	-	PAI of Layer 3 (L3, crown)	0	6	0	3	2	0.2	3	6	4.5	0.45
L2T(1)	-	Leaf-to-total area ratio of L1	0	1	0	1	0.5	0.05	0	1	0.5	0.05
L2T(3)	-	Leaf-to-total area ratio of L3	0	1	0	1	0.75	0.075	0	1	0.75	0.075
fCv(1)	-	Fractional cover of L1	0	1	0	1	0.5	0.05	0	1	0.5	0.05
fCv(3)	-	Fractional cover of L3	0	1	0	0.7	0.3	0.03	0.5	1	0.8	0.08
LIDFa(3)	-	Leaf inclination distribution function parameter a of L3 [a]	−1	1	−1	1	−0.2	0.1	−1	1	−0.2	0.1
WIDFa(3)	-	Wood inclination distribution function parameter a of L3	−1	1	−1	1	−0.2	0.1	−1	1	−0.2	0.1
HSl(3)	-	Hot spot size parameter of L3	0	0.1	0	0.1	0.05	0.005	0	0.1	0.05	0.005
zeta(3)	-	Tree shape factor of L3	0	2	0	2	1	0.1	0	2	1	0.1
Other												
H_w	m	Water depth	0	2	0	2	0.4	0.04	0	2	0.4	0.04
R_{aw_so}	-	Bidirectional reflectance of water surface	0	0.03	0	0.03	0.02	0.002	0	0.03	0.02	0.002

Note: Uniform or normal is the probability distribution of input factors, with Min and Max for minimum and maximum values and Mean and STD for mean value and standard deviation. [a] Equivalent to average inclination angle (AIA) by a = (45 − AIA)$*\pi^2/360$ [11].

In addition to the GSA methods and ranges of inputs, the probability distributions of inputs might also impact the sensitivity values [32]. The uniform (UNI) and normal (NRM) probability distributions were both adopted for the input factors under SPS and DEN. The mean values for NRM were selected according to [28], and corresponding standard deviation values were set as 10% of the mean values, except for LIDFa(3) and WIDFa(3). NRM indicated that a priori information on the mangrove forest was available and the variability in canopy structure was small, while UNI indicated a larger variation in canopy structure. Under GEN, only a uniform distribution was applied as it might not be suitable to use a nominal value to represent such a wide range of PAI(3).

2.3.2. Correlated PAI(3) and fCv(3)

In this case, the linear relationship between PAI and fCv of mangroves from [28] was applied to generate samples for correlated PAI(3) and fCv(3). First, the ranges of PAI(3) and fCv(3) were set as 1–5 and 0.1–0.9. Hence, their values were within 0–6 and 0–1 after adding random variations. Then random samples of all 16 factors were generated as in the GEN case. Second, PAI(3) or fCv(3) was modified to make them correlated. For VBSA, fCv(3) values were recalculated according to PAI(3) values and the linear relationship, and subsequently, random variations within ± 0.1 were added (Figure 3b). In the case of PAWN, the same methodology was used to compute unconditional CDF and conditional CDFs for most factors except fCv(3). When fCv(3) was fixed for computing the corresponding conditional CDFs, PAI(3) was recalculated based on fCv(3), and then a random variation within ± 0.8 was given. Finally, sensitivity indices for these two sample datasets were computed using VBSA and PAWN, respectively. Correlated inputs were only applied to GEN as a demonstration because the relationship between PAI and fCv could be different for various sites and species.

2.4. Vegetation Indices

VIs have been broadly applied for mapping the biophysical properties of mangroves such as LAI [2,3,31], but the impact of woody material, fCv and background water was rarely reported. Although a large number of VIs were available [33] (https://www.indexdatabase.de/, accessed on 9 December 2019), this study did not intend to investigate all possible choices. Instead, the VIs in Table 2 were simulated based on the spectral response function of the S2A multispectral instrument to assess their sensitivities to the input factors such as PAI, fCv, L2T ratio and H_w. These indices were either widely applied for terrestrial vegetation such as NDVI [34] and EVI [35] or in particular, proposed for aquatic vegetation such as NDAVI [36]. Some of the VIs for aquatic vegetation included shortwave infrared (SWIR) bands to address the existence of background water, e.g., RGVI [37] and WFI [38].

Table 2. Definitions and sources of the adopted vegetation indices in this study.

Indices	Descriptions	References
$NDVI = \frac{\rho_{NIR} - \rho_R}{\rho_{NIR} + \rho_R}$	Normalised Difference Vegetation Index	[34]
$SAVI = (1 + L)\frac{\rho_{NIR} - \rho_R}{\rho_{NIR} + \rho_R + L}$	Soil Adjusted Vegetation Index (L: 0–1)	[39]
$EVI = G\frac{\rho_{NIR} - \rho_R}{\rho_{NIR} + C_1\rho_R - C_2\rho_B + L}$	Enhanced Vegetation Index (G = 2.5, L = 1, C1 = 6, C2 = 7.5)	[35]
$NDWI = \frac{\rho_G - \rho_{NIR}}{\rho_G + \rho_{NIR}}$	Normalised Difference Water Index	[40]
$MNDWI = \frac{\rho_G - \rho_{SWIR1}}{\rho_G + \rho_{SWIR1}}$	Modified Normalised Difference Water Index	[41]
$NDAVI = \frac{\rho_{NIR} - \rho_B}{\rho_{NIR} + \rho_B}$	Normalised Difference Aquatic Vegetation Index	[36]
$WAVI = (1 + L)\frac{\rho_{NIR} - \rho_B}{\rho_{NIR} + \rho_B + L}$	Water Adjusted Vegetation Index (L: 0–1)	[21]
$WFI = \frac{\rho_{NIR} - \rho_R}{\rho_{SWIR2}}$	Wetland Forest Index	[38]
$MDI1 = \frac{\rho_{NIR} - \rho_{SWIR1}}{\rho_{SWIR1}}$ $MDI2 = \frac{\rho_{NIR} - \rho_{SWIR2}}{\rho_{SWIR2}}$	Mangrove Discrimination Index using SWIR 1 or SWIR2	[38]
$LSWI = \frac{\rho_{NIR} - \rho_{SWIR1}}{\rho_{NIR} + \rho_{SWIR1}}$	Land Surface Water Index	[42]
$RGVI = 1 - \frac{\rho_B + \rho_R}{\rho_{NIR} + \rho_{SWIR1} + \rho_{SWIR2}}$	Rice Growth Vegetation Index	[37]

2.5. Sensitivity Analysis Methods

2.5.1. Variance-Based Sensitivity Analysis

VBSA employed variance to represent sensitivity and attributed the variance of the model output to the variances of input factors. For a model with k independent input factors and a scalar output Y: $Y = f(X_1, X_2, \ldots, X_k)$, the variance ($V(\cdot)$) of Y can be decomposed as [14,15]:

$$V(Y) = \sum_i V_i + \sum_i \sum_{j>i} V_{ij} + \ldots + V_{12\ldots k} \quad (1)$$

The first order indices S_i could be represented by the expected reduction in output variance if X_i could be fixed [15]:

$$S_i = \frac{V_i}{V(Y)} = \frac{V_{X_i}(E_{\mathbf{X}_{\sim i}}(Y|X_i))}{V(Y)} \quad (2)$$

where $E(\cdot)$ was the expectation operator and subscripts X_i or $\mathbf{X}_{\sim i}$ indicated that the operation ($E(\cdot)$ or $V(\cdot)$) was taken over X_i or all factors but X_i. The total order indices S_i^T could be written as [15]:

$$S_i^T = \frac{E_{\mathbf{X}_{\sim i}}(V_{X_i}(Y|\mathbf{X}_{\sim i}))}{V(Y)} = 1 - \frac{V_{\mathbf{X}_{\sim i}}(E_{X_i}(Y|\mathbf{X}_{\sim i}))}{V(Y)} \quad (3)$$

Likewise, $V_{\mathbf{X}_{\sim i}}(E_{X_i}(Y|\mathbf{X}_{\sim i}))$ was the expected reduction in output variance if all input factors but X_i could be fixed. Other partial variances in Equation (1) corresponded to the interactions among input factors, and more details could be found in [15]. Both first and total order indices were broadly used, but this study only focused on total order indices. Due to numerical errors, this approach might produce negative sensitivity indices, usually corresponding to noninfluential factors [7]. Hence, negative values were reset to zero.

Open-source software was available for VBSA, such as SimLab [43] and the SAFE toolbox [44] (https://www.safetoolbox.info/info-and-documentation/, accessed on 7 May 2019). The SAFE toolbox for MATLAB® was employed in this study as it also provided a DBSA method. Trial tests showed that the computed sensitivity indices in VBSA could be affected by the number of samples, especially for NRM. Multiple sample sizes from 16,000 (1000 × 16) to 128,000 (8000 × 16) with an increment of 16,000 were tested for SPS_NRM and DEN_NRM. Finally, 112,000 samples were generated for each VBSA case as the computed sensitivity indices of reflectance spectra were consistent and stable.

2.5.2. Density-Based Sensitivity Analysis

The PAWN method [25] in the SAFE toolbox was adopted for DBSA in this study. It employed output CDFs instead of PDFs to reduce computational costs and to avoid tuning parameters, e.g., the bin width for calculating PDF [25]. The distance between unconditional and conditional output CDFs was taken as a measure of sensitivity. First, N_u samples were randomly generated for a model of k input factors $Y = f(X_1, X_2, \ldots X_k)$, to evaluate the empirical unconditional CDF of output $F_Y(Y)$. Second, n samples were generated for each input factor X_i and acted as conditional points. At each conditional point, N_c random samples were generated over $\mathbf{X}_{\sim i}$ to obtain the empirical conditional CDF of output $F_{Y|X_i}(Y)$. The Kolmogorov–Smirnov (KS) statistic was applied as a measure of the distance between unconditional and conditional CDFs (as cited in [25]):

$$KS(X_i) = \max_Y \left| F_Y(Y) - F_{Y|X_i}(Y) \right| \quad (4)$$

In other words, it was the maximum divergence between conditional and unconditional CDFs for a conditional point of X_i. Considering all the conditional points of an input factor X_i, another statistic, e.g., the maximum or the median, was computed over all conditional points of X_i to obtain the PAWN sensitivity index P_i^T [25]:

$$P_i^T = \underset{X_i}{\text{statistic}}[KS(X_i)] \quad (5)$$

By its definition, P_i^T would be between 0 and 1, and higher values suggested a greater influence on the output [25]. Taking a result for S2A NDVI as an example, the unconditional and conditional CDFs and the KS statistics for all conditioning points of fCv(3) are in Figure 4. As the KS curves had various patterns, derived P_i^T values using the maximum or the median statistic could be different and thus lead to divergent conclusions. The maximum statistic was adopted because it could detect factors that had influences on the output for at least one conditioning point [32]. The number of samples in a PAWN case was 66,000 in this study, with 2000 samples used to calculate the unconditional CDF, and 20 conditioning points for each of the 16 input factors and 200 samples used per conditioning point. More conditioning points or samples did not change the computed sensitivity indices remarkably.

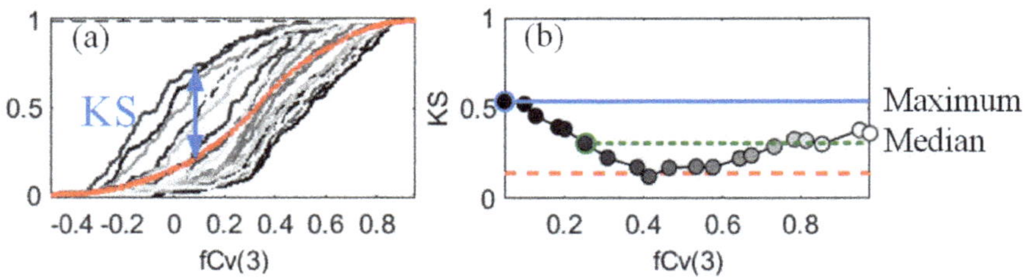

Figure 4. Unconditional (red) and conditional (grey) cumulative distribution functions (**a**) and corresponding Kolmogorov–Smirnov (KS) statistics (**b**) of the normalised difference vegetation index with regard to the fractional cover of the mangrove crown (fCv(3)). The PAWN index can be different if the maximum (blue) or the median (green) metric is adopted (**b**).

3. Results

The normalised sensitivity indices (NSIs) for reflectance spectra and sensitivity indices (SIs) for S2A bands and VIs from VBSA and PAWN were presented as stacked bar plots (Figures 5 and 6). As VIs had various scales, they were not normalised to avoid confusion, especially for VBSA indices that sometimes only one or two factors achieved positive sensitivity values, e.g., MDI2 in Figure 6c.

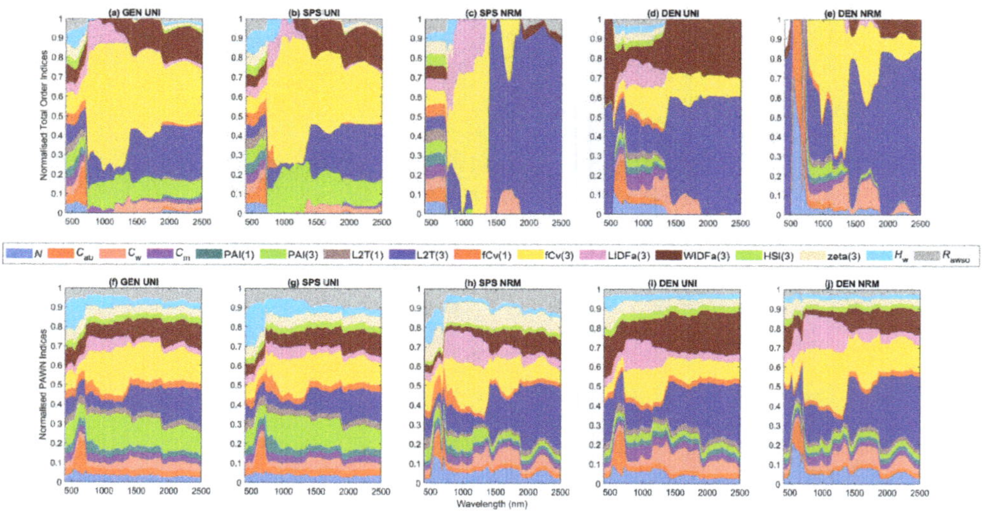

Figure 5. Stacked bar plots for normalised sensitivity indices (NSIs) of reflectance spectra in VBSA (**a–e**) and PAWN (**f–j**) for a general (GEN) canopy with uniformly sampled inputs (UNI) (**a**,**f**), a sparse (SPS) canopy with UNI (**b**,**g**), SPS with normally sampled inputs (NRM) (**c**,**h**), a dense (DEN) canopy with UNI (**d**,**i**) and DEN with NRM (**e**,**j**). Each x value corresponds to a wavelength. The NSI value is represented by the height of a bar and marked in the y-axis. Each colour corresponds to a factor defined in Table 1.

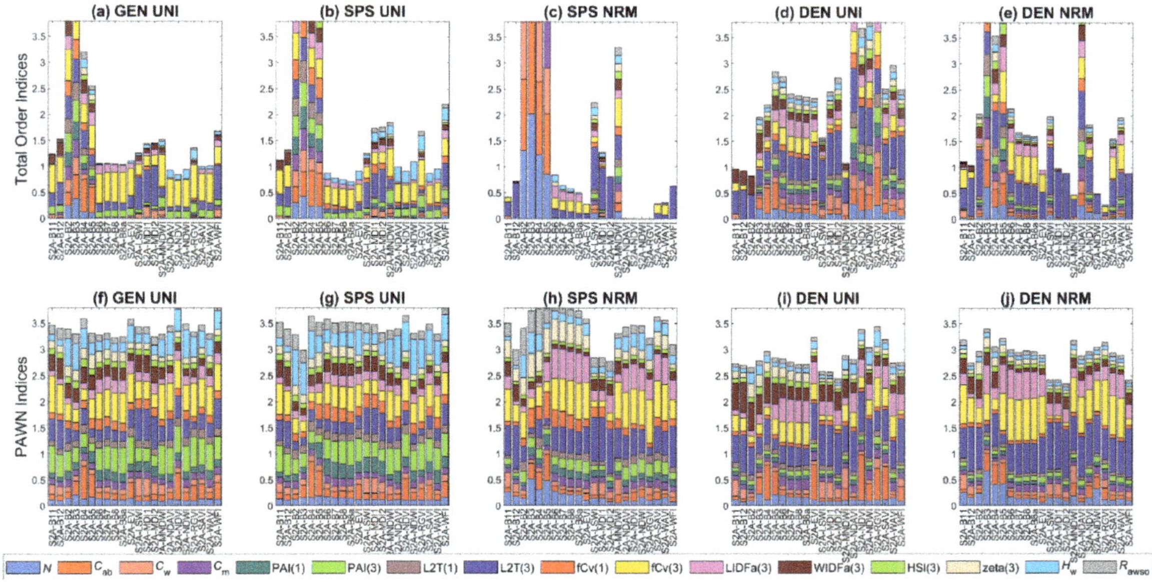

Figure 6. Stacked bar plots for sensitivity indices (SIs) of Sentinel-2A (S2A) band reflectance and vegetation indices (VIs) in VBSA (**a–e**) and PAWN (**f–j**) for a general (GEN) canopy with uniformly sampled inputs (UNI) (**a**,**f**), a sparse (SPS) canopy with UNI (**b**,**g**), SPS with normally sampled inputs (NRM) (**c**,**h**), a dense (DEN) canopy with UNI (**d**,**i**) and DEN with NRM (**e**,**j**). Each x value corresponds to a S2A band reflectance or a VI. The SI value is represented by the height of a bar and marked in the y-axis. The scale of the y-axis is restricted from 0 to 3.8 to show sufficient details. Each colour corresponds to a factor defined in Table 1.

3.1. General Scenario

For GEN, two large areas in blue and yellow in Figure 5a could be easily recognised, indicating that L2T(3) and fCv(3) had significant influences on canopy reflectance across nearly the whole visible to SWIR regions in VBSA. Exceptions were the boundaries of chlorophyll absorption bands around 600 nm and 710 nm where the impact of C_{ab} rose and fell abruptly and the NIR bands where the influence of L2T(3) dropped. WIDFa(3) was the next most influential factor, especially in the SWIR, followed by PAI(3) across NIR and SWIR bands. Comparatively, PAWN NSIs were slightly different from VBSA NSIs in that PAI(3) had higher NSI values while the NSI values of L2T(3) and fCv(3) were lower, especially in SWIR (Figure 5f). It is worth noting that H_w had higher NSI values in visible bands compared with its NSI values in other regions.

Since PAWN SIs were always between 0 and 1 while VBSA SIs could have various scales and negative values were reset to zero, the NSI patterns of VIs in the two methods could appear distinct. The stacked PAWN SIs were around 3.5 while stacked VBSA SIs varied from lower than 1 to higher than 6. The scale of the y-axis in Figure 6 was limited to 0–3.8 to maintain sufficient details. Once they were normalised, the length of each bar could be stretched or compressed. Thus, the relative length of the bar for a specific factor might look different in stacked bar plots of SIs and NSIs. In PAWN, multiple VIs had similarly high SI values of fCv(3) such as S2A-B11, MNDWI and RGVI (Figure 6f). A challenge with the PAWN indices was that all factors obtained a SI value between 0 and 1. Hence, the influential factors might look less outstanding than they did in VBSA, even if its SI value was high, e.g., the NSI of fCv(3) for MNDWI was 0.21, but its SI value was 0.68.

The output PDFs in VBSA were shown in Figure 7. Those outputs that had high NSI or SI values to fCv(3) in VBSA, such as S2A-B8a, MNDWI, NDAVI and RGVI, exhibited various PDFs. Their SIs with confidence intervals were obtained via bootstrapping and further examined. Although the PDF of RGVI was remarkably skewed, all factors had narrow confidence intervals (around 0.1 or lower) and the most influential factor fCv(3) was outstanding (not shown). The confidence intervals were also narrow for MNDWI and NDAVI but were relatively large for S2A-B8a (around 0.25 or higher, Figure 8a). In the case of PAWN, the KS plots for MNDWI with regard to input factors in Figure 8b could provide more insights. KS values of fCv(3) would increase if fCv(3) was approaching 0 or 1 from 0.4 and for the latter case, KS values would become stable after fCv(3) reached about 0.9. When L2T(3) moved towards the two ends 0 or 1, its KS values slightly increased but were not obvious. As for PAI(3) and H_w, their KS values only grew noticeably if their values decreased to zero, e.g., lower than 1 for PAI(3). This implied that PAI(3) was more influential when its value was low.

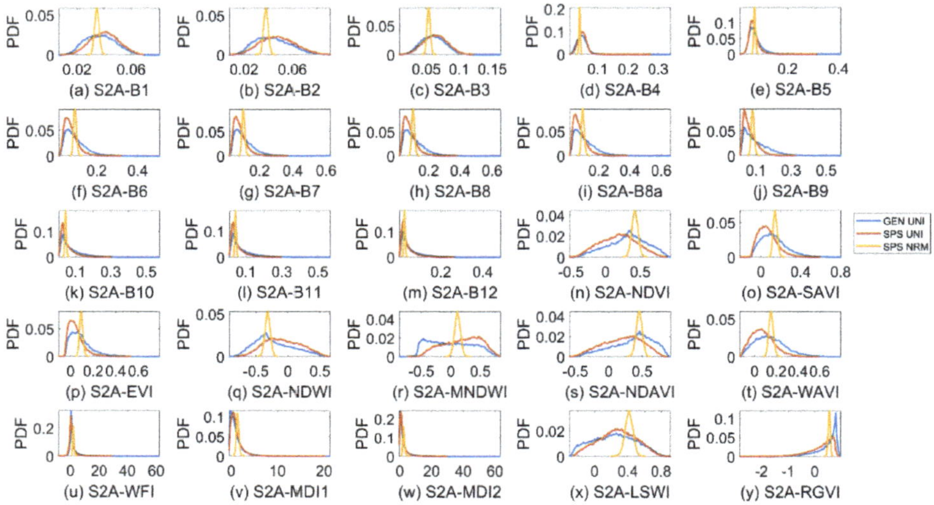

Figure 7. Probability density function (PDF) of Sentinel-2A (S2A) band reflectance and vegetation indices for a general (GEN) canopy with uniformly sampled inputs (UNI) (blue curves), a sparse (SPS) canopy with UNI (red curves) and SPS with normally sampled inputs (NRM) (yellow curves). The bin size for SPS_NRM is one-third of that for GEN_UNI and SPS_UNI to show details of the latter two.

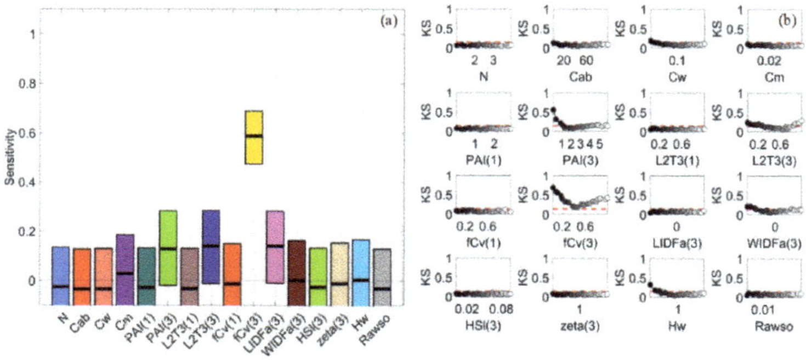

Figure 8. Mean sensitivity indices and confidence intervals estimated via bootstrapping for S2A-B8a in VBSA (**a**). Kolmogorov–Smirnov (KS) statistics for MNDWI with regard to input factors in PAWN (**b**).

3.2. Sparse Mangroves—Uniform Input Probability Distributions

The NSIs of reflectance spectra for SPS_UNI had many similarities with those for GEN_UNI, but there was an increased impact of H_w in visible bands (Figure 5b,g). In the NIR region, the NSI values of PAI(1) and fCv(1) slightly increased in PAWN, but only the NSI values of fCv(1) increased in VBSA. This implied that the understory and water background had a larger impact on SPS, which was intuitive. Therefore, more attention was paid to H_W in this section.

S2A-B3 achieved the highest SI value of 0.87 to H_W, but all factors had large confidence intervals (around 0.4 or higher, Figure 9a). Although H_W could still be distinguished, its bounds were already overlapped with those of other factors. This was somewhat unexpected as the S2A-B3 values were quite normally distributed and far from skewed (Figure 7c). Moreover, a larger sample size of up to 128,000 did not significantly reduce the confidence intervals. For RGVI and NDAVI, H_W, fCv(3) and PAI(3) were all identified as influential, but their SI values slightly varied and their confidence intervals might overlap. For NDWI, the SI of fCv(3) was slightly higher than that of H_W, i.e., 0.360 vs. 0.348. Their confidence intervals were narrow (lower than 0.1) but still overlapping. Ranking them further might not be reliable because of the overlapping confidence intervals [8,45]. Therefore, a high SI itself in VBSA might not guarantee a solid link between the output and an input factor. It was better to check the output distributions and confidence intervals as well, as suggested by [32].

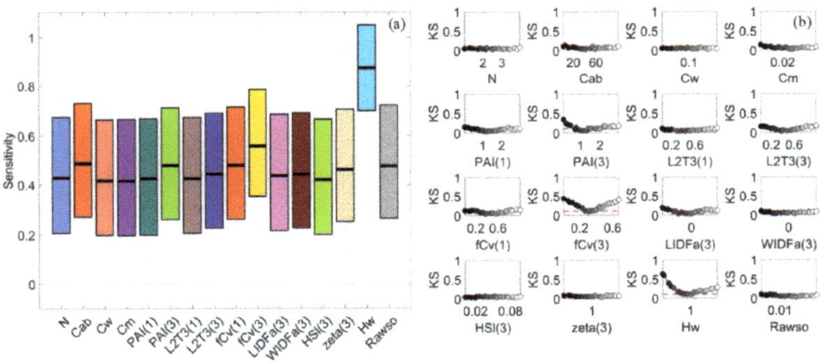

Figure 9. Mean sensitivity indices and confidence intervals estimated via bootstrapping for S2A-B3 in VBSA (**a**). Kolmogorov–Smirnov (KS) statistics for NDWI with regard to input factors in PAWN (**b**).

By comparison, S2A-B3 had similar SI patterns in PAWN, with the SI of H_W higher than 0.6 and those of other factors lower than 0.3, but the confidence intervals of all factors were much narrower (around 0.1 or lower). For NDWI, H_W, fCv(3) and PAI(3) had SI values of 0.685, 0.414 and 0.404, separately. All factors had narrow confidence intervals (lower than 0.1) in PAWN, and there was no overlapping between them and other factors. As for the KS plot Figure 9b, PAI(3) and fCv(3) presented similar patterns with those in Figure 8b, except that the KS values of fCv(3) continued rising with increasing fCv(3) since its maximum value was 0.7 rather than 1 in GEN. H_W had a two-way impact on NDWI and S2A-B3. This might be attributed to the different contributions of absorption and scattering by the water body and its components, both of which could vary with water depth. In contrast, only low H_W values could make a difference for VIs based on SWIR bands, e.g., MNDWI in Figure 8b, as the absorption of water in SWIR was strong.

3.3. Sparse Mangroves—Normal Input Probability Distributions

This case corresponded to small variations in the biophysical and biochemical properties of mangrove plots and the environmental factors. As shown in Figure 5b,c, there was a dramatic change in the NSI patterns of reflectance spectra in VBSA. fCv(3) and LIDFa(3) became dominant in NIR, while the most influential factor was L2T(3) in SWIR. Additionally, the impact of other factors was minor in these regions. Consequently, only a limited number of factors obtained positive SIs for multiple VIs such as the EVI and NDVI (Figure 6c). In contrast, the NSI patterns in PAWN also changed but more smoothly (Figure 5h). fCv(3) and L2T(3) were also the most influential factors in NIR and SWIR, separately. In addition, the influences of fCv(1) and LIDFa(3) increased in NIR and the impact of tree shape factor zeta(3) also increased.

As shown in Figure 7, the output PDFs in VBSA for SPS_NRM were remarkably different from those for GEN_UNI and SPS_UNI. This indicated that input distributions could affect the output distributions. However, the normal-like output PDFs did not guarantee that the confidence intervals of SIs were narrow. Taking MNDWI as an example, the confidence intervals were around 0.3, and there was overlapping between influential factors fCv(3) and L2T(3) and other factors (Figure 10a). Comparatively, the confidence intervals for MNDWI (Figure 10b), MDI1 and LSWI in PAWN were basically within 0.1.

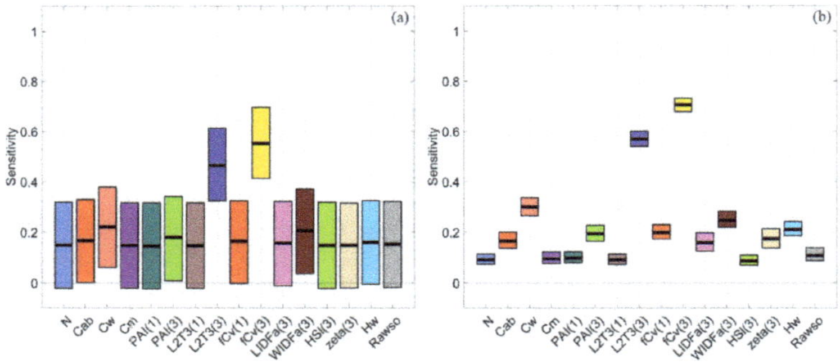

Figure 10. Mean sensitivity indices and confidence intervals estimated via bootstrapping for MNDWI in VBSA (**a**) and PAWN (**b**).

3.4. Dense Mangroves—Uniform Input Probability Distributions

When the canopy became dense, the impact of PAI(3) dropped quickly, and the impact of fCv(3) also declined noticeably, as shown by their shrunken areas in Figure 5d,i. H_w barely made a difference in this case. In the meanwhile, the influences of LIDFa(3) increased in NIR and C_w in SWIR. Moreover, the impact of L2T(3) and WIDFa(3) grew significantly, especially in SWIR. L2T(3) had the highest SI values over other factors for multiple VIs such as MDI2, LSWI, NDAVI and WFI (Figure 6d,i). This might be attributed to the distinct spectral responses of leaves and woody material, and the adopted reflectance values of woody material were higher than the reflectance of leaves in SWIR.

It was interesting to find that C_{ab} also had outstanding PAWN SI values for multiple VIs such as NDVI, EVI and S2A-B4 (Figure 6i). Comparatively, this was less significant for NDVI and EVI in VBSA (Figure 6d). The KS plots for NDVI in PAWN were presented in Figure 11b. L2T(3) had a two-way impact while the KS values of C_{ab} increased dramatically only when its values were lower than 20 $\mu g \cdot cm^{-2}$. This was because low C_{ab} values close to zero could remarkably increase the canopy reflectance in chlorophyll absorption bands [46]. The SI values of C_{ab} and L2T(3) for NDVI in PAWN were 0.879 and 0.771, respectively, with SI values of other factors lower than 0.4. The confidence intervals of all factors were within 0.1 in PAWN. In contrast, L2T(3) was the most influential factor (SI = 0.544) for NDVI in VBSA, followed by C_{ab} (SI = 0.207, Figure 11a). However, the confidence intervals were wide (around 0.3 or higher) and overlapping.

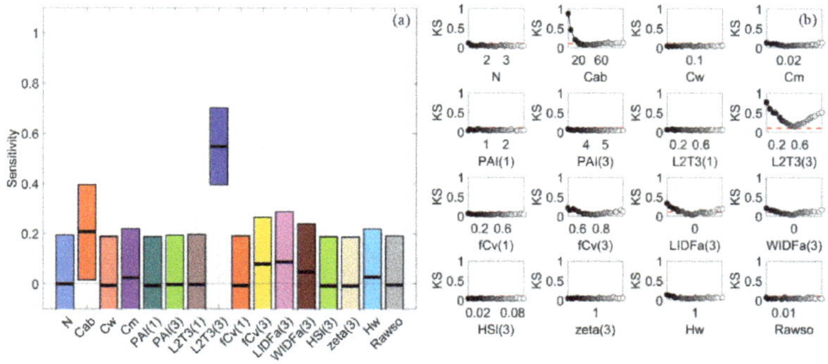

Figure 11. Mean sensitivity indices and confidence intervals estimated via bootstrapping for NDVI in VBSA (**a**). Kolmogorov–Smirnov (KS) statistics for NDVI with regard to input factors in PAWN (**b**).

3.5. Dense Mangroves—Normal Input Probability Distributions

Compared with the DEN_UNI case (Figure 5d,i), L2T(3) together with fCv(3) still had a dominant impact in NIR and SWIR for DEN_NRM, but L2T(3) was less influential in some NIR bands in VBSA (Figure 5e,j). In addition, LIDFa(3) also had noticeable influences in NIR. WIDFa(3) became less influential compared with that for DEN_UNI. Similar to SPS_NRM, only a limited number of factors had positive SI values in VBSA for multiple VIs such as EVI, MNDWI and WFI (Figure 6e) and the confidence intervals were generally wide (0.25 or higher) in VBSA.

LIDFa(3) was investigated in this section as other influential factors such as fCv(3) and PAI(3) had been covered in previous sections. As shown in Figure 6e,j, NIR bands were the most sensitive to LIDFa(3). Therefore, the KS statistics and confidence intervals for S2A-B8 in PAWN were presented in Figure 12. fCv(3), L2T(3) and LIDFa(3) had high SI values and were distinct from other factors. The KS plots of fCv(3) and L2T(3) in Figure 12b were similar to those in Figures 8b and 9b, indicating that the output CDF diverged from the unconditional CDF when the fCv(3) and L2T(3) values increased or decreased. The KS plot of LIDFa(3) presented a similar pattern to that of fCv(3). Moreover, the KS values were higher when LIDFa(3) moved towards −0.6 ($\theta_l = 67°$) than they were when LIDFa(3) was close to 0 ($\theta_l = 45°$), which implied that canopy reflectance in NIR changed more significantly if leaves were vertically distributed.

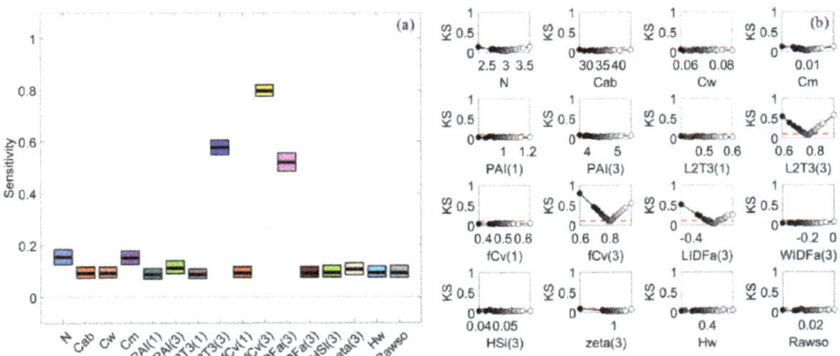

Figure 12. Mean sensitivity indices and confidence intervals estimated via bootstrapping for S2A-B8 (**a**) and Kolmogorov–Smirnov (KS) statistics for S2A-B8 with regard to input factors (**b**) in PAWN.

3.6. General Scenario with Correlated PAI(3) and fCv(3)

If PAI(3) and fCv(3) were correlated for a general scenario, both were recognised as influential factors in PAWN. Besides, similar NSIs of canopy reflectance could be obtained with the GEN_UNI with random PAI(3) and fCv(3) values (Figures 5f and 13b). In contrast,

VBSA only identified PAI(3) as being influential in NIR and SWIR while fCv(3) had virtually no impact (Figures 5a and 13a). The computed NSIs in VBSA in Figure 13a might be meaningless because there were correlated inputs. Thus, no further interpretation was given. This plot demonstrated that the correlated biophysical properties of mangroves changed the calculated sensitivity indices in VBSA.

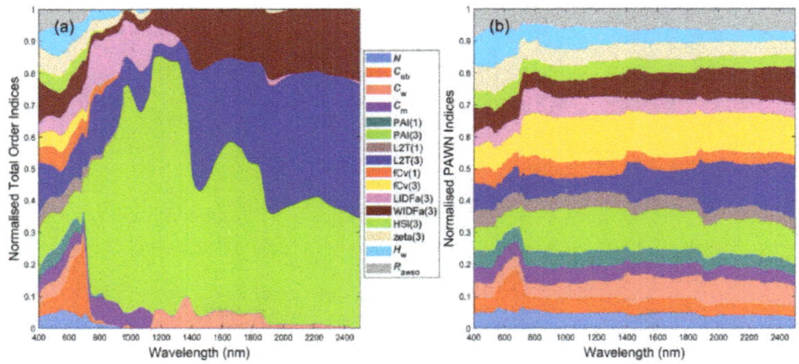

Figure 13. Stacked bar plots for normalised sensitivity indices of reflectance spectra in VBSA (**a**) and PAWN (**b**) for a general canopy with correlated PAI(3) and fCv(3). Each x value corresponds to an output, e.g., reflectance or VI, and each colour represents an input factor. The SI value is represented by the height of a bar and marked in the y-axis.

3.7. A Brief Summary

The results demonstrated the influences of SA methods and input configurations on the calculated SIs for reflectance of mangroves and the importance of examining the confidence intervals of calculated SIs, which might be ignored in previous SA studies on the RS of vegetation. The calculated NSI and SI values highlighted that canopy reflectance of mangroves was sensitive to fCv(3) and L2T(3) and PAI(3). The influences of H_w for a sparse mangrove canopy and inclination distributions of plant material and C_{ab} for a dense canopy might also be noteworthy. VIs with SWIR bands such as MNDWI and RGVI also had potential for mapping the fCv(3) and PAI(3) of mangroves and traditional VIs like EVI. Several VIs such as LSWI, MDI and WFI were sensitive to the L2T(3) of mangroves and might be helpful for estimating the LAI of mangroves from multispectral satellite images.

4. Discussion

4.1. Global Sensitivity Analysis Methods and Interpretations of the Results

The NSIs for canopy reflectance spectra of vegetation, as shown in Figure 5a–e, were frequently adopted in VBSA [19,20,22,47]. In addition, NSIs for VIs were also applied to evaluate the sensitivities of selected VIs to specific biophysical and biochemical properties of vegetation [22,47]. Stacked NSI plots for reflectance spectra were intuitive and straightforward for identifying the most influential factors and corresponding wavelength regions. However, SI values of various VIs could have distinct scales and only a limited number of factors might have non-negative SI values in VBSA, such as EVI and SAVI in Figure 6c. Therefore, a factor could obtain a high and dominant NSI value for a VI, e.g., the NSI of fCv(3) was 0.641 for EVI, while its absolute SI value was only 0.181 and not the highest compared with other VIs. Hence, the NSI values might mislead us to conclude that EVI was the most sensitive to fCv(3) in this case. Moreover, if a VI was only sensitive to a limited number of factors, the confidence intervals of these factors might be large and overlapping, and, thus, the results were not robust.

With regards to the output PDFs in VBSA, the results only showed that input distributions could affect output distributions but did not reveal any promising links between output PDFs and the confidence intervals of calculated SIs. A highly skewed output PDF such as RGVI in Figure 7y for SPS_UNI did not necessarily mean the SIs in VBSA had wide or overlapping confidence intervals. On the other hand, a normal-like output PDF such as S2A-B3 in Figure 7c for SPS_UNI did not guarantee robust SIs in VBSA (Figure 9a). In addition, different numbers of input samples in VBSA did not significantly change the output PDFs as long as they were large enough but could result in divergent SI values in this study, especially for normally distributed input samples. Comparatively, the confidence intervals of calculated SIs for S2A band reflectance and VIs were usually narrow in PAWN with 2000 samples for the unconditional CDF and 200 samples per conditional point by 20 conditional points for conditional CDFs. The KS plots could be used to identify the value ranges of inputs where conditional output CDFs diverged from the unconditional CDF, which implied that there were noticeable changes in the output values.

4.2. Differences between Sparse and Dense Mangrove Canopies

The results also demonstrated that input ranges and distributions could affect the computed SIs, particularly in VBSA. For the GEN_UNI case that PAI varied from 0 to 6, fCv(3), L2T(3) and PAI(3) were recognised as the most influential factors (Figure 5a,f). This partly coincided with the results of [20] that fCv and PAI were among the most influential factors. For the SPS_UNI case, the impact of fCv(3) became slightly less dominant in VBSA while the impact of PAI(3) increased (Figure 5a,b). The changes in the impact of PAI(3) and fCv(3) in PAWN were not significant (Figure 5f,g). Additionally, increased influences of H_w and fCv(1) in visible and NIR bands were identified by both methods. If the inputs were turned into normal distributions (SPS_NRM), the computed NSIs of reflectance spectra also changed remarkably. As shown in Figure 5c,h, PAI(3) became less influential while the impact of fCv(3) and L2T(3) increased in NIR and SWIR, respectively.

If the canopy was dense with uniformly distributed inputs (DEN_UNI), the NSI values of fCv(3) and PAI(3) decreased, particularly for PAI(3) (Figure 5d,i). Factors regarding the understory and background such as PAI(1) and H_w had negligible influence now. In the meanwhile, biophysical properties related to woody material, such as L2T(3) and WIDFa(3), had a dominant impact in SWIR. The impact of LIDFa(3) also grew remarkably in NIR. In addition, the influences of leaf parameters, including C_{ab}, C_w and C_m, increased by various degrees in different spectral regions. For the DEN_NRM case, the NSI patterns in PAWN (Figure 5j) were similar but slightly different from Figure 5i for DEN_UNI. The impact of fCv(3), L2T(3) and LIDFa(3) increased while that of WIDFa(3) decreased. In VBSA, fCv(3) and L2T(3) became the most influential factors, and the influences of LIDFa(3) and WIDFa(3) both decreased (Figure 5e).

In summary, PAI(3) had a larger impact under the sparse canopy while the influences of leaf parameters and inclination angles increased under a dense canopy. This agreed with the results in [19,22]. The new factors fCv(3) and L2T(3) were among the most influential factors under all examined scenarios except for the correlated case where fCv(3) was not identified as influential in VBSA. H_w mainly affected the reflectance of sparse canopies, but it was worth noting that the water body was not allowed to immerse the crown in this study. Otherwise, the infrared reflectance of submerged mangroves would reduce dramatically, which had been used to identify them from satellite images [4,48].

4.3. Potential Limitations and Suggestions

The structure of real mangrove forests was more complicated than the model could simulate. Simplifications in the model might result in overestimated contributions from fCv(3) because it was applied to weight the reflected radiation from mangroves directly. Additionally, the woody material was assumed to be randomly distributed and to share the same fractional cover with leaves in the adopted model [28]. Its reflectance spectra were averaged from field measurements over the lower parts (lower than 2 m) of stems [28], which might be different from the reflectance of branches and shoots in the crowns. Consequently, the contributions of L2T(3) and its SIs could also be overestimated.

Although there were potential divergences in the computed SI values of L2T(3), the impact of woody material on canopy reflectance were demonstrated [5,49]. In addition, caution for the inclination distributions of leaves and wood was still needed because they could be an obstacle when mapping the LAI of mangroves from satellite images. The leaf area ratio could be measured by classifying the point clouds acquired by laser scanning [50,51]. Average inclination angles could also be calculated from point clouds [52,53] or digital photos [54]. However, convenient and operational methods were still lacking [55], particularly when they needed to be measured together with PAI and fCv at many of the plots to be used as calibration or validation data.

For future studies, it might be a good start to map the fCv of mangroves together with PAI because field data of fCv were relatively easy to collect, e.g., by taking hemispherical photos or upward photos [56]. Moreover, the data could be used to assess the potential correlation between PAI and fCv, and further the impact of fCv and background water on PAI mapping. According to the GSA results in this study, multiple VIs may be suitable for mapping PAI and fCv of mangroves from S2A images. Besides traditional VIs such as EVI and NDVI, other VIs such as MNDWI and RGVI were also sensitive to fCv and PAI. The new VIs for mangroves or aquatic plants such as MDI, WFI and RGVI had a common characteristic that SWIR bands were incorporated. This study showed that the SWIR bands and these VIs could be helpful for mapping mangroves, but the optimal VI for mapping a biophysical property from a specific dataset might vary.

5. Conclusions

Based on a canopy reflectance model of mangroves, this study demonstrated that GSA methods (variance-based or density-based), input ranges and probability distributions (uniform or normal) could affect the computed sensitivity indices under the examined mangrove scenarios. Briefly, fCv(3) and L2T(3) were among the influential factors for the infrared reflectance of mangroves under the examined scenarios. PAI(3) was also influential for a sparse canopy but became less influential for a dense canopy. In contrast, inclination distributions of plant material and leaf parameters, e.g., leaf water content, could become more influential in infrared bands for a dense canopy. Moreover, the influence of water depth was noteworthy for a sparse canopy and maybe other scenarios if the water body could immerse the crown.

Since the results and conclusions can be different if a specific model, GSA method and input configuration are adopted, it may be essential to perform a tailored GSA according to the study area and available data. Based on the results, it is recommended that attention should be paid to the L2T and fCv as they may affect estimating the LAI or PAI of mangroves. A complete field dataset that includes the factor to be mapped such as LAI and if possible, other influential factors such as fCv, L2T ratio and inclination angles would be beneficial to further analyses and evaluations of their impact. This, in turn, could contribute to the development of a protocol on field data collection for mapping mangrove from remote sensing images. Considering the challenges in field data collection, it may be a good start to collect and map PAI and fCv of mangroves. For VI based mapping methods, more choices such as MNDWI and RGVI deserve an attempt if SWIR bands are available, but tests are still needed to determine the optimal VI for mapping a specific property based on the available field data and RS images.

Author Contributions: Conceptualization, C.N. and S.P.; Data curation, C.N.; Formal analysis, C.N.; Funding acquisition, C.N. and S.P.; Methodology, C.N.; Project administration, C.N. and S.P.; Resources, C.N. and S.P.; Software, C.N.; Supervision, S.P. and C.R.; Writing—original draft, C.N.; Writing—review and editing, S.P. and C.R. All authors have read and agreed to the published version of the manuscript.

Funding: Chunyue Niu was supported by an Australian Government Research Training Program (RTP) Scholarship and Remote Sensing Research Centre, School of Earth and Environmental Sciences, The University of Queensland.

Institutional Review Board Statement: Not applicable.

Informed Consent Statement: Not applicable.

Data Availability Statement: Not applicable.

Acknowledgments: We thank Francesca Pianosi, Fanny Sarrazin and Thorsten Wagener at the University of Bristol for sharing the SAFE toolbox. We are grateful to the anonymous reviewers for their valuable comments.

Conflicts of Interest: The authors declare no conflict of interest.

References

1. Kamal, M.; Phinn, S. Hyperspectral data for mangrove species mapping: A comparison of pixel-based and object-based approach. *Remote Sens.* **2011**, *3*, 2222–2242. [CrossRef]
2. Kovacs, J.M.; Wang, J.; Flores-Verdugo, F. Mapping mangrove leaf area index at the species level using IKONOS and LAI-2000 sensors for the Agua Brava Lagoon, Mexican Pacific. *Estuar. Coast. Shelf Sci.* **2005**, *62*, 377–384. [CrossRef]
3. Heenkenda, M.K.; Maier, S.W.; Joyce, K.E. Estimating Mangrove Biophysical Variables Using WorldView-2 Satellite Data: Rapid Creek, Northern Territory, Australia. *J. Imaging* **2016**, *2*, 24. [CrossRef]
4. Jia, M.; Wang, Z.; Wang, C.; Mao, D.; Zhang, Y. A new vegetation index to detect periodically submerged Mangrove forest using single-tide Sentinel-2 imagery. *Remote Sens.* **2019**, *11*, 2043. [CrossRef]
5. Widlowski, J.-L.; Côté, J.-F.; Béland, M. Abstract tree crowns in 3D radiative transfer models: Impact on simulated open-canopy reflectances. *Remote Sens. Environ.* **2014**, *142*, 155–175. [CrossRef]
6. Cárdenas, N.Y.; Joyce, K.E.; Maier, S.W. Monitoring mangrove forests: Are we taking full advantage of technology? *Int. J. Appl. Earth Obs. Geoinf.* **2017**, *63*, 1–14. [CrossRef]
7. Saltelli, A.; Ratto, M.; Andres, T.; Campolongo, F.; Cariboni, J.; Gatelli, D.; Saisana, M.; Tarantola, S. *Global Sensitivity Analysis: The Primer*; John Wiley & Sons: Hoboken, NJ, USA, 2008.
8. Pianosi, F.; Beven, K.; Freer, J.; Hall, J.W.; Rougier, J.; Stephenson, D.B.; Wagener, T. Sensitivity analysis of environmental models: A systematic review with practical workflow. *Environ. Model. Softw.* **2016**, *79*, 214–232. [CrossRef]
9. Jacquemoud, S.; Baret, F. PROSPECT: A model of leaf optical properties spectra. *Remote Sens. Environ.* **1990**, *34*, 75–91. [CrossRef]
10. Verhoef, W. Light scattering by leaf layers with application to canopy reflectance modeling: The SAIL model. *Remote Sens. Environ.* **1984**, *16*, 125–141. [CrossRef]
11. Verhoef, W. Theory of Radiative Transfer Models Applied in Optical Remote Sensing of Vegetation Canpioes. Ph.D. Thesis, Wageningen Agricultural University, Wageningen, The Netherlands, 1998.
12. Verhoef, W.; Bach, H. Coupled soil–leaf-canopy and atmosphere radiative transfer modeling to simulate hyperspectral multi-angular surface reflectance and TOA radiance data. *Remote Sens. Environ.* **2007**, *109*, 166–182. [CrossRef]
13. Jacquemoud, S.; Verhoef, W.; Baret, F.; Bacour, C.; Zarco-Tejada, P.J.; Asner, G.P.; François, C.; Ustin, S.L. PROSPECT+ SAIL models: A review of use for vegetation characterization. *Remote Sens. Environ.* **2009**, *113*, S56–S66. [CrossRef]
14. Sobol, I.M. Global sensitivity indices for nonlinear mathematical models and their Monte Carlo estimates. *Math. Comput. Simul.* **2001**, *55*, 271–280. [CrossRef]
15. Saltelli, A.; Annoni, P.; Azzini, I.; Campolongo, F.; Ratto, M.; Tarantola, S. Variance based sensitivity analysis of model output. Design and estimator for the total sensitivity index. *Comput. Phys. Commun.* **2010**, *181*, 259–270. [CrossRef]
16. Liu, J.; Pattey, E.; Jégo, G. Assessment of vegetation indices for regional crop green LAI estimation from Landsat images over multiple growing seasons. *Remote Sens. Environ.* **2012**, *123*, 347–358. [CrossRef]
17. Saltelli, A.; Tarantola, S.; Chan, K.-S. A quantitative model-independent method for global sensitivity analysis of model output. *Technometrics* **1999**, *41*, 39–56. [CrossRef]
18. Campbell, G. Derivation of an angle density function for canopies with ellipsoidal leaf angle distributions. *Agric. For. Meteorol.* **1990**, *49*, 173–176. [CrossRef]
19. Xiao, Y.; Zhao, W.; Zhou, D.; Gong, H. Sensitivity analysis of vegetation reflectance to biochemical and biophysical variables at leaf, canopy, and regional scales. *IEEE Trans. Geosci. Remote Sens.* **2014**, *52*, 4014–4024. [CrossRef]
20. Mousivand, A.; Menenti, M.; Gorte, B.; Verhoef, W. Global sensitivity analysis of the spectral radiance of a soil–vegetation system. *Remote Sens. Environ.* **2014**, *145*, 131–144. [CrossRef]
21. Villa, P.; Mousivand, A.; Bresciani, M. Aquatic vegetation indices assessment through radiative transfer modeling and linear mixture simulation. *Int. J. Appl. Earth Obs. Geoinf.* **2014**, *30*, 113–127. [CrossRef]
22. Zhou, G.; Ma, Z.; Sathyendranath, S.; Platt, T.; Jiang, C.; Sun, K. Canopy Reflectance Modeling of Aquatic Vegetation for Algorithm Development: Global Sensitivity Analysis. *Remote Sens.* **2018**, *10*, 837. [CrossRef]
23. Zhou, G.; Niu, C.; Xu, W.; Yang, W.; Wang, J.; Zhao, H. Canopy modeling of aquatic vegetation: A radiative transfer approach. *Remote Sens. Environ.* **2015**, *163*, 186–205. [CrossRef]
24. Saltelli, A. Sensitivity analysis for importance assessment. *Risk Anal.* **2002**, *22*, 579–590. [CrossRef]
25. Pianosi, F.; Wagener, T. A simple and efficient method for global sensitivity analysis based on cumulative distribution functions. *Environ. Model. Softw.* **2015**, *67*, 1–11. [CrossRef]
26. Borgonovo, E.; Tarantola, S. Moment independent and variance-based sensitivity analysis with correlations: An application to the stability of a chemical reactor. *Int. J. Chem. Kinet.* **2008**, *40*, 687–698. [CrossRef]
27. Borgonovo, E.; Plischke, E. Sensitivity analysis: A review of recent advances. *Eur. J. Oper. Res.* **2016**, *248*, 869–887. [CrossRef]

28. Niu, C. Canopy Reflectance Modelling for Mapping Coastal Wetland Vegetation of South East Queensland, Australia. Ph.D. Thesis, The University of Queensland, Brisbane, Australia, 2021.
29. Kearney, M.S.; Stutzer, D.; Turpie, K.; Stevenson, J.C. The effects of tidal inundation on the reflectance characteristics of coastal marsh vegetation. *J. Coast. Res.* **2009**, *25*, 1177–1186. [CrossRef]
30. McKay, M.D.; Beckman, R.J.; Conover, W.J. Comparison of three methods for selecting values of input variables in the analysis of output from a computer code. *Technometrics* **1979**, *21*, 239–245.
31. Kamal, M.; Phinn, S.; Johansen, K. Assessment of multi-resolution image data for mangrove leaf area index mapping. *Remote Sens. Environ.* **2016**, *176*, 242–254. [CrossRef]
32. Noacco, V.; Sarrazin, F.; Pianosi, F.; Wagener, T. Matlab/R workflows to assess critical choices in Global Sensitivity Analysis using the SAFE toolbox. *MethodsX* **2019**, *6*, 2258–2280. [CrossRef]
33. Henrich, V.; Götze, E.; Jung, A.; Sandow, C.; Thürkow, D.; Gläßer, C. Development of an Online indices-database: Motivation, concept and implementation. In Proceedings of the 6th EARSeL Imaging Spectroscopy SIG Workshop Innovative Tool for Scientific and Commercial Environment Applications, Tel Aviv, Israel, 16–18 March 2009.
34. Rouse, J.W.; Haas, R.H.; Schell, J.A.; Deering, D.W.; Harlan, J.C. Monitoring the vernal advancement and retrogradation (green wave effect) of natural vegetation. In *NASA/GSFCT Type III Final Report*; Greenbelt, MD, USA, 1974.
35. Huete, A.R.; Liu, H.Q.; Batchily, K.V.; Van Leeuwen, W. A comparison of vegetation indices over a global set of TM images for EOS-MODIS. *Remote Sens. Environ.* **1997**, *59*, 440–451. [CrossRef]
36. Villa, P.; Laini, A.; Bresciani, M.; Bolpagni, R. A remote sensing approach to monitor the conservation status of lacustrine Phragmites australis beds. *Wetl. Ecol. Manag.* **2013**, *21*, 399–416. [CrossRef]
37. Nuarsa, I.W.; Nishio, F.; Hongo, C. Spectral characteristics and mapping of rice plants using multi-temporal Landsat data. *J. Agric. Sci.* **2011**, *3*, 54. [CrossRef]
38. Wang, D.; Wan, B.; Qiu, P.; Su, Y.; Guo, Q.; Wang, R.; Sun, F.; Wu, X. Evaluating the Performance of Sentinel-2, Landsat 8 and Pléiades-1 in Mapping Mangrove Extent and Species. *Remote Sens.* **2018**, *10*, 1468. [CrossRef]
39. Huete, A.R. A soil-adjusted vegetation index (SAVI). *Remote Sens. Environ.* **1988**, *25*, 295–309. [CrossRef]
40. McFeeters, S.K. The use of the Normalized Difference Water Index (NDWI) in the delineation of open water features. *Int. J. Remote Sens.* **1996**, *17*, 1425–1432. [CrossRef]
41. Xu, H. Modification of normalised difference water index (NDWI) to enhance open water features in remotely sensed imagery. *Int. J. Remote Sens.* **2006**, *27*, 3025–3033. [CrossRef]
42. Xiao, X.; Boles, S.; Liu, J.; Zhuang, D.; Frolking, S.; Li, C.; Salas, W.; Moore, B. Mapping paddy rice agriculture in southern China using multi-temporal MODIS images. *Remote Sens. Environ.* **2005**, *95*, 480–492. [CrossRef]
43. Tarantola, S.; Becker, W. SIMLAB Software for Uncertainty and Sensitivity Analysis. In *Handbook of Uncertainty Quantification*; Ghanem, R., Higdon, D., Owhadi, H., Eds.; Springer: New York, NY, USA, 2015.
44. Sarrazin, F.; Pianosi, F.; Wagener, T. An Introduction to the SAFE Matlab Toolbox With Practical Examples and Guidelines. In *Sensitivity Analysis in Earth Observation Modelling*; Petropoulos, G., Srivastava, P.K., Eds.; Elsevier: Amsterdam, The Netherlands, 2017; pp. 363–378.
45. Vanuytrecht, E.; Raes, D.; Willems, P. Global sensitivity analysis of yield output from the water productivity model. *Environ. Model. Softw.* **2014**, *51*, 323–332. [CrossRef]
46. Jacquemoud, S. Inversion of the PROSPECT+ SAIL canopy reflectance model from AVIRIS equivalent spectra: Theoretical study. *Remote Sens. Environ.* **1993**, *44*, 281–292. [CrossRef]
47. Morcillo-Pallarés, P.; Rivera-Caicedo, J.P.; Belda, S.; De Grave, C.; Burriel, H.; Moreno, J.; Verrelst, J. Quantifying the Robustness of Vegetation Indices through Global Sensitivity Analysis of Homogeneous and Forest Leaf-Canopy Radiative Transfer Models. *Remote Sens.* **2019**, *11*, 2418. [CrossRef]
48. Zhang, X.; Treitz, P.M.; Chen, D.; Quan, C.; Shi, L.; Li, X. Mapping mangrove forests using multi-tidal remotely-sensed data and a decision-tree-based procedure. *Int. J. Appl. Earth Obs. Geoinf.* **2017**, *62*, 201–214. [CrossRef]
49. Malenovský, Z.; Martin, E.; Homolová, L.; Gastellu-Etchegorry, J.-P.; Zurita-Milla, R.; Schaepman, M.E.; Pokorný, R.; Clevers, J.G.; Cudlín, P. Influence of woody elements of a Norway spruce canopy on nadir reflectance simulated by the DART model at very high spatial resolution. *Remote Sens. Environ.* **2008**, *112*, 1–18. [CrossRef]
50. Ma, L.; Zheng, G.; Eitel, J.U.; Magney, T.S.; Moskal, L.M. Determining woody-to-total area ratio using terrestrial laser scanning (TLS). *Agric. For. Meteorol.* **2016**, *228*, 217–228. [CrossRef]
51. Li, Z.; Strahler, A.; Schaaf, C.; Jupp, D.; Schaefer, M.; Olofsson, P. Seasonal change of leaf and woody area profiles in a midlatitude deciduous forest canopy from classified dual-wavelength terrestrial lidar point clouds. *Agric. For. Meteorol.* **2018**, *262*, 279–297. [CrossRef]
52. Bailey, B.N.; Mahaffee, W.F. Rapid measurement of the three-dimensional distribution of leaf orientation and the leaf angle probability density function using terrestrial LiDAR scanning. *Remote Sens. Environ.* **2017**, *194*, 63–76. [CrossRef]
53. Vicari, M.B.; Pisek, J.; Disney, M. New estimates of leaf angle distribution from terrestrial LiDAR: Comparison with measured and modelled estimates from nine broadleaf tree species. *Agric. For. Meteorol.* **2019**, *264*, 322–333. [CrossRef]
54. Qi, J.; Xie, D.; Li, L.; Zhang, W.; Mu, X.; Yan, G. Estimating Leaf Angle Distribution From Smartphone Photographs. *IEEE Geosci. Remote Sens. Lett.* **2019**, *16*, 1190–1194. [CrossRef]

55. Yan, G.; Hu, R.; Luo, J.; Weiss, M.; Jiang, H.; Mu, X.; Xie, D.; Zhang, W. Review of indirect optical measurements of leaf area index: Recent advances, challenges, and perspectives. *Agric. For. Meteorol.* **2019**, *265*, 390–411. [CrossRef]
56. Weiss, M.; Baret, F. *CAN_EYE V6.4.91 User Manual*; INRA-UMR: Avignon, France, 2017.

remote sensing

MDPI

Article

Mangrove Forest Cover and Phenology with Landsat Dense Time Series in Central Queensland, Australia

Debbie A. Chamberlain [1,2,*], Stuart R. Phinn [2] and Hugh P. Possingham [1,3]

1 Centre for Biodiversity and Conservation Science, School of Biological Sciences, The University of Queensland, St. Lucia, QLD 4072, Australia; h.possingham@uq.edu.au
2 Remote Sensing Research Centre, School of Earth and Environmental Sciences, The University of Queensland, St. Lucia, QLD 4072, Australia; s.phinn@uq.edu.au
3 Office of the Queensland Chief Scientist, Queensland Government, Brisbane, QLD 4000, Australia
* Correspondence: d.chamberlain@uq.edu.au; Tel.: +61-4-0182-8535

Abstract: Wetlands are one of the most biologically productive ecosystems. Wetland ecosystem services, ranging from provision of food security to climate change mitigation, are enormous, far outweighing those of dryland ecosystems per hectare. However, land use change and water regulation infrastructure have reduced connectivity in many river systems and with floodplain and estuarine wetlands. Mangrove forests are critical communities for carbon uptake and storage, pollution control and detoxification, and regulation of natural hazards. Although the clearing of mangroves in Australia is strictly regulated, Great Barrier Reef catchments have suffered landscape modifications and hydrological alterations that can kill mangroves. We used remote sensing datasets to investigate land cover change and both intra- and inter-annual seasonality in mangrove forests in a large estuarine region of Central Queensland, Australia, which encompasses a national park and Ramsar Wetland, and is adjacent to the Great Barrier Reef World Heritage site. We built a time series using spectral, auxiliary, and phenology variables with Landsat surface reflectance products, accessed in Google Earth Engine. Two land cover classes were generated (mangrove versus non-mangrove) in a Random Forest classification. Mangroves decreased by 1480 hectares (−2.31%) from 2009 to 2019. The overall classification accuracies and Kappa coefficient for 2008–2010 and 2018–2020 land cover maps were 95% and 95%, respectively. Using an NDVI-based time series we examined intra- and inter-annual seasonality with linear and harmonic regression models, and second with TIMESAT metrics of mangrove forests in three sections of our study region. Our findings suggest a relationship between mangrove growth phenology along with precipitation anomalies and severe tropical cyclone occurrence over the time series. The detection of responses to extreme events is important to improve understanding of the connections between climate, extreme weather events, and biodiversity in estuarine and mangrove ecosystems.

Keywords: Landsat; mangrove forests; time series; Google Earth Engine; random forests; phenology; TIMESAT; climate; monitoring; Great Barrier Reef

Citation: Chamberlain, D.A.; Phinn, S.R.; Possingham, H.P. Mangrove Forest Cover and Phenology with Landsat Dense Time Series in Central Queensland, Australia. *Remote Sens.* **2021**, *13*, 3032. https://doi.org/10.3390/rs13153032

Academic Editor: Brigitte Leblon

Received: 8 May 2021
Accepted: 28 July 2021
Published: 2 August 2021

1. Introduction

Coastal and estuarine ecosystems are valued for their ecosystem services, particularly coastal protection, as a nursery for fish, and carbon sequestration [1,2]. Mangrove forests are an important part of tropical estuarine ecosystems and considered integral to the emerging blue carbon economy. Blue carbon consists of carbon that is stored, sequestered, or released from coastal vegetation ecosystems [3]. Despite this, human interference leading to widespread degradation and deforestation has caused a decline in mangrove cover and biomass. Carbon emissions from global mangrove loss is estimated to be 2391 Tg $CO_{2\,eq}$ by the end of the century [4]. For reference, emissions from the global transportation sector are projected to rise to 11,900 Tg $CO_{2\,eq}$ by 2100 [5]. Although the rate of mangrove loss has decreased substantially since the 1990s, from ~2% to <0.4% per year [6] increasing human pressure along the coast has synergistic effects with climate change and threatens the distribution and abundance of mangroves. As coastal forests with unique adaptations to saline conditions, mangroves form a characteristic vegetation zone along sheltered bays, tidal inlets and estuaries in the tropics and subtropics [7]. The major threats to mangroves are climate change, clearing (through urban areas, ports, aquaculture, and industry development), timber harvesting, dieback, changes in hydrology (e.g., the restriction or alteration of flows), and pollution [8–10]. Understanding mangrove ecosystems and mapping their extent is critical to meeting the UN sustainable development indicator 6.6.1: "change in the extent of water-related ecosystems over time". Indicator 6.6.1 is used in determining progress toward meeting Sustainable Development Goal 6 (SDG 6), which is to "ensure availability and sustainable management of water and sanitation for all" [11]. Forest degradation mapping at the canopy scale supports SDG 6, and is primarily based on time series of vegetation indices obtained from optical sensors on satellites, e.g., medium spatial resolution Landsat satellites [12].

A key requirement for understanding the impact of human activities on climate, ecosystem function, and biodiversity is assessing the rate of change in tropical forest cover [13,14]. Many studies have analysed change dynamics in forest and wetland ecosystems through remote sensing data over time [15,16]; however, image sequences are often incomplete and of limited temporal range [17,18]. In tropical regions, many observations each year are required to obtain a complete cloud-free set of images for a region of interest [13,19]. Landsat-based classification with supervised machine learning techniques for land cover, along with spectral transformations using Vegetation Indices (VIs), are an efficient method for accurate mapping and monitoring of tropical forests and mangroves [20–22]. Random Forests (RF) are an ensemble learning technique for land cover classification, robust to outliers and noise, and computationally undemanding than other classification methods (e.g., boosting, support vector machines) for practical applications with a lot of data [23–25]. Given the reported high accuracy of RF in land cover classification of forests and wetlands, RF have been employed to map the extent of mangrove forests in both the sub-tropics and tropics [26–28].

The most commonly used VI is the normalized difference vegetation index (NDVI) [29], a broadband green-sensitive (photosynthetically active) index for detecting seasonal and inter-annual variations in canopy vigour and growth in relation to climate parameters and relating these variations to the capacity of the canopy to photosynthesize [30]. In accordance, phenological variations are changes in the rate of growth as indicated by changes in photosynthesis that can be detected by satellites through spectral transformations of VIs. Indeed, the variation from season to season in the satellite spectral data as measured by remote sensing and vegetation phenology, is an essential approach for understanding the role of vegetation in climate change [31]. Xiao et al. [32] challenged the idea that habitats in tropical environments do not show variation from season to season and they identified changes in growth rate driven by variation in rainfall. With dense time series, the suitability of phenology-based smoothing techniques such as harmonic regression and Savitsky-Golay have recently been demonstrated to map tropical forest disturbance and degradation [33,34]. Furthermore, to capture the temporal variation of vegetation growth cycles the localised climate at each time series location should be extracted. Localised precipitation patterns will influence the phenophases, and if the spatial distribution of weather stations is insufficient, these subtle variations in climate will not be recorded [35]. Finally, it has been shown that under a comparable climatic environment, genotypes of species distributed across a study region may vary enough to express a different phenophase [36].

Few studies have used VIs as phenological variables in classification approaches to maximise spectral contrast among vegetative communities and discriminate wetlands and mangroves from other vegetation types (e.g., grasses, crops) [16,37]. Li et al. [38] investigated wetland dynamics with greenness trends under intense land cover change on the coast of China. Lamb et al. [39] mapped estuarine emergent wetlands based on seasonal structural change, and [40] identified salt marsh seasonal trend decomposition in the

Chesapeake and Delaware Bays, USA. Wu et al. [16] extracted phenological differences in the wetland vegetation of Chongming Island, China with time series harmonic regression. As these analyses were computationally demanding, the researchers used time series of the Landsat archive linked with wetland vegetation phenology features as a Google Earth Engine (GEE)-managed pipeline. The GEE platform enables processing and classification applications for large-scale geospatial mapping and monitoring of land cover and land use change [22]. Nonetheless, change dynamics and degradation in mangrove forests integrating phenology-based dense time series has rarely been investigated with remote sensing approaches [15]. In this paper, we use the term 'phenology' in a general context to describe the cyclic vegetation activity of mangroves over time. Here, we apply machine learning classification procedures to a time series of mangrove cover and examine climate drivers of mangrove growth dynamics across Central Queensland, Australia. To our knowledge, this is the first study to examine mangrove phenology across different sites on the east coast of Australia. Harmonic regression captures cyclic behaviour on an intra- and inter-annual basis, and the Savitzky-Golay filter captures subtle dynamics during the vegetative growing season. We used both approaches to fully examine the seasonal variability in mangrove forests in our study region. The study region is a large estuarine area adjacent to the Great Barrier Reef, Australia. Although management of Great Barrier Reef catchments is important to maintain the ecological integrity of the reef, monitoring of estuarine areas receives little attention. Moreover, monitoring of Ramsar Wetland sites on the east coast is absent. The objectives of the study are: (1) to quantify how coastal mangrove forests have spatially and temporally changed in a decadal period (2009–2019) within estuarine catchments of the Great Barrier Reef; (2) to investigate the seasonality in tropical mangrove forests through a dense time series of satellite images, exploring the inter-annual and seasonal responses and variations of mangrove phenology to 21 years of rainfall variability; (3) to inform regional coastal planning, conservation efforts, and policy-makers.

2. Materials and Methods

2.1. Study Area

Our study area is located within the northeast coast drainage division of the Central Queensland coast encompassing the Mackay Whitsunday Natural Resource Region and part of the Isaac Region, Central Queensland, Australia (Figure 1). Cape Palmerston National Park is positioned in the Ince Bay Receiving Waters adjacent to the World Heritage listed Great Barrier Reef. The Shoalwater and Corio Bays Area Ramsar site is located south of Cape Palmerston National Park. A large proportion of the marine waters in the Ramsar site are included in marine parks (Commonwealth and Queensland), including the Great Barrier Reef Marine Park (Commonwealth) and Great Barrier Reef Coast Marine Park (State) [41]. The Ramsar wetland supports a broad range of natural values including nationally/internationally threatened wetland species, significant species diversity and large populations of waterbirds, green turtles, dugong, and fish that use the site for vital life history functions such as roosting, nesting, feeding, and breeding. Cape Palmerston National Park is listed as a category II protected area on the International Union for Conservation of Nature (IUCN) World Database on Protected Areas and covers 7200 hectares [42]. The Shoalwater and Corio Bays Area Ramsar site covers 239,100 hectares [43].

The primary intensive land use in the region is the cultivation of sugar cane, making up 18% of the catchment area, with Mackay being the largest sugar-producing region in Australia [44]. Grazing is also an important land use, accounting for 42% [45]. The region's estuaries directly support several commercial fisheries, e.g., East Coast Inshore Fin Fish Fishery, East Coast Otter Trawl Fishery, and Line Fishery (Reef) [46]. Additionally, recreational fishing is a considerable activity in the region, with 25% of the population participating in fishing for recreation, far greater than the state average of 15% [47]. Mangroves and associated communities cover 87,994 hectares of tidal land in the region, with nine wetland areas recognized as nationally important [48], and one Ramsar wetland recognized as internationally important. The Ramsar wetland site is subtropical with species from both temperate and tropical regions. Within the high rainfall areas of the Central Queensland coast bioregion, estuarine wetlands are about equally dominated by saltpan and samphire flats along the high intertidal area; yellow and orange mangroves (*Ceriops tagal* and *Bruguiera* spp.) dominate along the mid-intertidal area; and the stilted mangrove (*Rhizophora stylosa*) dominates in the lower intertidal area [48]. The river mangrove *Aegicera corniculatum* occupies an intermediate-upstream estuarine position while the ubiquitous grey mangrove *Avicennia marina* inhabits a wide range of tidal inundation levels [49]. Twenty-three tree and shrub species of mangroves are present [50]. Mean maximum temperature is between 22.7 °C in July and 30.4 °C in January, while the mean minimum temperature is between 11.4 °C in July and 23.2 °C in January. The dry season is May-November, the wet season occurs December-April. Mean annual rainfall is 1528.4 mm, while the wettest month (344.8 mm) is recorded in February. The study area is located between latitude 20°22′–22°58′S and longitude 147°19′–150°46′E (Figure 1).

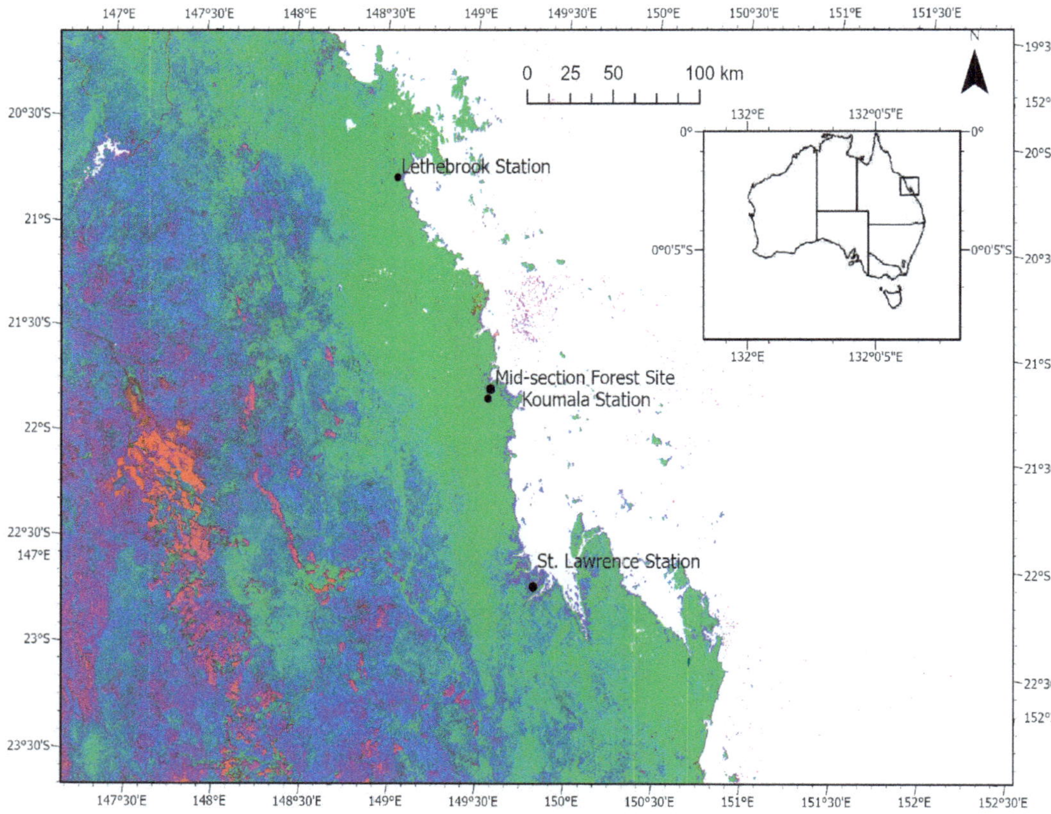

Figure 1. Study site with weather stations at Lethebrook, Koumala, and St. Lawrence-Bowen to Shoalwater and Corio Bays Area Ramsar Wetland, Central Queensland-Landsat fractional cover image visualised by using the Band 1-red (bare ground and rock), Band 2-green (vegetation), Band 3-blue (non-green vegetation indicative of drier habitats with less vegetative cover), provided by [51].

2.2. Image Classification

Medium-resolution satellite imagery is suitable for mapping mangrove and wetland areas on a regional scale [52]. There are two reasons for selecting Landsat imagery: (1) it is acquired at regular intervals, and (2) it is freely available from USGS. Data processing and analysis was performed in Google Earth Engine (GEE). We used a Landsat time series to understand how mangroves have changed over time, identify areas of loss/gain and understand patterns of change in our study region. Landsat data were derived from the United States Geological Survey (USGS) Landsat 7 and 8 Collections, atmospherically corrected surface reflectance products [53]. In regions that are severely cloud-affected, such as the tropical coastal zone, it is necessary to accumulate enough Landsat data to create a cloud-free composite, therefore we chose three-year decadal windows, 2008–2010 and 2018–2020. We created a composite on a per pixel per band basis using the median reducer in GEE [27]. Topographic variables were included in the GEE image stack as a 1 arc-second (approximately 30 m) digital elevation model, the Shuttle Radar Topography Mission (SRTM), sourced from the National Aeronautics and Space Administration Agency (NASA) Jet Propulsion Lab [54]. The SRTM provides elevation and slope data subset to include only low elevation coastal areas where mangroves may be present. We masked to elevations in our study region that are less than 65 m. above mean sea level. Two distinct data sources served as reference datasets: Global Mangrove Watch (GMW) Project [55] providing the 2009 reference data; and Geoscience Australia's Digital Earth Australia platform [56] providing the 2019 reference data. For cloud/shadow removal we filtered the Landsat composite using the quality assessment (QA) band [57] and the GEE median reducer. We calculated four VIs [58,59]: the Simple Ratio (SR) indicates healthy vegetation; the Normalized Difference Vegetation Index (NDVI), the most widely used VI and useful in understanding vegetation density and assessing changes in plant health; the Modified Normalized Difference Moisture Index (MNDWI) assesses vegetation water content; Green Chlorophyll Vegetation Index (GCVI) estimates green leaf biomass. In GEE, we applied masks to areas of low elevation and high NDVI and MNDWI. Additional masking allows us to focus on areas that are more likely to have mangroves. The auxiliary variables in the analysis were the topographic variables derived from the digital elevation model. The phenology variable was the median NDVI calculated from the time series in the three-year decadal windows, 2008–2010 and 2018–2020. The phenology variable was included to separate mangrove trees from other vegetation types (e.g., grass and crops) based on differences in seasonality [60].

Since this study seeks to evaluate predictor variables that tell us the extent of mangrove forest, only two land cover classes were generated (mangrove versus non-mangrove that includes bare soil, sand beach, water, crops, pasture, and native vegetation). We sought a supervised classification approach that is robust to normal distribution departures, can overcome limitations such as overfitting, and uses different subsets of the same training dataset [61]. Machine learning allows us to use samples of areas with and without mangroves to detect mangrove forests across a region. RF is a supervised, nonparametric, and ensemble tree-based machine learning algorithm, which has increasing use in mangrove mapping with Landsat images [62]. RF has been found to outperform comparative classification models in projection accuracy and computation cost [63,64]. Based on the predictors (Landsat bands), the trees will vote for each pixel to detect mangrove vs. non-mangrove with the most supported value assigned to each pixel. We used 100 decision trees with five predictors including the vegetation indices NDVI, MNDWI, SR, and GCVI. In GEE, the RF classifier chooses five predictors based on a random selection from seven bands in Landsat 8: 'B5', 'B6', 'B4', and the VIs. In the same way, the RF classifier chooses five predictors based on a random selection from seven bands in Landsat 7: 'B4', 'B5', 'B3', and the VIs. The training samples were collected by creating geometries of the location of mangroves using the VIs and elevation masks with Landsat false-colour composites in GEE. The total number of sample points (pixels) was 19,500. In our analysis, 80% of the sample points were randomly split and used to train the classifier. The decision trees were fully grown without pruning using a sample (with replacements) of 20% of the training data, also randomly split, to test the accuracy and validate the RF classifier [65,66]. For accuracy assessment a confusion matrix was calculated based on the sample data and classification result giving the evaluation metrics, user's and producer's accuracy, overall accuracy, and Kappa coefficient. Moreover, the mangrove forest maps were visually inspected and compared with high-resolution mangrove canopy cover maps from Digital Earth Australia to evaluate the performance of our method [56]. Considering that the Digital Earth Australia maps quantify the change in the extent of mangrove forest over the period 1987 to 2016 at an annual, national scale, there is concurrence with the classification maps in our study that demonstrate a regional and local approach.

2.3. Mangrove Phenology

All available images (i.e., 2012 to 2020) for our study region (Bowen to Shoalwater and Corio Bays Area Ramsar Wetland) were acquired from the Landsat 8 surface reflectance data with a spatial resolution of 30 m and temporal resolution of 16 days and pre-processed in GEE using functions including masking clouds/cloud shadows. Images were stacked together to build a time series to be used in the models.

We used linear and harmonic regression models to analyse the seasonal parameter trends in our time series. First, we fitted an ordinary least squares model to the time series of all Landsat 8 statistical coefficients with the linear regression reducer in GEE. Second, we detrended the data to reveal underlying fluctuations in the time series. Third, we fitted Fourier-style harmonic regression equations including a trend term to each pixel and spectral component to identify phenological behaviours [37]. Harmonic regression is a mathematical technique largely focused on NDVI that is used to decompose a complex, static signal into a series of individual cosine waves (terms) and an additive term [31]. The time series is approximated as a trigonometric polynomial, with coefficients added to the harmonic model to get the amplitude (A) and phase (φ):

$$p_t = \beta_0 + \beta_1 t + A\cos(2\pi\omega t - \varphi) + e_t = \beta_0 + \beta_1 t + \beta_2 \cos(2\pi\omega t) + \beta_3 \sin(2\pi\omega t) + e_t \quad (1)$$

where p_t is the value of the harmonised NDVI within a pixel obtained from the satellite data at time t, β_0 is a constant component of the regression that designates the start of greenness at each pixel, β_1 is a linear coefficient (slope/linear trend), β_2 is the cosine term coefficient, β_3 is the sine term coefficient, e_t is a random error, and ω is the angular frequency. We set $\omega = 1$ (one cycle per unit time for an annual cycle) to fit the model to the time series.

TIMESAT Analysis

The program TIMESAT is an approach for extracting phenological parameters from remote sensing data and was partly developed from the premise that changes in vegetation phenology may be an indication of climate change [67]. The method is based on nonlinear least squares fits to the upper envelope of the NDVI data [68]. First, we applied a median filter and interpolated the missing values. Second, we used the Savitzky-Golay model and seasonal trend decomposition for a yearly growing season in tropical mangrove forests. The Savitzky-Golay filter follows within-season variations and therefore captures subtle dynamics during the growing season [69]. The result is a smoothed curve adapted to the upper envelope of the NDVI values [68]. To analyse the dense time series (1999–2020), we first created a harmonise function between L5, L7, and L8 sensors. There are small, but potentially significant differences between the spectral characteristics of Landsat ETM+ and OLI, depending on the application. It is advisable to harmonise for a long time series that spans Landsat TM, ETM+, and OLI sensors [31]. Following the harmonise function, we added the NDVI band to the collection. We extracted 13 phenological metrics for each season (21 years) from the TIMESAT program with Landsat dense time series at our study region. TIMESAT provides phenology parameters such as: start of season (SOS) depicted as initiation of a consistent upward trend in the NDVI time series, or the beginning of measurable photosynthesis in the mangrove canopy; end of season (EOS) identified as the end of consistent downward trend in the time series, or the end of measurable photosynthesis in the mangrove canopy; season length (LOS) or duration of photosynthetic activity (time of SOS to EOS); peak NDVI value (the maximum recorded NDVI value); base value (the average of the minimum values left and right of the season); amplitude (the range between the maximum NDVI and the base level or the maximum increase in canopy photosynthetic activity above the baseline); peak time (POS: time of the middle of the season, or time of maximum photosynthesis in the canopy); left derivative (rate of increase at the beginning of season or growth rate); right derivative (rate of decrease at the end of season or rate of senescence [70–72]. Mangrove productivity over the region was estimated using the amplitude, peak NDVI, left and right derivates, large integral, and small integral. The large integral is the function describing the season from the start to the end (an estimate of the total vegetation productivity in the annual cycle). The small integral is the difference between the function describing the season and the base level from season start to end. It is a measure of the productivity within the growing season [70,73]. Further, we used the Seasonal-Trend Decomposition by LOESS (STL) method to remove spikes and outliers in the time series and produce a smoother trendline [74].

Relationships between TIMESAT seasonality parameters were examined with linear regression models by the increasing/decreasing patterns (regression slope) and goodness-of-fit (adjusted coefficient of determination, Adjusted R^2). To characterise the geographic distribution of environmental drivers of mangrove phenology, we applied climate data for 1999–2020 to the time series. Monthly rain gauge precipitation data interpolated from ground station measurements were obtained from the Australian Bureau of Meteorology online climate database (www.bom.gov.au, accessed on 1 March 2021). For the northern section of our study region, rainfall data were primarily gathered from adjacent Lethebrook Station (station number: 33041) with missing data supplemented from Bowen Station (station number: 33264). For the mid-section of our study region rainfall data were primarily gathered from the adjacent Koumala Station (station number: 33038) with missing data supplemented from Plane Creek Station (station number: 33059). For the southern section of our study region rainfall data were collected from two adjacent stations at St. Lawrence (station numbers: 33065 and 33210). As rainfall amount differs across the region, trendlines of the precipitation data were superimposed on the interpolated and filtered NDVI from the three locations across our study site: northern at Repulse Bay, mid-section at Rocky Dam Creek/Cape Palmerston National Park, and southern at Herbert Creek bordering Shoalwater and Corio Bays Area Ramsar Wetland. The peak of the season (POS) in relation to precipitation trends were compared between sites.

RF classification, linear and harmonic regression modelling and dense time series data processing was done with Google Earth Engine. The TIMESAT program was used for phenometric extraction (Figure 2). Statistical analysis was done with the R statistical environment [75] (r-project.org/).

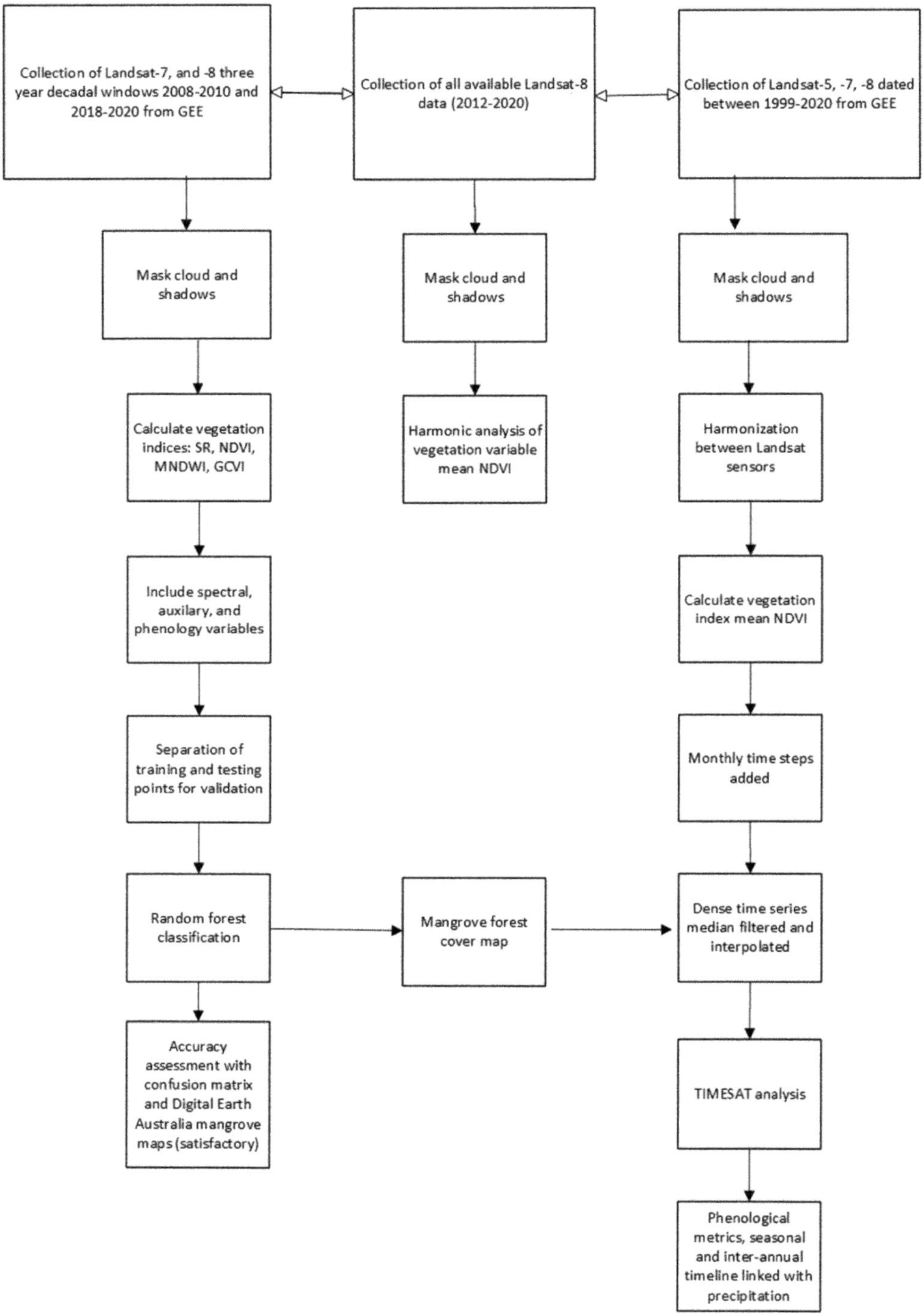

Figure 2. Workflow used for land cover classification and phenological analysis.

3. Results

The aim of this research is to quantify how coastal mangrove forests have spatially and temporally changed in a decadal period (2009–2019) within estuarine catchments of the Great Barrier Reef and to explore the seasonality, climate responses, and variability of mangroves through a dense time series of satellite images in Central Queensland, Australia. Both biotic and abiotic factors constrain the phenological trajectory of a mangrove forest, e.g., regional climate, seawater and soil salinity, latitude, and local vegetation interactions, and these can vary across different locations and years. Sea surface temperatures in the Australian region have warmed by around 1 °C since 1910, with the Great Barrier Reef warming by 0.8 °C in the same period. Moreover, observations show that there has been an increase in the intensity of heavy rainfall events in Australia. (Figures S1 and S2 in the Supplementary Materials). Against the background of climate change, our study examines the phenodynamics of mangrove forests across different sites on the east coast of Australia.

3.1. Image Classification

The mangrove forest extent derived from the RF classification shows a decrease of 1480 hectares (−2.31%) over the decadal window, 2009–2019 (Table 1). In ArcGIS, we generated a Landsat 8 Operational Land Imager (OLI) true-colour mosaic of our study region, overlaying the mosaic with the mangrove change map from GEE (Figure 3). The overall classification accuracy and Kappa coefficient for the 2009 and 2019 land cover maps were 95% and 95%, respectively (Table 2). The high accuracies suggest that the RF classifier is an efficient model for projecting mangrove forest cover dynamics on both training and test datasets. The mangrove forest maps satisfactorily aligned with the high-resolution mangrove canopy cover continental-scale maps from Digital Earth Australia [56].

Table 1. Mangrove forest extent derived from Random Forest classification with Landsat NDVI images 2009 and 2019 (Bowen to Shoalwater and Corio Bays Area Ramsar Wetland, Central Queensland).

Mangrove Extent 2009 (Ha)	Mangrove Extent 2019 (Ha)	Decrease from 2009 to 2019 (Ha)	Percent Area Decrease from 2009 to 2019 (%)
63,990	62,510	−1480	−2.31

Table 2. Accuracy assessment from Random Forest classification 2009 and 2019 Landsat decadal time series analysis (Bowen to Shoalwater and Corio Bays Area Ramsar Wetland, Central Queensland).

Land Class Name	Producer's Accuracy (%)		User's Accuracy (%)	
	2009	2019	2009	2019
Mangrove forest	0.99	0.96	0.98	0.95
Non-Mangrove Forest	0.95	0.99	0.96	0.99
Overall accuracy	0.98	0.98		
Kappa coefficient	0.95	0.95		

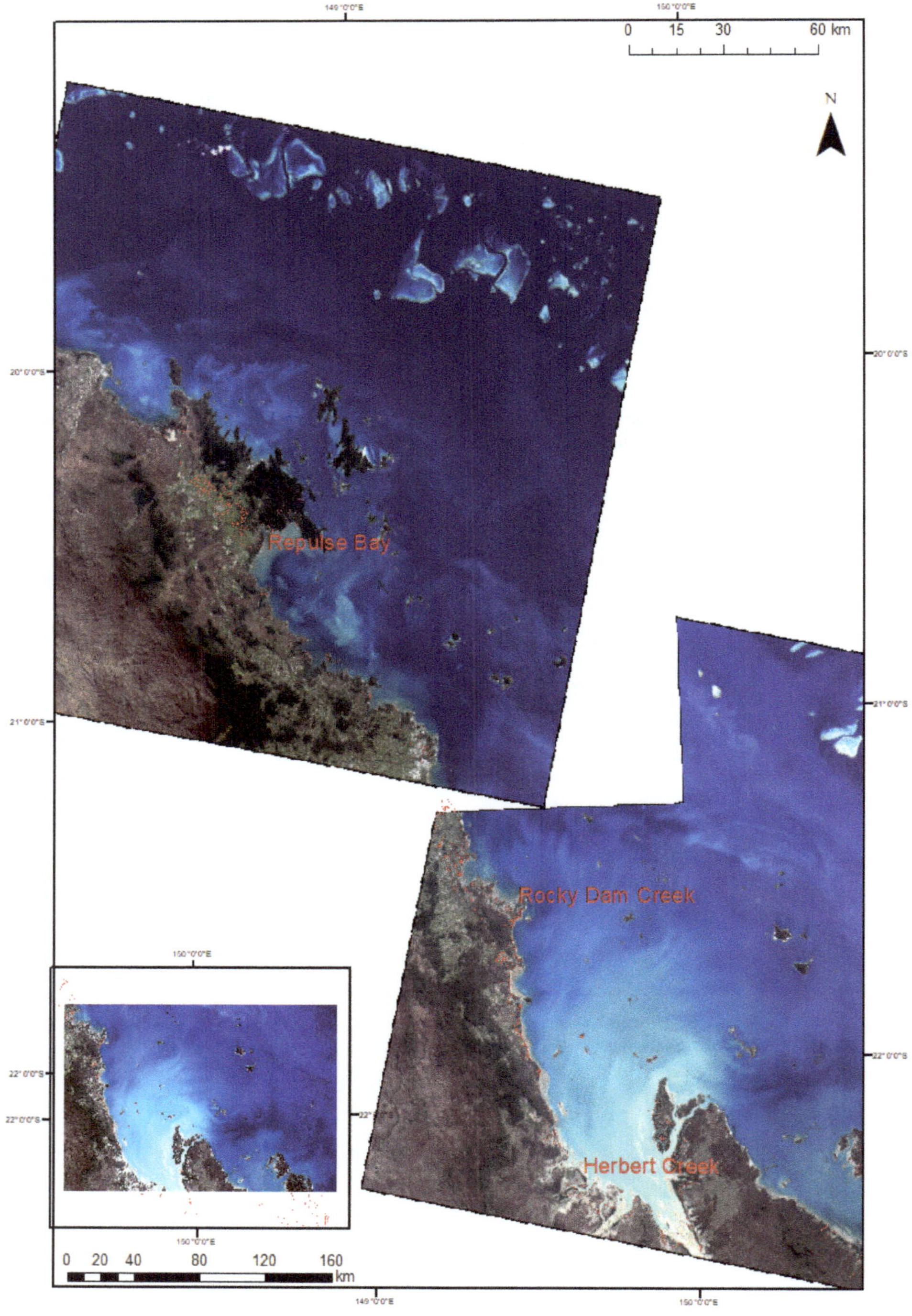

Figure 3. Landsat Operational Land Imager (OLI) true-colour mosaic (images captured 16 July 2020 and 9 August 2020) overlaid with the change map of mangrove forest extent (decline) in red from the Random Forest classification and showing the three sites input to GEE for time series construction and subsequent phenology analysis: northern site, Repulse Bay; mid-section, Rocky Dam Creek/Cape Palmerston National Park; and southern site, Herbert Creek, study region is Bowen to Shoalwater and Corio Bays Area Ramsar Wetland.

3.2. Mangrove Phenology

The detrended time series (2012–2020) shows peaks and troughs in the time series but has missing data caused by cloud masking of pixels in GEE (Figure S3 in the Supplementary Materials). Nevertheless, there is a decrease in mean NDVI across the time series with strong seasonality. Values of the peak growth and photosynthetic activity range from 0.72 to 0.9 and values of the minimum photosynthetic activity range from 0.62 to 0.82. The predicted values of the mangrove harmonic regression model were compared with the actual mean NDVI values and show a general trend in the time series (Figure 4). The seasonal trend indicates that high values were attained towards the end of the growing season (generally in winter) and low values were realized in late spring (October to November with the exception of 2018 where the lowest value was in summer, i.e., December, and 2019 where the lowest value was in early spring, i.e., September, Figure 5). We suggest that under tropical conditions, frequent cloud cover and high temperatures in spring and summer may limit the ability of mangroves to photosynthesize hence greater NDVI values occur in winter. This outcome agrees with field studies conducted by [49] with *Avicennia marina* in tropical Australia. We identified that tropical mangroves on the east coast of Australia display a disproportional bimodal seasonality with two periods of high mean NDVI and two periods of low NDVI values annually through the Landsat time series (1999–2020) (Figures 6 and 7).

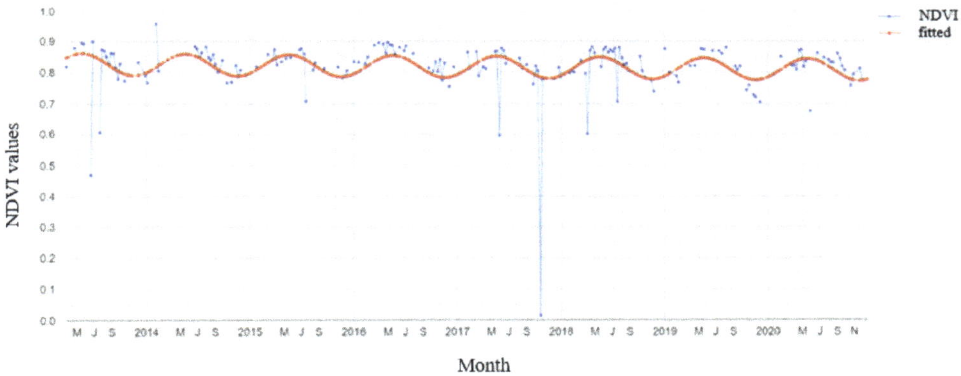

Figure 4. Harmonic regression model original and fitted values of all available (2012–2020) Landsat 8 imagery with mangrove NDVI. The red line is the fitted NDVI values and the blue dotted line is the original NDVI values for mangrove forest in the mid-section of our study region. The site is located directly adjacent to Koumala Station, Figure 1, Bowen to Shoalwater and Corio Bays Area Ramsar Wetland, Central Queensland.

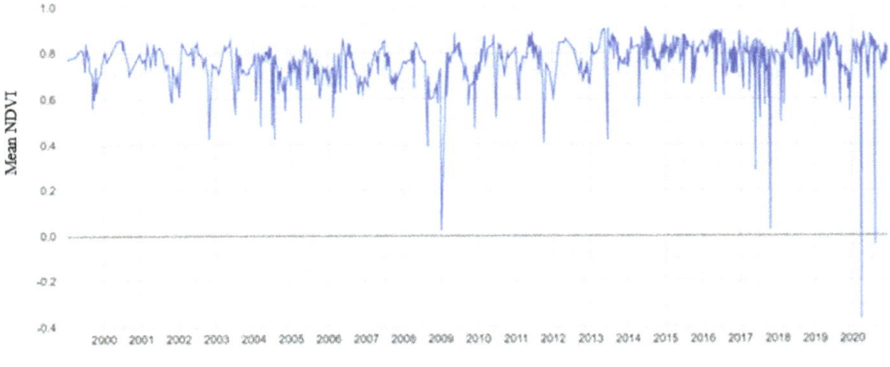

Figure 5. Mean NDVI temporal pattern across the dense Landsat time series (1999–2020) for mangrove forest in the mid-section of our study region. The site is located directly adjacent to Koumala Station, Figure 1, Bowen to Shoalwater and Corio Bays Area Ramsar Wetland, Central Queensland.

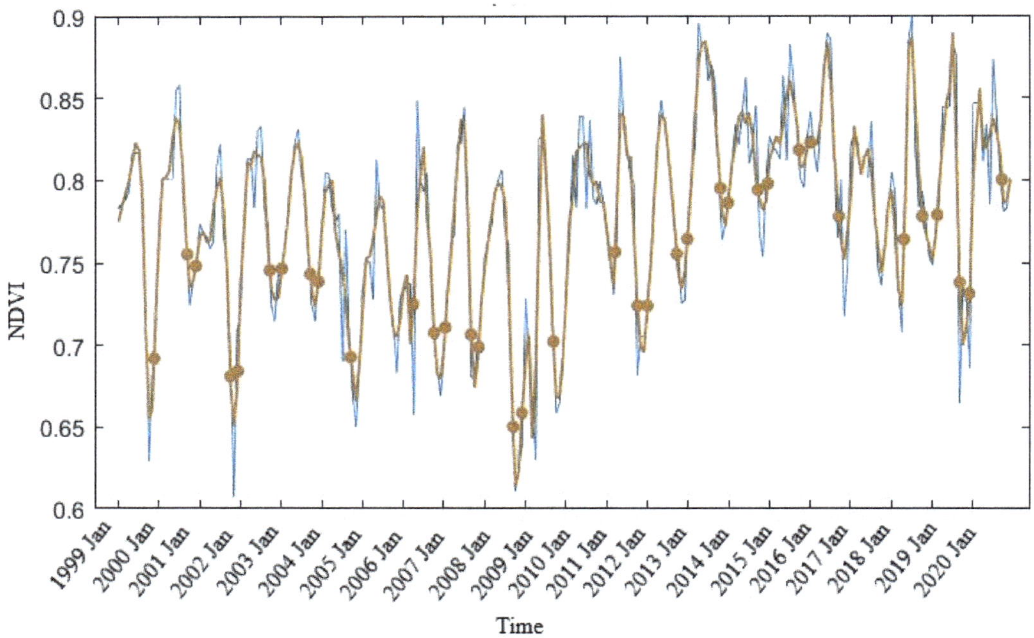

Figure 6. Mangrove forest intra-annual smoothing phenology with the Savitzky-Golay filter in TIMESAT using Landsat NDVI dense time-series data 1999–2020, blue line indicates the decadal NDVI data, brown line indicates the smoothed time series fitted data, brown dots display the start and end of the growing season for mangrove forest in the mid-section of our study region, Figure 1, Bowen to Shoalwater and Corio Bays Area Ramsar Wetland, Central Queensland.

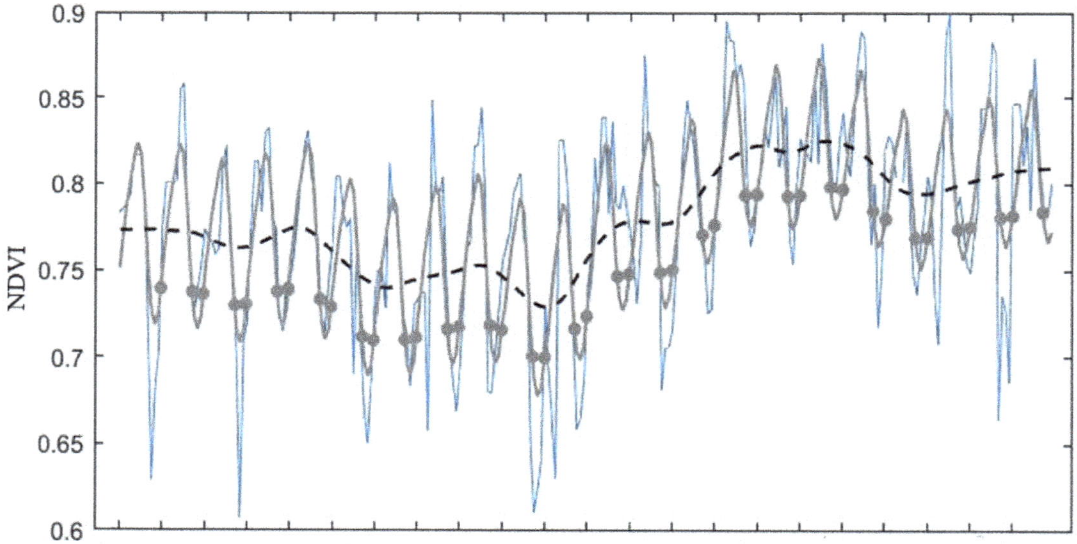

Figure 7. Mangrove forest seasonal trend decomposition in TIMESAT with Landsat NDVI data and dense time-series 1999–2020, black line indicates the seasonal trend, grey dots display the start and end of the growing season for mangrove forest in the mid-section our study region, Figure 1, Bowen to Shoalwater and Corio Bays Area Ramsar Wetland, Central Queensland.

TIMESAT Analysis

In the TIMESAT program we used the Savitzky-Golay algorithm and extracted 13 attributes describing the spectro-temporal profile of mangroves for each season (Table 3). The filtering process preserved the trend of the original NDVI data and the fitted curve depicted the seasonal variation in forest characteristics; it, therefore, provides reliable data for the extraction of phenological information. For the mid-section of our study region (Rocky Dam Creek/Cape Palmerston National Park, Figure 3) we found the SOS to be December, EOS to be September/October, and the LOS to be 10 months. POS varied through the timeseries (February to July): 1999–2003 in June/July; 2004 in February; 2005–2006 in May; 2007–2008 in June/July; 2009–2010 in May/June; 2011–2012 in May; 2013 in April; 2014–2020 in June/July (Figure 6). There is a marked upward movement in the seasonal trend decomposition of NDVI data from the summer of 2009 to 2020 (Figure 6). Notwithstanding the length of the season gradually becoming shorter, the long integral (total productivity in the annual cycle) increased (Figure S4 in the Supplementary Materials). The trend in the peak NDVI value (maximum recorded NDVI value in the time series) increased from the summer of 2009 (Figure 7 and Figure S5 in the Supplementary Materials). In evergreen areas the small integral (productivity within the growing season) may be small even if the total vegetation production is large. The small integral and amplitude (difference between the maximal value and the base level) both declined from 2009 in our study region due to a positive change in the base value increasing the amount of green vegetation for this period (Figures S6 and S7 in the Supplementary Materials). We found a significantly high correlation ($p < 0.001$) between the phenology metrics from TIMESAT through the time series particularly season length and duration of the peak value (Pearson correlation coefficient r value = −0.91, Adjusted $R^2 = 0.8182$), and season (year) and season length (Pearson correlation coefficient r value = −0.91, Adjusted $R^2 = 0.8179$) indicating a dynamic trend in the cyclical behaviour of tropical mangrove forests (Table 4).

Except for the northern region, Central Queensland is largely decoupled from southern monsoonal influences and exhibits highly variable phenology that is rainfall pulse driven. Monthly rainfall was 0–800 mm at our study site, Rocky Dam Creek/Cape Palmerston National Park (Figure 3). Australia's millennium drought of 2001–2007 is reflected in the rainfall pattern and the reduced mean NDVI trendline (Figure 8). Considerable decreases in NDVI occurred in 2001, 2002, 2004, and 2008 that are consistent with the trend of monthly rainfall. In addition to the general trend, phenological characteristics and rainfall patterns correspond with major climate events such as the EL Niño-Southern Oscillation (ENSO). The increase in precipitation from the summer of 2009 in the mid-section of our study region produced a concomitant increase in photosynthetically active growth and biomass production in mangrove forests at this site. The extreme and extended rainfall event from La Niño that occurred over Queensland in the summer of 2010/2011 was reflected in the net seasonal photosynthetic rates for this period. Further, a strong La Niño event is apparent in the Southern Oscillation Index (SOI, the difference in surface air pressure measured between Tahiti and Darwin, northern Australia) that spikes in the 2010/11 summer (Figure S8 in Supplementary Materials). In addition, a severe tropical cyclone (Tropical Cyclone Yasi, Category 5) brought an intense rainfall event in February 2011 to the entire region [76]. A relatively high rainfall pattern continued until mid-Autumn of 2017 when a severe tropical cyclone (Tropical Cyclone Debbie, Category 4) impacted the coast and destroyed coastal vegetation [77]. Recovery from the cyclone is evident in 2018 when the mean NDVI reached 0.9 at Rocky Dam Creek/Cape Palmerston National Park (Figure 8).

Table 3. Savitsky-Golay filter with phenometrics extracted from TIMESAT for the Landsat dense time series of mangrove forest in the mid-section of our study region, Startt.: start of season is the beginning of measurable photosynthesis in the mangrove canopy (month of the year); Endt.: end of season is the end of measurable photosynthesis in the mangrove canopy (month of the year); Length: duration of photosynthetic activity (days); Baseval.: the average of the minimum values left and right of the season (NDVI); Peakt.: time of maximum photosynthesis in the canopy (days); Peakval.: the maximum recorded NDVI value (NDVI); Ampl.: the maximum increase in canopy photosynthetic activity above the baseline (NDVI); L.deriv.: growth rate (day/NDVI); R.deriv.: rate of senescence (day/NDVI); L.integral: total productivity (day/NDVI); S.integral: seasonal productivity (day/NDVI); Startval.: value at start of season (NDVI); Endval.: value at end of season (NDVI), Bowen to Shoalwater and Corio Bays Area Ramsar Wetland, Central Queensland.

Seas.	Startt.	Endt.	Length	Baseval.	Peakt.	Peakval.	Ampl.	L.deriv.	R.deriv.	L.integral	S.integral	Startval.	Endval.
1	11.9	21.12	9.222	0.721	17.34	0.8539	0.1329	0.03167	0.03633	9.555	0.9028	0.7173	0.7779
2	26.35	33.71	7.36	0.7283	30.53	0.8131	0.08484	0.01231	0.04226	7.002	0.4479	0.7697	0.7208
3	36.6	46	9.405	0.7184	41.26	0.8337	0.1153	0.0316	0.02131	9.411	0.7901	0.7249	0.7581
4	49.65	57.38	7.728	0.7438	53.56	0.8246	0.08079	0.02252	0.02102	7.876	0.438	0.7563	0.7637
5	59.89	69.77	9.878	0.7227	63.97	0.8039	0.08119	0.01741	0.01763	9.234	0.5616	0.7596	0.7184
6	72.56	84.11	11.55	0.7069	77.04	0.7975	0.0906	0.02542	0.01493	10.56	0.6589	0.7171	0.733
7	86	94.3	8.3	0.701	90.26	0.8121	0.1111	0.02122	0.03094	8.354	0.6427	0.7359	0.7106
8	97.41	105.2	7.776	0.6941	102.1	0.8428	0.1487	0.02885	0.04959	7.722	0.7812	0.7167	0.7309
9	108.1	117.4	9.345	0.6692	113.2	0.8055	0.1363	0.02137	0.04906	8.29	0.9292	0.7234	0.6694
10	119.7	129.4	9.732	0.6548	126.1	0.8275	0.1727	0.02239	0.0459	8.826	0.9689	0.6738	0.7048
11	132.3	144	11.73	0.7202	137.4	0.8444	0.1243	0.03738	0.01137	10.41	1.043	0.7082	0.7818
12	146.9	154.2	7.284	0.7394	150.4	0.8503	0.111	0.02247	0.03217	7.939	0.5453	0.783	0.7401
13	157.4	164.9	7.516	0.7331	161.3	0.8385	0.1054	0.03491	0.02762	7.17	0.5718	0.7378	0.7706
14	169.2	178.5	9.321	0.7749	173.6	0.8965	0.1216	0.0348	0.02082	9.346	0.8217	0.7822	0.8163
15	181.6	190.5	8.946	0.8042	186.1	0.8529	0.04878	0.01479	0.01081	9.162	0.3164	0.8076	0.8202
16	193.8	201.7	7.847	0.8168	198.1	0.8744	0.05757	0.01329	0.01511	8.472	0.3035	0.8245	0.8322
17	205.6	213.2	7.618	0.8043	209.8	0.8898	0.08543	0.01478	0.03051	8.493	0.4493	0.8352	0.8076
18	215.2	224.5	9.297	0.7786	220.8	0.8359	0.05734	0.007241	0.01802	8.956	0.3909	0.7968	0.7833
19	227.8	238.2	10.41	0.7684	235	0.8755	0.1071	0.03052	0.03727	10.58	0.5864	0.7912	0.7884
20	242.6	249.4	6.795	0.7457	246.2	0.9123	0.1666	0.03448	0.05107	7.449	0.7378	0.7958	0.7623
21	251.6	261.5	9.971	0.7633	254.9	0.8783	0.115	0.03907	0.008088	9.942	0.7833	0.7555	0.817

Table 4. Response and explanatory variables used in the linear regression models with Pearson correlation coefficient r values and adjusted R^2 at significantly high values ($p < 0.001$).

Response Variable	Explanatory Variable Pearson Correlation Coefficient r	Adjusted R^2
Season (year)	Length of season −0.91	0.818
Season (year)	Seasonal amplitude −0.78	0.597
Season (year)	Rate of senescence −0.9	0.793
Season (year)	Peak photosynthetic activity 0.7	0.462
Season (year)	Seasonal productivity −0.88	0.775
Length of season	Period of peak −0.91 photosynthetic activity	0.812
Length of season	Rate of senescence −0.9	0.721
Growth rate	Seasonal amplitude 0.87	0.755
Seasonal productivity	Rate of senescence 0.89	0.771

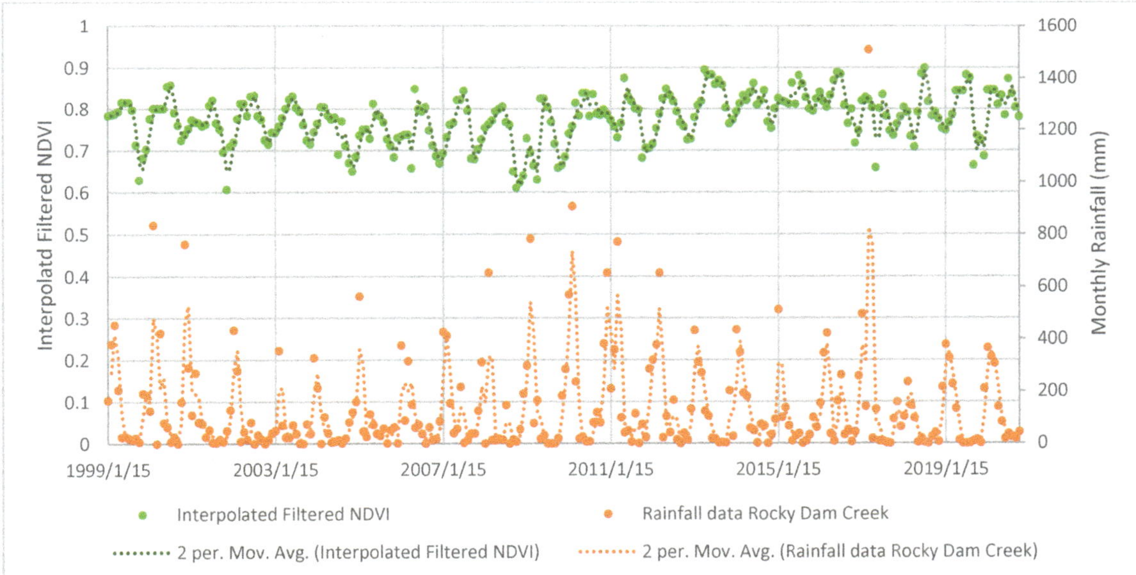

Figure 8. Interpolated, filtered NDVI and rainfall trendlines for the mid-section of our study region, Rocky Dam Creek/Cape Palmerston National Park, rainfall data gathered from Koumala Station (station number: 33038).

In comparison to the mid-section, the northern section of our study region, Repulse Bay, displays a generally higher level of mean NDVI and less temporal variability (Figure 3 and Figure S9 in the Supplementary Materials). Monthly rainfall was 0–900 mm across the time series. The POS occurred in diverse months (May to August), exemplifying a different climate pattern at this site. The northern site exhibited a NDVI decrease in 2002 similar to our mid-section site; however, the decrease occurred in the winter of 2002 (at the northern site it was early summer) and was less pronounced (Figure 9). From 2003–2020 a steady increase in NDVI activity was apparent. High rainfall peaks occurred from 2007–2012, followed by 6 years of moderate rainfall, escalating to a monsoonal trough event in January 2019, then falling to moderate rainfall conditions in 2020. A tropical cyclone (Tropical Cyclone Jasmine, Category 2) affected the northern section in February 2012. A substantial decrease in NDVI was apparent in the winter of 2017 that was not reflected in the rainfall regime but possibly occurred from the influence of cloud cover.

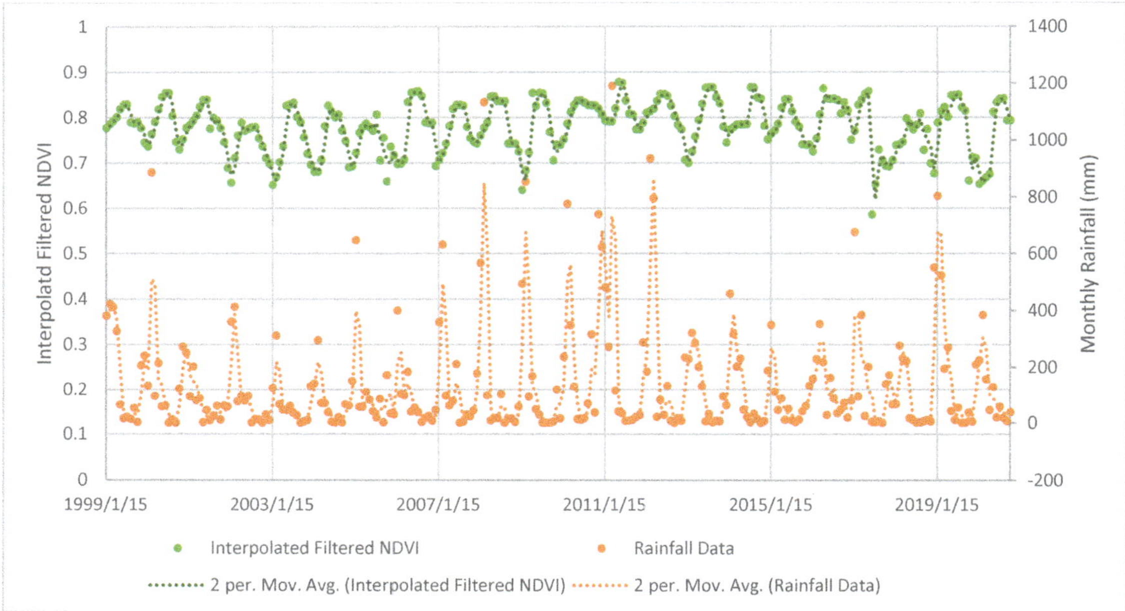

Figure 9. Interpolated, filtered NDVI and rainfall trendlines for the northern section of our study region, Repulse Bay, rainfall data gathered from Lethebrook Station (station number: 33041).

The southern section of our study region (Herbert Creek bordering Shoalwater and Corio Bays Area Ramsar Wetland) exhibited drier conditions than the other sections (Figure 3 and Figure S10 in the Supplementary Materials). A shorter season is apparent with POS occurring in May to July. The NDVI trendline occurred at a lower level than the other sites, but with less temporal variability. Monthly rainfall was 0–390 mm across the time series. Substantial rainfall years in 2010–2011 corresponded with the La Niño pattern of precipitation throughout Queensland but with less intensity (Figure 10). A large decrease in NDVI occurred in 2007 after prolonged low rainfall conditions. The NDVI trendline at this site reflects the El Niño year of 2006/2007 with a low rainfall level and the La Niña year of 2007/2008 with a high rainfall level. POS consistent NDVI years over the 3 sections of our study region were in 2010 with POS in May/June, 2015 with POS in June, and 2019 with POS in June/July.

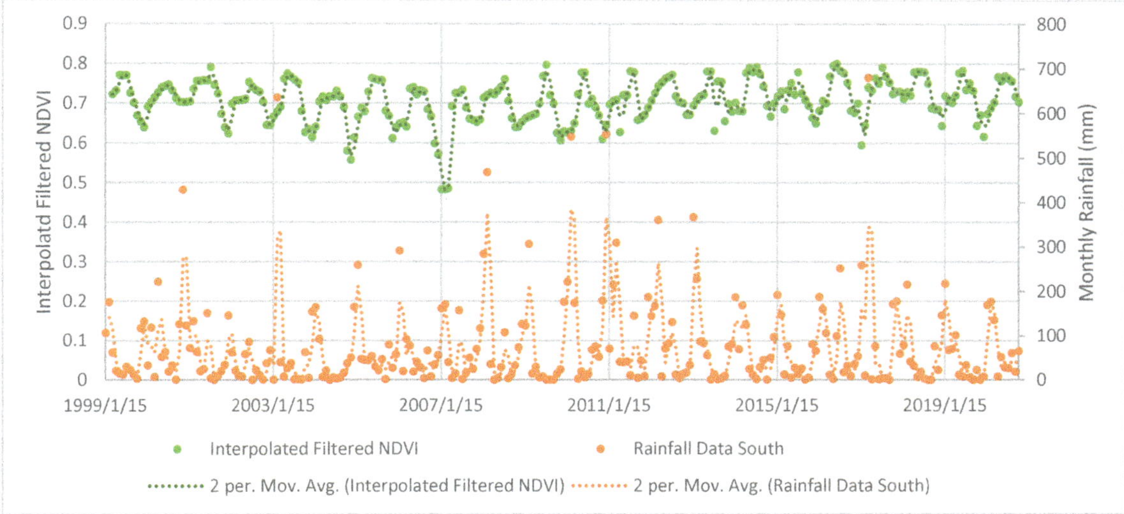

Figure 10. Interpolated, filtered NDVI and rainfall trendlines for the southern section of our study region, Herbert Creek bordering Shoalwater and Corio Bays Area Ramsar Wetland, rainfall data gathered from St. Lawrence Station (station number: 33065).

4. Discussion

4.1. Change Dynamics in Mangrove Forests

With one of the richest flora in the world comprising thirty-nine species, old-world tropical mangroves found in the Indo-Pacific including Australia are notable for their attributes of species diversity, abundance, and succession and they are therefore considered to be the most dominant and important mangroves globally [78,79]. We examined decadal changes in mangrove forest cover in Great Barrier Reef catchments using a machine learning approach with the online cloud-computing platform, GEE. With GEE, we compiled a dense time series of NDVI-based Landsat imagery, inputted the time series to the TIMESAT program and investigated the phenodynamics of mangroves at different locations in our study region. This work has relevance to the maintenance of biodiversity, conservation of estuarine wetlands and mangroves, and the role of coastal forests in climate change. To our knowledge, this is the first study to examine and compare the phenology of tropical mangroves with remote sensing time series across different sites adjacent to the Great Barrier Reef, on the east coast of Australia.

We demonstrated that the mangrove extent derived from the RF classification in our study area decreased by 1480 hectares (−2.31%) over the three-year decadal window, 2009–2019. Our findings corroborate a previous study of mangrove deforestation and degradation that described a total of 11,285 hectares lost worldwide from episodic disturbance (climatic events such as dieback, disturbance mortality, climatic extremes, sea level variability) between 2000–2016 with 8004 hectares from Australia [80]. Global annual mangrove loss rates through deforestation, determined from remote sensing datasets, are estimated to be between 0.26% and 0.66% [81]. Decoupling human-induced mangrove mortality such as degradation, from declines due to episodic disturbance events is inherently difficult to achieve [82]. Moreover, spatial inconsistencies can be present as tropical coastal zones are affected by clouds and atmospheric contamination due to their proximity to oceans and the climate regime [83]. Remote sensing-based time series approaches for change detection require a series of images captured continuously throughout a time period. Thus, there is a need for substantially extra and regular image acquisitions over the area of interest [84]. As with previous studies on mangrove change dynamics [26,85,86], we found that the GEE platform provided efficient data preparation and processing capability for classification and land cover mapping with spectral, auxiliary, and phenology variables.

VIs, particularly NDVI are often used as variables in research related to regional remote sensing forest assessments and shown to be linked not only to canopy structure and leaf area index (the area of single sides leaves per area of soil) but also to canopy photosynthesis [29]. The contrast between strong near-infrared reflectance and low red reflectance of green vegetation is the basis for most VIs. NDVI, and related VIs are functional variants of the simple SR index [87]. We considered four VIs as predictor variables in our classification, SR, NDVI, MNDWI, and GCVI. VIs are commonly used in land cover classification as they increase the classification accuracy, for example: [15] selected NDVI and [88] chose SR to quantify mangrove changes over the Western Arabian Gulf and United Arab Emirates, respectively; [26] generated MNDWI for mapping mangrove extent in China. A major limitation of NDVI however, is that it saturates when mapping high amounts of biomass and cannot distinguish dense vegetation [87]. Another essential point is that NDVI is sensitive to soil, atmospheric effects, and aerosols [29,59]. Furthermore, in open mangrove canopies tidal stage can produce NDVI variations when there is no change in the vegetation amount or condition. Whereas GCVI has been developed as a proxy for chlorophyll content with Landsat surface reflectance data in crops [89], to our knowledge no previous authors have incorporated GCVI for mapping tropical forests or wetlands.

4.2. *Mangrove Phenology*

Our multitemporal analysis based on mean-NDVI Landsat time series (from a representative region in the study area, that is, the mid-section for approximately each month in the study period 1999–2019) revealed a strong seasonality in mangrove forests. Specifically, we found a high growth in winter and a trough of growth in late spring to summer. Such cyclical effect in seasonality has been proposed for mangroves in other tropical regions [31] and can be explained by the fact that local climate properties affect the resource availability of water through seasonal influences of tides, rainfall, and river flows and with variation in the strength of ENSO cycles [90]. Considering all available Landsat 8 imagery we demonstrated that the Fourier-style harmonic regression method provides an explicit depiction of temporal shifts in a region with substantial climate variability. Although rarely met in practice, a stringent requirement of investigating continuous intra-annual dynamics of vegetation phenology, is to acquire error-free remotely sensed data at regular time instants [91], and we successfully achieved this goal. The start, length, peak, and end of season are phenological markers that indicate the photosynthetic activity over the growing season. Specifically, we found SOS (start of the season) to be December, EOS (end of the season) to be September/October, and the LOS (length of the season) to be 10 months, regulated by the precipitation pattern in the study region. We suggest that synchronization of new leaf growth with dry season (winter) shedding, shifts canopy composition toward younger, higher radiometric green-efficient leaves, explaining obvious seasonal increases in ecosystem photosynthesis in our study region [92,93]. For instance, studies conducted on *Xylocarpus* spp. phenology-related litter fall in Australia demonstrate that leaf shedding occurred in the mid to late dry season on the tropical east coast [94] and *Avicennia marina* displayed low leafing and shedding rates throughout the year with a seasonal increase in winter [49]. The phenology of *Rhizophora stylosa* in both Northern and Eastern Australia was documented by [95] who detected lower photosynthetic values during the wet season and higher values during the dry season with Landsat imagery, in accordance with our study. Tropical regions experience variable climate conditions with frequent cloud cover and high temperatures during spring and summer months that may limit the capacity of mangroves to photosynthesize. Consequently, greater NDVI values occur in the winter months under cooler temperatures and more radiance. Conversely, a few studies have reported lower spectral index values during the dry season and higher values during the wet season [96,97].

The mismatch that we uncovered between the gradual contraction in the season length, and the increase in seasonal productivity can be explained by mangrove response to the resource availability of water and perhaps nutrients in tropical, coastal soils. Higher rainfall conditions as occurred from the summer of 2009 following six years of drought, produced an associated increase in photosynthetically active growth and biomass production in mangrove forests. High rainfall increases mangrove growth rates by helping to supply sediments and nutrients thereby stimulating productivity [97]. According to [98] water availability influences several ecological processes including vegetative cover, species composition, and community biomass of tidal wetlands. Freshwater is essential for mangroves to maintain the turgidity of cells and tissues, mechanisms of cell augmentation, and stomata regulation [99]. Consequently, in drought years resource-use strategies become geared towards maximum survival as opposed to rapid biomass gain.

TIMESAT Analysis

Our TIMESAT-based approach with the adaptive Savitzky-Golay filter generated smoothed data for each time step, and seasonality parameters for each identified growing season over the time series. To provide a clear and interpretable vegetation signal we chose to implement the Savitzky-Golay filter as it iteratively tightens the search window in order to capture very rapid increase or decrease in the data and follows local variations in the seasonal curve more closely [100]. Our findings that a disproportionately bimodal pattern is evident in the time series e g., late summer and winter peaks, confirm studies in phenology-based litterfall conducted on the east coast of Australia with *Avicennia marina* [101], which postulated that bimodality in leaf production is interrelated with position in the inter-tidal zone and frequent tidal inundation. More importantly, abiotic factors such as hydrology and photoperiod and biotic factors such as the timing of the reproductive cycle, appear to be linked to mangrove vegetative phenology [102].

The seasonal dynamics derived using TIMESAT, with mangrove NDVI time series as input, showed differences in the spatio-temporal patterns of growth dynamics. While this is the case, the three different sections of our study region are analogous in that all sections are tropical estuarine ecosystems with the same mangrove species such as the stilted mangrove *Rhizophora stylosa*, the grey mangrove *Avicennia marina*, the yellow mangrove *Ceriops tagal*, and the river mangrove *Aegiceras corniculatum*. Macroclimatic drivers such as rainfall and temperature regimes are recognized for their ability to limit and influence processes within coastal-estuarine wetlands [103]. The premise that rainfall has the greater importance in affecting biodiversity, biomass, and phenology of tropical mangrove vegetation [49] is supported by our study. For example, our observation that peaks and troughs in the NDVI trendline correspond with the trend of monthly rainfall and are magnified by climate events occurred at all three study sections, albeit with differential affects. Accordingly, we found the northern site that was also the wettest had the highest NDVI values and the southern site that was the driest had the lowest NDVI values. The site with the greatest variability in the trendline was the mid-section (Rocky Dam Creek/Cape Palmerston National Park). The mid-section was also the site with moderate monthly rainfall indicating that mangrove growth variability responds to localised patterns.

We also found that the La Niña event of 2010–2011 and its associated rainfall changes impacted vegetation growth, increasing NDVI values at all section sites. Another essential point is that the Southern Oscillation Index spikes in La Niño years, particularly 2010–2011. Indeed, a study conducted at Corio Bay in the Ramsar wetland (a generally euhaline site) in the El Niño year of 2006–2007 reported salinity reached exceedingly high values (up to 40 parts per thousand) towards the mouth. Conversely, at Corio Bay in the La Niña year of 2007–2008 salinities down to inordinately low values (23 parts per thousand) were recorded, which were associated with flooding of the adjacent Fitzoy River and local catchment run off [43]. Notably, our analysis shows that extreme climate-related salinity values such as these, are reflected in the mangrove NDVI trendline where a trough occurs in 2006–2007 followed by rise in 2007–2008 at our southern section (Herbert Creek bordering Shoalwater and Corio Bays Area Ramsar Wetland). Our results are particularly concerning for tropical mangrove forests as they highlight contemporary research on the intensification of ENSO in the coming decades with climate change [10,82].

Episodic disturbance events such as tropical cyclones and climatic extremes are purported to be responsible for approximately 80% of global mangrove loss from natural causes since 1990 [82]. In addition to coastal damage from strong winds [104], impacts from tropical cyclones include high rainfall that exacerbates the hydrological and hydrodynamic conditions [105]. Moreover, the anomalies in rainfall induce lower salinity, greater turbidity, as well as higher dissolved nutrients and phytoplankton biomass in the coastal waters when compared with typical conditions [106]. The frequency and intensity of cyclones and storms has increased as a consequence of greater sea-surface temperature, with further escalation predicted under most climate change scenarios [107,108]. Greater cyclone frequency has compounding effects on mangrove forests and may extend or halt mangrove recovery [107]. Our findings show decreases in the NDVI values associated with the occurrence of tropical cyclones in each section of our study region. Tropical Cyclone Yasi impacted all three sites in 2011, Tropical Cyclone Jasmine destroyed the northern section site in 2012, and Tropical Cyclone Debbie devastated the mid-section and southern section sites in 2017, which was the highest precipitation event over the time series.

4.3. Limitations of the Study

Remote sensing data and tools are fundamental methods for measuring land cover, but there are key drawbacks in the change detection of wetlands such as mangrove forests. The need for assessing accuracy of a map generated from any remotely sensed product has become a universal requirement and an integral part of any classification project. The user community needs to know the accuracy of the classified image data being presented [109] otherwise the classification lacks robust validation which could have serious implications for some users and may lessen their confidence in remote sensing as a source of land cover data. Accuracy assessment (1) allows self-evaluation and to learn from mistakes in the classification process; (2) provides quantitative comparison of the classified maps; (3) ensures greater reliability of the resulting maps/spatial information. Therefore, we presented robust validation methods that detail the user's and producer's accuracies of change with Kappa coefficient. Furthermore, we compared the mangrove forest classified maps with Digital Earth Australia high-resolution imagery (at the continental-scale) for visual refinement and to evaluate the performance of our method [56]. The second key drawback is that there can be a similarity of spectral reflectance in the NIR and SWIR regions between mangrove forest and the mixture of other natural vegetation and water at high tide compared to that obtained at low tide [26,86,110]. The third drawback is that quantification is difficult in long-term studies because inconsistencies and gaps in data collection due to observation error are common [111]. The fourth drawback is that temperate climates use simple spring indicators and measures related to 1 January that are not appropriate for tropical phenology because of the circularity of the data. Thus, tropical phenological approaches necessitate taking account of data circularity, be flexible, quantitative, and attribute confidence to conclusions [112]. Finally, a key drawback is that in similar climate conditions, genotypes of mangrove species spread across a study region may vary enough to exemplify different phenophases. In addition, although mangroves are highly specialized and occupy unique and heterogeneous suites of habitats within Australia, they are not a genetic entity but an ecological one [98,112,113]. Consequently, our study has accessed nearby weather stations to our section sites such that localised and subtle climate variations (known to influence phenophases) in location are recorded, and this has reflected in the timeline signals.

4.4. Implications for the Conservation of Mangrove Forests

Growing evidence of mangrove's ability to sequester and store carbon has brought attention to their conservation value from international climate change policy and decision makers [114]. Although estuarine monitoring is almost completely absent at Shoalwater and Corio Bay Areas Ramsar Wetland, it is critical to establish a monitoring program to provide baseline assessments of vegetation seasonality in relation to weather patterns and detect responses to extreme events. The Ramsar Convention on Wetlands is an intergovernmental treaty with Parties to the Convention from 171 countries. The purpose of the Ramsar Convention is to promote wetland conservation and wise use to ensure that the benefits of wetlands contribute towards meeting the UN Sustainable Development goals, Aichi Biodiversity Targets, Paris Agreement on Climate Change, and other related international commitments [115]. The United Nations' 2030 Agenda for Sustainable Development together with SDGs require the monitoring and mapping of mangroves through the sustainable development indicator 6.6.1: "change in the extent of water-related ecosystems over time" [11]. Monitoring of mangrove forests is critical to preserve the ecosystem services they provide such as pollution control, detoxification, and regulation of natural hazards. Hydrologic conditions, land use, and management are important drivers of coastal vegetation communities with many land use changes developing over decades. So far, studies on tropical forests and mangrove cover with phenodynamics have been limited by the unavailability of suitable remote sensing imagery. Dense time series of satellite images has allowed a comprehensive understanding of mangrove phenology and its relation to climate factors across different sites. Our results are of relevance for conservation strategies of tropical coastal wetlands in Australia that account for foundation species and their seasonal responses to climate drivers.

5. Conclusions

In this paper, we demonstrated an operational method to (1) map mangrove forests in Central Queensland, Australia using an NDVI time series, which was built in GEE from Landsat imagery, and (2) examine seasonal and inter-annual changes in mangrove phenology extracted with the TIMESAT program, and linked to natural and extreme climate variability. Surprisingly, estuarine monitoring has been absent in the coastal ecosystems of Shoalwater and Corio Bay Ramsar Wetland. Mangrove forests in this southern section of our study region are bounded by an internationally recognized Ramsar Wetland Treaty and therefore changes in their ecology are important for Australia's reporting commitments to the Ramsar Convention. Policy guidelines of the Convention underline that urgent action is needed to raise awareness of the benefits of wetlands, with greater safeguards for their survival. We propose that a regulated monitoring program be instigated so that information can be gathered on the change dynamics and seasonality of estuarine communities and mangroves under threat from episodic disturbances.

Monitoring is essential, ideally before and particularly after an extreme event, to improve our knowledge of the connections among climate, weather, and biodiversity. Our objective to acquire long-term and spatially extensive data was due to the highly variable nature of biological data, and the need to examine periodical events at different locations. The detection of responses to extreme events is important to improve understanding of the connections between climate, extreme weather events, and biodiversity. Such knowledge is essential to inform conservation management attempts to mitigate the impacts of extreme events with ongoing climate change.

Supplementary Materials: The following are available online at https://www.mdpi.com/article/10.3390/rs13153032/s1, Figure S1: Annual sea surface temperature anomaly for the Great Barrier Reef (1999–2020) from Australian Bureau of Meteorology using NOAA ERSST V5 gridded data, Figure S2: Annual rainfall anomaly for Queensland (1900–2020) from Australian Bureau of Meteorology, Figure S3: Linear regression detrended time series of all available Landsat 8 imagery with NDVI for the study region, Bowen to Shoalwater and Corio Bays Area Ramsar Wetland, Central Queensland, Figure S4: Length of season and long integral (total productivity in the annual cycle) in the time series for our study region, Figure S5: The peak NDVI value (maximum recorded NDVI value in the time series) and short integral (the difference between the function describing the season and the base level from season start to end) in our study region, Figure S6: Amplitude (the range between the maximum NDVI and the base level or the maximum increase in canopy photosynthetic activity above the baseline) and the short integral (the difference between the function describing the season and the base level from season start to end) in our study region, Figure S7: Season year (1999–2020) with seasonal metrics from the Savitzky-Golay filter in TIMESAT, Ampl. (amplitude), L.deriv. (growth rate). R.deriv. (rate of senescence), Figure S8: Southern Oscillation Index and rainfall trendline for the mid-section of our study region, rainfall data gathered from Koumala Station (station number: 33038), Figure S9: Mangrove forest seasonal trend decomposition in TIMESAT with Landsat NDVI data and dense time-series 1999–2020, northern section of our study region, Repulse Bay, black line indicates the seasonal trend, grey dots display the start and end of the growing season, Central Queensland, Figure S10: Mangrove forest seasonal trend decomposition in TIMESAT with Landsat NDVI data and dense time-series 1999–2020, southern section of our study region, Herbert Creek bordering Shoalwater and Corio Bays Area Ramsar Wetland, black line indicates the seasonal trend, grey dots display the start and end of the growing season, Central Queensland.

Author Contributions: S.R.P. and D.A.C. jointly conceived and designed this study. H.P.P. provided input into the contents of the paper, suggestions for analysis and editing of the draft and final versions. D.A.C. processed and analysed the data and led the writing of the paper with all authors making contributions. All authors have read and agreed to the published version of the manuscript.

Funding: This research received no external funding.

Institutional Review Board Statement: Not applicable.

Informed Consent Statement: Not applicable.

Conflicts of Interest: The authors declare no conflict of interest.

References

1. Lee, S.Y.; Primavera, J.H.; Dahdouh-Guebas, F.; McKee, K.; Bosire, J.O.; Cannicci, S.; Diele, K.; Fromard, F.; Koedam, N.; Marchand, C.; et al. Ecological role and services of tropical mangrove ecosystems: A reassessment. *Glob. Ecol. Biogeogr.* **2014**, *23*, 726–743. [CrossRef]
2. Sheaves, M.; Johnston, R.; Baker, R. Use of mangroves by fish: New insights from in-forest videos. *Mar. Ecol. Prog. Ser.* **2016**, *549*, 167–182. [CrossRef]
3. Macreadie, P.I.; Anton, A.; Raven, J.A.; Beaumont, N.; Connolly, R.M.; Friess, D.A.; Kelleway, J.J.; Kennedy, H.; Kuwae, T.; Lavery, P.S.; et al. The future of Blue Carbon science. *Nat. Commun.* **2019**, *10*, 3998. [CrossRef] [PubMed]
4. Adame, M.F.; Connolly, R.M.; Turschwell, M.P.; Lovelock, C.E.; Fatoyinbo, T.; Lagomasino, D.; Goldberg, L.A.; Holdorf, J.; Friess, D.A.; Sasmito, S.D.; et al. Future carbon emissions from global mangrove forest loss. *Glob. Chang. Biol.* **2021**. [CrossRef]
5. The International Council on Clean Transportation. *Vision 2050: A Strategy to Decarbonize the Global Transport Sector by Mid-Century*; The International Council on Clean Transportation: Washington, DC, USA, 2020; pp. 1–25.
6. Friess, D.A.; Yando, E.S.; Abuchahla, G.M.O.; Adams, J.B.; Cannicci, S.; Canty, S.W.J.; Cavanaugh, K.C.; Connolly, R.M.; Cormier, N.; Dahdouh-Guebas, F.; et al. Mangroves give cause for conservation optimism, for now. *Curr. Biol.* **2020**, *30*, R153–R154. [CrossRef] [PubMed]
7. Saintilan, N.; Rogers, K.; Kelleway, J.J.; Ens, E.; Sloane, D.R. Climate change impacts on the coastal wetlands of Australia. *Wetlands* **2018**, *38*, 1145–1154. [CrossRef]
8. Spalding, M.D.; Ruffo, S.; Lacambra, C.; Meliane, I.; Hale, L.Z.; Shepard, C.C.; Beck, M.W. The role of ecosystems in coastal protection: Adapting to climate change and coastal hazards. *Ocean Coast. Manag.* **2014**, *90*, 50–57. [CrossRef]
9. Fang, X.; Hou, X.; Li, X.; Hou, W.; Nakaoka, M.; Yu, X. Ecological connectivity between land and sea: A review. *Ecol. Res.* **2018**, *33*, 51–61. [CrossRef]
10. Lovelock, C.E.; Feller, I.C.; Reef, R.; Hickey, S.; Ball, M.C. Mangrove dieback during fluctuating sea levels. *Sci. Rep.* **2017**, *7*, 1680. [CrossRef]
11. UNGA: United Nations General Assembly. *Work of the Statistical Commission Pertaining to the 2030 Agenda for Sustainable Development: Resolution/Adopted by the General Assembly (A/RES/71/313)*; UNGA: New York, NY, USA, 2017.
12. de Beurs, K.M.; Henebry, G.M. Spatio-temporal statistical methods for modelling land surface phenology. In *Phenological Research: Methods for Environmental and Climate Change Analysis*; Hudson, I.L., Keatley, M.R., Eds.; Springer: New York, NY, USA, 2010; pp. 177–208.
13. Hostert, P.; Griffiths, P.; van der Linden, S.; Pflugmacher, D. Time series analyses in a new era of optical satellite data. In *Remote Sensing Time Series*; Springer International Publishing: Cham, Switzerland, 2015; pp. 25–41. [CrossRef]
14. Keenan, R.J. Climate change impacts and adaptation in forest management: A review. *Ann. For. Sci.* **2015**, *72*, 145–167. [CrossRef]
15. Li, W.; El-Askary, H.; Qurban, M.A.; Li, J.; ManiKandan, K.P.; Piechota, T. Using multi-indices approach to quantify mangrove changes over the Western Arabian Gulf along Saudi Arabia coast. *Ecol. Indic.* **2019**, *102*, 734–745. [CrossRef]

16. Wu, N.; Shi, R.; Zhuo, W.; Zhang, C.; Zhou, B.; Xia, Z.; Tao, Z.; Gao, W.; Tian, B. A classification of tidal flat wetland vegetation combining phenological features with Google Earth Engine. *Remote Sens.* **2021**, *13*, 443. [CrossRef]
17. Keatley, M.R.; Hudson, I.L. Introduction and overview. In *Phenological Research: Methods for Environmental and Climate Change Analysis*; Hudson, I.L., Keatley, M.R., Eds.; Springer: New York, NY, USA, 2010; pp. 1–22. [CrossRef]
18. Younes, N.; Joyce, K.E.; Maier, S.W. All models of satellite-derived phenology are wrong, but some are useful: A case study from northern Australia. *Int. J. Appl. Earth Obs. Geoinf.* **2021**, *97*, 102285. [CrossRef]
19. Jimenez Galo, A.J. Monitoring of tropical forest cover with remote sensing. In *Tropical Forestry Handbook*; Pancel, L., Köhl, M., Eds.; Springer: Berlin, Germany, 2016; pp. 1–19. [CrossRef]
20. DeVries, B.; Pratihast, A.K.; Verbesselt, J.; Kooistra, L.; Herold, M. Characterizing forest change using community-based monitoring data and Landsat time series. *PLoS ONE* **2016**, *11*, e0147121. [CrossRef] [PubMed]
21. Duarte, E.; Barrera, J.A.; Dube, F.; Casco, F.; Hernández, A.J.; Zagal, E. Monitoring approach for tropical coniferous forest degradation using remote sensing and field data. *Remote Sens.* **2020**, *12*, 2531. [CrossRef]
22. Souza, C.M.; Shimbo, J.Z.; Rosa, M.R.; Parente, L.L.; Alencar, A.A.; Rudorff, B.F.T.; Hasenack, H.; Matsumoto, M.; Ferreira, G.L.; Souza-Filho, P.W.M.; et al. Reconstructing three decades of land use and land cover changes in Brazilian biomes with Landsat archive and Earth Engine. *Remote Sens.* **2020**, *12*, 2735. [CrossRef]
23. Criminisi, A. Decision Forests: A unified framework for classification, regression, density estimation, manifold learning and semi-supervised learning. *Found. Trends Comput. Graph. Vis.* **2011**, *7*, 81–227. [CrossRef]
24. Hastie, T.; Tibshirani, R.; Friedman, J. *The Elements of Statistical Learning: Data Mining, Inference, and Prediction*, 2nd ed.; Springer: New York, NY, USA, 2017.
25. Pham, L.T.H.; Brabyn, L. Monitoring mangrove biomass change in Vietnam using SPOT images and an object-based approach combined with machine learning algorithms. *ISPRS J. Photogramm. Remote Sens.* **2017**, *128*, 86–97. [CrossRef]
26. Hu, L.; Li, W.; Xu, B. Monitoring mangrove forest change in China from 1990 to 2015 using Landsat-derived spectral-temporal variability metrics. *Int. J. Appl. Earth Obs. Geoinf.* **2018**, *73*, 88–98. [CrossRef]
27. Diniz, C.; Cortinhas, L.; Nerino, G.; Rodrigues, J.; Sadeck, L.; Adami, M.; Souza-Filho, P. Brazilian mangrove status: Three decades of satellite data analysis. *Remote Sens.* **2019**, *11*, 808. [CrossRef]
28. Bihamta Toosi, N.; Soffianian, A.R.; Fakheran, S.; Pourmanafi, S.; Ginzler, C.; Waser, T.L. Land cover classification in mangrove ecosystems based on VHR satellite data and machine learning—an upscaling approach. *Remote Sens.* **2020**, *12*, 2884. [CrossRef]
29. Xue, J.; Su, B. Significant remote sensing vegetation indices: A review of developments and applications. *J. Sens.* **2017**, *2017*, 1353691. [CrossRef]
30. Grace, J.; Nichol, C.; Disney, M.; Lewis, P.; Quaife, T.; Bowyer, P. Can we measure terrestrial photosynthesis from space directly, using spectral reflectance and fluorescence? *Glob. Chang. Biol.* **2007**, *13*, 1484–1497. [CrossRef]
31. Venkatappa, M.; Sasaki, N.; Shrestha, R.P.; Tripathi, N.K.; Ma, H.O. Determination of vegetation thresholds for assessing land use and land use changes in Cambodia using the Google Earth Engine cloud-computing platform. *Remote Sens.* **2019**, *11*, 1514. [CrossRef]
32. Xiao, X.; Zhang, J.; Yan, H.; Wu, W.; Biradar, C. Land surface phenology. In *Phenology of Ecosystem Processes: Applications in Global Change Research*; Noormets, A., Ed.; Springer: New York, NY, USA, 2009; pp. 247–270. [CrossRef]
33. Priyadarshi, N.; Chowdary, V.M.; Srivastava, Y.K.; Das, I.C.; Jha, C.S. Reconstruction of time series MODIS EVI data using de-noising algorithms. *Geocarto Int.* **2018**, *33*, 1095–1113. [CrossRef]
34. Shamim Hasan Mandal, M.; Kamruzzaman, M.; Hosaka, T. Elucidating the phenology of the Sundarbans mangrove forest using 18-year time series of MODIS vegetation indices. *Tropics* **2020**, *29*, 41–55. [CrossRef]
35. Sadinski, W.; Gallant, A.L.; Roth, M.; Brown, J.; Senay, G.; Brininger, W.; Jones, P.M.; Stoker, J. Multi-year data from satellite- and ground-based sensors show details and scale matter in assessing climate's effects on wetland surface water, amphibians, and landscape conditions. *PLoS ONE* **2018**, *13*, e0201951. [CrossRef]
36. Kelly, N. Accounting for correlated error structure within phenological data: A case study of trend analysis of snowdrop flowering. In *Phenological Research: Methods for Environmental and Climate Change Analysis*; Hudson, L.L., Keatley, M.R., Eds.; Springer: New York, NY, USA, 2010; pp. 271–298. [CrossRef]
37. Adams, B.; Iverson, L.; Matthews, S.; Peters, M.; Prasad, A.; Hix, D.M. Mapping forest composition with Landsat time series: An evaluation of seasonal composites and harmonic regression. *Remote Sens.* **2020**, *12*, 610. [CrossRef]
38. Li, D.; Lu, D.; Wu, M.; Shao, X.; Wei, J. Examining land cover and greenness dynamics in Hangzhou Bay in 1985–2016 using Landsat time-series data. *Remote Sens.* **2017**, *10*, 32. [CrossRef]
39. Lamb, B.T.; Tzortziou, M.A.; McDonald, K.C. Evaluation of approaches for mapping tidal wetlands of the Chesapeake and Delaware Bays. *Remote Sens.* **2019**, *11*, 366. [CrossRef]
40. Campbell, A.D.; Wang, Y. Salt marsh monitoring along the mid-Atlantic coast by Google Earth Engine enabled time series. *PLoS ONE* **2020**, *15*, e0229605. [CrossRef]
41. Department of Agriculture Water and the Environment. Directory of Important Wetlands in Australia-Information Sheet, Sarina Inlet-Ince Bay Aggregation-QLD053. Available online: https://www.environment.gov.au/cgi-bin/wetlands/report.pl (accessed on 1 March 2021).
42. Chamberlain, D.; Phinn, S.; Possingham, H. Remote sensing of mangroves and estuarine communities in Central Queensland, Australia. *Remote Sens.* **2020**, *12*, 197. [CrossRef]

43. Ronan, M. *Ramsar Information Sheet: Shoalwater and Corio Bays Area, Australia*; Queensland Department of Environment and Heritage Protection: Brisbane, QLD, Australia, 2018.
44. Folkers, A.; Rohde, K.; Delaney, K.; Flett, I. *Mackay Whitsunday Water Quality Improvement Plan 2014–2021*; Reef Catchments: Mackay, QLD, Australia, 2014.
45. Reef Catchments Limited. *Natural Resource Management Plan*; Mackay Whitsunday Isaac: Mackay, QLD, Australia, 2014.
46. Pascoe, S.; Innes, J.; Tobin, R.; Stoeckle, N.; Paredes, S.; Dauth, K. *Beyond GVP: The Value of Inshore Commercial Fisheries to Fishers and Consumers in Regional Communities on Queensland's East Coast*; FRDC Project No 2013-301: Deakin, VIC, Australia, 2016.
47. Webley, J.; McInnes, K.; Teixeira, D.; Lawson, A.; Quinn, R. *Statewide Recreational Fishing Survey 2013-14*; Department of Agriculture and Fisheries: Brisbane, QLD, Australia, 2015.
48. Reef Catchments. *State of the Region Report*; Mackay Whitsunday Isaac: Mackay, QLD, Australia, 2013.
49. Duke, N.C. *Australia's Mangroves: The Authoritative Guide to Australia's Mangrove Plants*; University of Queensland: Brisbane, QLD, Australia, 2006.
50. Duke, N.C.; Larkum, A.W.D. Mangroves and seagrasses. In *The Great Barrier Reef: Biology, environment and Management*, 2nd ed.; Hutchings, P., Kingsford, M., Hoegh-Guldberg, O., Eds.; CSIRO: Clayton South, VIC, Australia, 2019; pp. 219–228.
51. AusCover. Seasonal Fractional Vegetation Cover for Queensland Derived from USGS Landsat Images. Available online: http://data.auscover.org.au/xwiki/bin/view/Product+pages/Landsat+Seasonal+Fractional+Cover (accessed on 1 March 2021).
52. Kuenzer, C.; Bluemel, A.; Gebhardt, S.; Quoc, T.V.; Dech, S. Remote sensing of mangrove ecosystems: A review. *Remote Sens.* **2011**, *3*, 878–928. [CrossRef]
53. Vermote, E.; Justice, C.; Claverie, M.; Franch, B. Preliminary analysis of the performance of the Landsat 8/OLI land surface reflectance product. *Remote Sens. Environ.* **2016**, *185*, 46–56. [CrossRef]
54. Farr, T.G.; Rosen, P.A.; Caro, E.; Crippen, R.; Duren, R.; Hensley, S.; Kobrick, M.; Paller, M.; Rodriguez, E.; Roth, L.; et al. The Shuttle Radar Topography Mission. *Rev. Geophys.* **2007**, *45*, 1–33. [CrossRef]
55. Bunting, P.; Rosenqvist, A.; Lucas, R.; Rebelo, L.-M.; Hilarides, L.; Thomas, N.; Hardy, A.; Itoh, T.; Shimada, M.; Finlayson, M. The global mangrove watch-a new 2010 global baseline of mangrove extent. *Remote Sens.* **2018**, *10*, 1669. [CrossRef]
56. Lymburner, L.; Bunting, P.; Lucas, R.; Scarth, P.; Alam, I.; Phillips, C.; Ticehurst, C.; Held, A. Mapping the multi-decadal mangrove dynamics of the Australian coastline. *Remote Sens. Environ.* **2020**, *238*, 111185. [CrossRef]
57. USGS. *Landsat Collection 1 Level 1 Product Definition*; United States Geological Survey: Vermillion, SD, USA, 2017; pp. 1–24.
58. Wu, C.; Niu, Z.; Gao, S. The potential of the satellite derived green chlorophyll index for estimating midday light use efficiency in maize, coniferous forest and grassland. *Ecol. Indic.* **2012**, *14*, 66–73. [CrossRef]
59. Pastor-Guzman, J.; Atkinson, P.; Dash, J.; Rioja-Nieto, R. Spatiotemporal variation in mangrove chlorophyll concentration using Landsat 8. *Remote Sens.* **2015**, *7*, 14530–14558. [CrossRef]
60. Karlson, M.; Ostwald, M.; Reese, H.; Sanou, J.; Tankoano, B.; Mattsson, E. Mapping tree canopy cover and aboveground biomass in Sudano-Sahelian Woodlands using Landsat 8 and Random Forest. *Remote Sens.* **2015**, *7*, 10017–10041. [CrossRef]
61. Berhane, T.M.; Lane, C.R.; Wu, Q.; Autrey, B.C.; Anenkhonov, O.A.; Chepinoga, V.V.; Liu, H. Decision-Tree, rule-based, and Random Forest classification of high-resolution multispectral imagery for wetland mapping and inventory. *Remote Sens.* **2018**, *10*, 580. [CrossRef]
62. Pham, T.D.; Xia, J.; Ha, N.T.; Bui, D.T.; Le, N.N.; Tekeuchi, W. A review of remote sensing approaches for monitoring blue carbon ecosystems: Mangroves, seagrasses and salt marshes during 2010-2018. *Sensors* **2019**, *19*, 1933. [CrossRef]
63. Belgiu, M.; Drăguţ, L. Random forest in remote sensing: A review of applications and future directions. *ISPRS J. Photogramm. Remote Sens.* **2016**, *114*, 24–31. [CrossRef]
64. Liu, B.; Gao, L.; Li, B.; Marcos-Martinez, R.; Bryan, B.A. Nonparametric machine learning for mapping forest cover and exploring influential factors. *Landsc. Ecol.* **2020**, *35*, 1683–1699. [CrossRef]
65. Pelletier, C.; Valero, S.; Inglada, J.; Champion, N.; Dedieu, G. Assessing the robustness of Random Forests to map land cover with high resolution satellite image time series over large areas. *Remote Sens. Environ.* **2016**, *187*, 156–168. [CrossRef]
66. Maxwell, A.E.; Warner, T.A.; Fang, F. Implementation of machine-learning classification in remote sensing: An applied review. *Int. J. Remote Sens.* **2018**, *39*, 2784–2817. [CrossRef]
67. Eklundh, L.; Jonsson, P. TIMESAT for processing time-series data from satellite sensors for land surface monitoring. In *Multitemporal Remote Sensing: Methods and Applications*; Ban, Y., Ed.; Springer International Publishing: Cham, Switzerland, 2016; Volume 20, pp. 177–194.
68. Jönsson, P.; Eklundh, L. TIMESAT—A program for analyzing time-series of satellite sensor data. *Comput. Geosci.* **2004**, *30*, 833–845. [CrossRef]
69. Stanimirova, R.; Cai, Z.; Melaas, E.K.; Gray, J.M.; Eklundh, L.; Jönsson, P.; Friedl, M.A. An empirical assessment of the MODIS land cover dynamics and TIMESAT land surface phenology algorithms. *Remote Sens.* **2019**, *11*, 2201. [CrossRef]
70. Valderrama-Landeros, L.; España-Boquera, M.L.; Baret, F.; Sánchez-Vargas, N.; Sáenz-Romero, C. Capacity of phenological data derived from Cyclopes Lai for the year 2000 to distinguish land cover types in the State of MichoacÁn, Mexico. *Revista Chapingo Serie Ciencias Forestales Y Del Ambiente* **2014**, *20*, 261–276. [CrossRef]
71. Hentze, K.; Thonfeld, F.; Menz, G. Evaluating crop area mapping from MODIS time-series as an assessment tool for Zimbabwe's "Fast Track Land Reform Programme". *PLoS ONE* **2016**, *11*, e0156630. [CrossRef]

72. Kong, F.; Li, X.; Wang, H.; Xie, D.; Li, X.; Bai, Y. Land cover classification based on fused data from GF-1 and MODIS NDVI time series. *Remote Sens.* **2016**, *8*, 741. [CrossRef]
73. Davis, C.L.; Hoffman, M.T.; Roberts, W. Long-term trends in vegetation phenology and productivity over Namaqualand using the GIMMS AVHRR NDVI3g data from 1982 to 2011. *S. Afr. J. Bot.* **2017**, *111*, 76–85. [CrossRef]
74. Ghosh, S.; Mishra, D. Analyzing the long-term phenological trends of salt marsh ecosystem across Coastal Louisiana. *Remote Sens.* **2017**, *9*, 1340. [CrossRef]
75. R Core Team. R: A Language and Environment for Statistical Computing. 2020. Available online: https://www.R-project.org/ (accessed on 1 March 2021).
76. Bureau of Meteorology. Tropical Cyclone Reports. Available online: http://www.bom.gov.au/cyclone/tropical-cyclone-knowledge-centre/history/past-tropical-cyclones/ (accessed on 1 March 2021).
77. Commonwealth of Australia. *Reef 2050 Long Term Sustainability Plan 2018*; Commonwealth of Australia: Canberra, Australia, 2018.
78. Bahuguna, A.; Chauhan, H.B.; Sen Sarma, K.; Bhattacharya, S.; Ashutosh, S.; Pandey, C.N.; Thangaradjou, T.; Gnanppazham, L.; Selvam, V.; Nayak, S.R. Mangrove inventory of India at community level. *Natl. Acad. Sci. Lett.* **2013**, *36*, 67–77. [CrossRef]
79. Semeniuk, V.; Cresswell, I.D. Australian mangroves: Anthropogenic impacts by industry, agriculture, ports, and urbanisation. In *Threats to Mangrove Forests: Hazards, Vulnerability, and Management*; Makowski, C., Finkl, C.W., Eds.; Springer International Publishing: Cham, Switzerland, 2018; pp. 173–197. [CrossRef]
80. Goldberg, L.; Lagomasino, D.; Thomas, N.; Fatoyinbo, T. Global declines in human-driven mangrove loss. *Glob. Chang. Biol.* **2020**, *26*, 5844–5855. [CrossRef] [PubMed]
81. Hamilton, S.E.; Casey, D. Creation of a high spatio-temporal resolution global database of continuous mangrove forest cover for the 21st century (CGMFC-21). *Glob. Ecol. Biogeogr.* **2016**, *25*, 729–738. [CrossRef]
82. Sippo, J.Z.; Lovelock, C.E.; Santos, I.R.; Sanders, C.J.; Maher, D.T. Mangrove mortality in a changing climate: An overview. *Estuar. Coast. Shelf Sci.* **2018**, *215*, 241–249. [CrossRef]
83. Hamunyela, E.; Verbesselt, J.; Herold, M. Using spatial context to improve early detection of deforestation from Landsat time series. *Remote Sens. Environ.* **2016**, *172*, 126–138. [CrossRef]
84. Hirschmugl, M.; Gallaun, H.; Dees, M.; Datta, P.; Deutscher, J.; Koutsias, N.; Schardt, M. Methods for mapping forest disturbance and degradation from optical Earth observation data: A review. *Curr. For. Rep.* **2017**, *3*, 32–45. [CrossRef]
85. Pimple, U.; Simonetti, D.; Sitthi, A.; Pungkul, S.; Leadprathom, K.; Skupek, H.; Som-ard, J.; Gond, V.; Towprayoon, S. Google Earth Engine based three decadal Landsat imagery analysis for mapping of mangrove forests and its surroundings in the Trat Province of Thailand. *J. Commun.* **2018**, *6*, 247–264. [CrossRef]
86. Hu, L.; Xu, N.; Liang, J.; Li, Z.; Chen, L.; Zhao, F. Advancing the mapping of mangrove forests at national-scale using Sentinel-1 and Sentinel-2 time-series data with Google Earth Engine: A case study in China. *Remote Sens.* **2020**, *12*, 3120. [CrossRef]
87. Glenn, E.P.; Huete, A.R.; Nagler, P.L.; Nelson, S.G. Relationship between remotely-sensed vegetation indices, canopy attributes and plant physiological processes: What vegetation indices can and cannot tell us about landscape. *Sensors* **2008**, *8*, 2136–2160. [CrossRef]
88. Elmahdy, S.I.; Ali, T.A.; Mohamed, M.M.; Howari, F.M.; Abouleish, M.; Simonet, D. Spatiotemporal mapping and monitoring of mangrove forests changes from 1990 to 2019 in the Northern Emirates, UAE using Random Forest, Kernel Logistic Regression and Naive Bayes Tree Models. *Front. Environ. Sci.* **2020**, *8*, 102. [CrossRef]
89. Gitelson, A.A.; Peng, Y.; Masek, J.G.; Rundquist, D.C.; Verma, S.; Suyker, A.; Baker, J.M.; Hatfield, J.L.; Meyers, T. Remote estimation of crop gross primary production with Landsat data. *Remote Sens. Environ.* **2012**, *121*, 404–414. [CrossRef]
90. Lovelock, C.E.; Adame, M.F.; Bennion, V.; Hayes, M.; Reef, R.; Santini, N.; Cahoon, D.R. Sea level and turbidity controls on mangrove soil surface elevation change. *Estuar. Coast. Shelf Sci.* **2015**, *153*, 1–9. [CrossRef]
91. Carrao, H.; Gonalves, P.; Caetano, M. A nonlinear harmonic model for fitting satellite image time series: Analysis and prediction of land cover dynamics. *IEEE Trans. Geosci. Remote Sens.* **2010**, *48*, 1919–1930. [CrossRef]
92. Bi, J.; Knyazikhin, Y.; Choi, S.; Park, T.; Barichivich, J.; Ciais, P.; Fu, R.; Ganguly, S.; Hall, F.; Hilker, T.; et al. Sunlight mediated seasonality in canopy structure and photosynthetic activity of Amazonian rainforests. *Environ. Res. Lett.* **2015**, *10*. [CrossRef]
93. Wu, J.; Albert, L.P.; Lopes, A.P.; Restrepo-Coupe, N.; Hayek, M.; Wiedemann, K.T.; Guan, K.; Stark, S.C.; Christoffersen, B.; Prohaska, N.; et al. Leaf development and demography explain photosynthetic seasonality in Amazon evergreen forests. *Science* **2016**, *351*, 972–976. [CrossRef]
94. Robertson, A.I.; Dixon, P.; Zagorskis, I. Phenology and litter production in the mangrove genus Xylocarpus along rainfall and temperature gradients in tropical Australia. *Mar. Freshw. Res.* **2021**, *72*, 551–562. [CrossRef]
95. Younes, N.; Northfield, T.D.; Joyce, K.E.; Maier, S.W.; Duke, N.C.; Lymburner, L. A novel approach to modelling mangrove phenology from satellite images: A case study from Northern Australia. *Remote Sens.* **2020**, *12*, 4008. [CrossRef]
96. Pastor-Guzman, J.; Dash, J.; Atkinson, P.M. Remote sensing of mangrove forest phenology and its environmental drivers. *Remote Sens. Environ.* **2018**, *205*, 71–84. [CrossRef]
97. Songsom, V.; Koedsin, W.; Ritchie, R.J.; Huete, A. Mangrove phenology and environmental drivers derived from remote sensing in Southern Thailand. *Remote Sens.* **2019**, *11*, 955. [CrossRef]
98. Duke, N.C.; Field, C.; Mackenzie, J.R.; Meynecke, J.-O.; Wood, A.L. Rainfall and its possible hysteresis effect on the proportional cover of tropical tidal-wetland mangroves and saltmarsh–saltpans. *Mar. Freshw. Res.* **2019**, *70*, 1047–1055. [CrossRef]
99. Santini, N.S.; Reef, R.; Lockington, D.A.; Lovelock, C.E. The use of fresh and saline water sources by the mangrove Avicennia marina. *Hydrobiologia* **2014**, *745*, 59–68. [CrossRef]

100. Eklundh, L.; Jönsson, P. TIMESAT: A software package for time-series processing and assessment of vegetation dynamics. In *Remote Sensing Time Series: Revealing Land Surface Dynamics*; Kuenzer, C., Dech, S., Wagner, W., Eds.; Springer International Publishing: Cham, Switzerland, 2015; pp. 141–158. [CrossRef]
101. Coupland, G.T.; Paling, E.I.; McGuinness, K.A. Vegetative and reproductive phenologies of four mangrove species from northern Australia. *Aust. J. Bot.* **2005**, *53*, 109–117. [CrossRef]
102. Morton, D.C.; Nagol, J.; Carabajal, C.C.; Rosette, J.; Palace, M.; Cook, B.D.; Vermote, E.F.; Harding, D.J.; North, P.R. Amazon forests maintain consistent canopy structure and greenness during the dry season. *Nature* **2014**, *506*, 221–224. [CrossRef] [PubMed]
103. Osland, M.J.; Enwright, N.M.; Day, R.H.; Gabler, C.A.; Stagg, C.L.; Grace, J.B. Beyond just sea-level rise: Considering macroclimatic drivers within coastal wetland vulnerability assessments to climate change. *Glob. Chang. Biol.* **2016**, *22*, 1–11. [CrossRef] [PubMed]
104. Feng, Y.; Negrón-Juárez, R.I.; Chambers, J.Q. Remote sensing and statistical analysis of the effects of hurricane María on the forests of Puerto Rico. *Remote Sens. Environ.* **2020**, *247*, 111940. [CrossRef]
105. Das, C.S.; Mallick, D.; Mandal, R.N. Mangrove forests in changing climate: A global overview. *J. Indian Soc. Coast. Agric. Res.* **2020**, *38*, 104–124.
106. Rossi, S.; Soares, M.d.O. Effects of El NiÑo on the coastal ecosystems and their related services. *Mercator* **2017**, *16*, 1–16. [CrossRef]
107. Asbridge, E.; Lucas, R.; Ticehurst, C.; Bunting, P. Mangrove response to environmental change in Australia's Gulf of Carpentaria. *Ecol. Evol.* **2016**, *6*, 3523–3539. [CrossRef]
108. Hoque, M.A.-A.; Phinn, S.; Roelfsema, C. A systematic review of tropical cyclone disaster management research using remote sensing and spatial analysis. *Ocean Coast. Manag.* **2017**, *146*, 109–120. [CrossRef]
109. Foody, G.M. Assessing the accuracy of land cover change with imperfect ground reference data. *Remote Sens. Environ.* **2010**, *114*, 2271–2285. [CrossRef]
110. Younes Cárdenas, N.; Joyce, K.E.; Maier, S.W. Monitoring mangrove forests: Are we taking full advantage of technology? *Int. J. Appl. Earth Obs. Geoinf.* **2017**, *63*, 1–14. [CrossRef]
111. Bush, E.R.; Bunnefeld, N.; Dimoto, E.; Dikangadissi, J.-T.; Jeffery, K.; Tutin, C.; White, L.; Abernethy, K.A. Towards effective monitoring of tropical phenology: Maximizing returns and reducing uncertainty in long-term studies. *Biotropica* **2018**, *50*, 455–464. [CrossRef]
112. Bush, E.R.; Abernethy, K.A.; Jeffery, K.; Tutin, C.; White, L.; Dimoto, E.; Dikangadissi, J.T.; Jump, A.S.; Bunnefeld, N.; Freckleton, R. Fourier analysis to detect phenological cycles using long-term tropical field data and simulations. *Methods Ecol. Evol.* **2017**, *8*, 530–540. [CrossRef]
113. Cresswell, I.D.; Semeniuk, V. Australian mangroves: Their distribution and protection. In *Threats to Mangrove Forests: Hazards, Vulnerability, and Management*; Makowski, C., Finkl, C.W., Eds.; Springer International Publishing: Cham, Switzerland, 2018; pp. 3–22. [CrossRef]
114. Friess, D.A.; Rogers, K.; Lovelock, C.E.; Krauss, K.W.; Hamilton, S.E.; Lee, S.Y.; Lucas, R.; Primavera, J.; Rajkaran, A.; Shi, S. The state of the world's mangrove forests: Past, present, and future. *Annu. Rev. Environ. Resour.* **2019**, *44*, 89–115. [CrossRef]
115. Ramsar Convention on Wetlands. *Global Wetland Outlook: State of the World's Wetlands and Their Services to People*; Ramsar Convention Secretariat: Gland, Switzerland, 2018.

remote sensing

MDPI

Article

Spatio-Temporal Simulation of Mangrove Forests under Different Scenarios: A Case Study of Mangrove Protected Areas, Hainan Island, China

Bin Zhu [1,2], Jingjuan Liao [1,3,*] and Guozhuang Shen [1]

[1] International Research Center of Big Data for Sustainable Development Goals, Aerospace Information Research Institute, Chinese Academy of Sciences, Beijing 100094, China; zhubin@aircas.ac.cn (B.Z.); shengz@radi.ac.cn (G.S.)
[2] University of Chinese Academy of Sciences, Beijing 100049, China
[3] Key Laboratory of Earth Observation Hainan Province, Hainan 572029, China
* Correspondence: liaojj@aircas.ac.cn; Tel.: +86-10-82178160

Citation: Zhu, B.; Liao, J.; Shen, G. Spatio-Temporal Simulation of Mangrove Forests under Different Scenarios: A Case Study of Mangrove Protected Areas, Hainan Island, China. *Remote Sens.* **2021**, *13*, 4059. https://doi.org/10.3390/rs13204059

Academic Editor: Chandra Giri

Received: 1 September 2021
Accepted: 5 October 2021
Published: 11 October 2021

Publisher's Note: MDPI stays neutral with regard to jurisdictional claims in published maps and institutional affiliations.

Abstract: Mangrove forests are important woody plant communities that grow in the intertidal zone between land and sea. They provide important social, ecological and economic services to coastal areas. In recent years, the growth environment of mangrove forests has been threatened. Mangrove forests have become one of the most endangered ecosystems in the world. To better protect mangrove forests, effective monitoring methods are essential. In this study, a spatio-temporal simulation method for mangrove forests was proposed in the mangrove protected areas of Hainan Island, China. This method compared the simulation accuracy of different models in terms of spatial characteristics, evaluated the applicability of driving factors in mangrove simulation and predicted the future spatio-temporal distribution and change trends of mangrove forests under different scenarios. The simulation results of different models showed that AutoRF (random forest with spatial autocorrelation) performs best in spatial characteristic simulation. Driving factors such as the Enhanced Vegetation Index (EVI), various location indices and the spatial autocorrelation factor can significantly improve the accuracy of mangrove simulations. The prediction results for Hainan Island showed that the mangrove area increased slowly under a natural growth scenario (NGS), decreased significantly under an economic development scenario (EDS) and increased significantly under a mangrove protection scenario (MPS) with 4460, 2704 and 5456 ha respectively by 2037. The contraction of mangrove forests is closely related to the expansion of aquaculture ponds, building land and cultivated land. Mangrove contraction is more severe in marginal or fragmented areas. The expansion of mangrove forests is due to the contraction of aquaculture ponds, cultivated land and other forests. The areas around existing mangrove forests and on both sides of the riverbank are typical areas prone to mangrove expansion. The MPS should be the most suitable development direction for the future, as it can reasonably balance economic development with mangrove protection.

Keywords: mangrove forests; Hainan Island; CLUE-S; spatio-temporal simulation; future change trends

1. Introduction

Mangrove forests are important woody plant communities that grow in the intertidal zone between land and sea. They are widely distributed in tropical and subtropical regions of the world between 30° N and 30° S latitude [1]. Mangrove forests can sequestrate carbon, mitigate climate change, maintain marine and land biodiversity, purify water, protect coastlines and coastal infrastructure and export economic products, thus providing important social, ecological and economic services for surrounding areas [2–10]. Mangrove forests are constantly threatened due to natural and anthropogenic factors such as extreme weather events, sea-level rise, aquaculture and urban development, making them one of the most endangered ecosystems in the world [11–14]. Currently, the world's mangrove forests have declined by about 40% compared to the middle of the last century [15]. Throughout the 1990s, the annual loss rate of mangrove forests was 1%, which was about twice that of terrestrial forests during the same period [16]. In recent years, with the continuous advancement of protection measures around the world, the loss rate of mangrove forests has slowed, but still remains at 0.26–0.66% per year [17]. If the loss rate of mangrove forests continues, about 40% of the world's mangrove forests will completely disappear by the end of this century. The services and products provided by mangrove forests will be significantly reduced or lost, which will have a negative impact on human survival and development [1]. Therefore, the protection of mangrove forests is urgent and has reached a broad international consensus. Effective mangrove monitoring methods are very important, which can provide a theoretical basis and decision support for mangrove protection, restoration and utilization.

Mangrove forests grow in the mudflats between land and sea. Traditional field survey methods have difficulty in obtaining accurate, comprehensive and timely mangrove data. In recent years, the rapid development of remote sensing technology has provided a new means for better extraction, analysis and prediction of mangrove forests. At present, mangrove monitoring mainly focuses on mangrove change analysis and time series analysis. Mangrove change analysis is a statistical analysis of the area, extent, conversion and landscape pattern of mangrove forests [8,13,18–29]. Although this method can reflect spatial distribution, area change and the causes of changes in mangrove forests, it only statistically analyzes the overall changes, and cannot accurately reveal the spatial trends of each section. Mangrove time series analysis is based on vegetation indices, with short time intervals and long time series. This method explores mangrove trends by linear regression analysis, Theil-Sen median trend analysis, the Mann-Kendall test, the Hurst exponent and so forth. It can reflect the change trends and future trends of mangrove forests, and further reveal the causes and effects of these trends [30–33]. Although this method accurately reflects the spatial trends of each section, it only analyzes the changes of historical data and cannot predict how long the sustainable trends will continue [34], which poses a challenge for the future continuous monitoring of mangrove forests.

Therefore, it is crucial to conduct spatio-temporal simulation of mangrove forests to predict the future spatio-temporal distribution of mangrove forests under different scenarios. Land-use change models are the key to spatio-temporal simulation, which can be divided into two categories, non-spatial models and spatial models. Non-spatial models were developed earlier, such as the SAhelian Land-Use model (SALU) [35], linear programming model [36,37], system dynamics model [38,39] and Markov chain model [40]. These models consider only quantity changes and do not measure location changes. Spatial models consist of micromodels (bottom-up) and macromodels (top-down), which consider both quantity and location changes. Micromodels first configure land-use changes according to demands at the microscale and then aggregate the results to the macroscale, with examples including cellular automation (CA) [41,42], SAMBA [43], the Future Land Use Simulation (FLUS) [44] and so forth. Macromodels first configure land-use changes at the macroscale, and then allocate demands to the microscale layer by layer; applications include the Land Use Planning and Analysis System (LUPAS) [45], the Conversion of Land Use and its Effects at Small region extent (CLUE-S) [46] and so forth. At present, few studies address the spatio-temporal simulation of mangrove forests, and these are mainly based on CA-Markov models. This model uses a Markov chain model to calculate the quantity changes, and then brings it into a CA model to simulate location changes. Mukhopadhyay et al. [47] utilized this model to predict the mangrove species of Bangladesh Sundarbans in 2025, 2055 and 2105. Bozkaya et al. [48] compared the CA-Markov and St. Markov models and predicted the distribution of mangrove forests along the northwest coast of Turkey in 2030. DasGupta et al. [49] used Multi-Layer Perceptron-Markov Chain Analysis (MLP-MCA) to predict the distribution of mangrove forests in Sundarbans, India in 2030, under four development scenarios. Tajbakhsh et al. [50] developed a hybrid model (CA-Markov-ANN) to predict the distribution of mangrove forests in Qeshm Island, Southern Iran in 2025.

However, the above studies for mangrove forests still have some shortcomings. First, previous studies are mainly based on the CA-Markov model. The CLUE-S model has been widely used in spatio-temporal simulation in different regions [51–53]. For long-term simulation with non-stationary change patterns, the CLUE-S model has better stability compared to the CA-Markov model. The statistical significance of the probability values in the CA-Markov model is affected when the number of area changes is small [54]. The CLUE-S model is more accurate in the simulation of the land-use spatial patterns compared to the CA-Markov model [55]. Second, the simulation accuracy of spatial characteristics needs to be improved. Spatial characteristics are key to the CLUE-S model to assess the potential for land-use changes, which was calculated from land-use data and driving factors by logistic regression [51,56], autologistic regression [52,57,58], NE-Logistic regression [56], artificial neural networks [51] and random forest (RF) [53,59], and so forth. Some studies have shown that machine learning methods such as RF have higher simulation accuracy compared to traditional logistic regression [53], but no systematic comparisons have been made. Third, driving factors need to be further refined and selected. Most driving factors in the previous studies are commonly used indices such as elevation, slope and location indices. Vegetation indices such as NDVI and its change trend can well reflect the change trends of vegetation [60], which can provide a reference for simulation and prediction. The spatial autocorrelation factor was proposed to address the spatial autocorrelation effect inherent to spatial statistical analysis [61]. Traditional logistic regression methods can significantly improve the simulation accuracy along with it [57], but machine learning methods have not yet considered it.

Mangrove protected areas in Hainan Island were selected as the study area to explore the spatio-temporal simulation of mangrove forests under different scenarios. The objectives of this study were as follows: (1) compare the simulation accuracy of different models in terms of spatial characteristics and evaluate the applicability of driving factors in mangrove simulation; (2) set different development scenarios and predict the future spatio-temporal distribution of mangrove forests; and (3) analyze the future change trends of mangrove forests.

2. Materials

2.1. Study Area

Hainan Island is rich in mangrove species, with 26 species of true mangrove, 12 species of semi-mangrove and more than 40 species of mangrove associates [62], containing almost all of the mangrove species found in China. For more than half a century, the mangrove area in Hainan Island has experienced a developmental process from a sharp decline to a slow increase. In the 1950s, the mangrove area of Hainan Island accounted for about a quarter of China's mangrove area, reaching 9992 ha [63]. Then, these mangrove forests suffered serious damage due to excessive economic development, such as marine aquaculture, land reclamation and mangrove deforestation. In the 1980s, the mangrove area was reduced by half, to reach 4836 ha [64]. By 2010, the mangrove area in Hainan Island had decreased to 3576 ha [21]. In recent years, as the government began to vigorously implement restrictive measures such as converting cultivated land to wetlands and converting fishponds to wetlands, the public awareness of protecting mangroves has increased significantly. The mangrove area of Hainan Island has responded with a slow increase, recovering to 4278 ha in 2017 [65]. At present, Hainan Island has established 10 mangrove reserves, including one national reserve, two provincial reserves and seven local reserves. Seven of these mangrove reserves were selected for this study (Figure 1 and Table S1). Dongzhaigang National Mangrove Reserve is the best preserved and most abundant mangrove reserve in China. Qinglangang Nature Mangrove Reserve consists of three regions: Huiwen, Puqian and Guannan (a1, a2 and a3 in Figure 1), containing the largest number of mangrove species in China. Dongfang Nature Reserve was established in 2006. Mangrove forests within it provide a good habitat for the international endangered species Platalea minor. The remaining four local protected reserves are located in the northern part of Hainan Island. The study area was selected within the 2 km buffer zone of the reserve boundary [27].

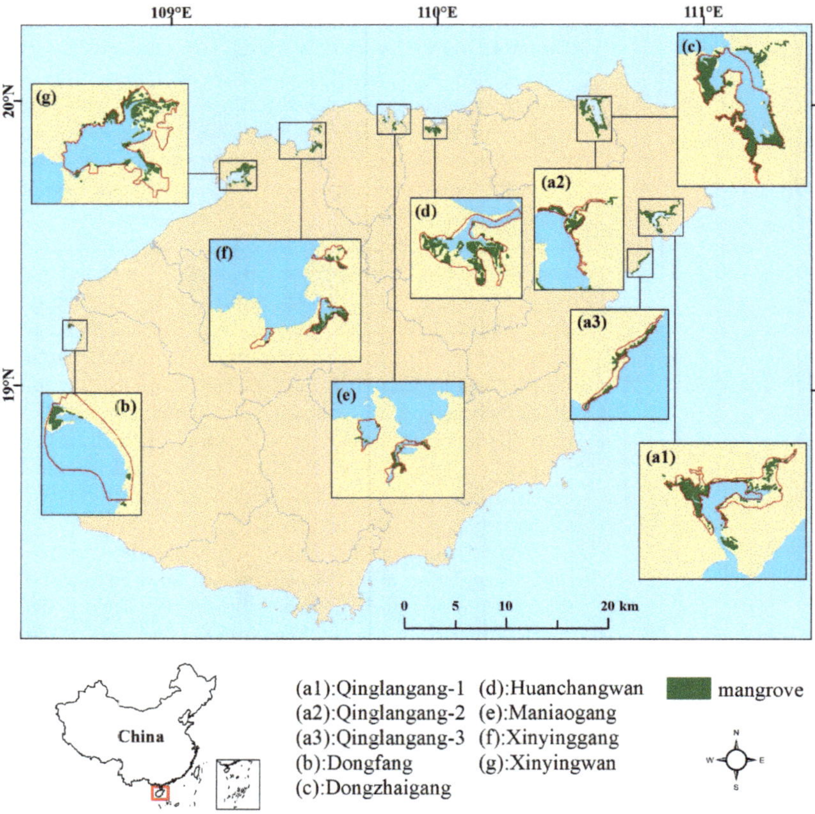

Figure 1. Location of selected reserves (**a1–g**) in this study (red lines are the boundary of reserves; green areas indicate the appearance of mangrove forests in 2017).

2.2. Data Sources

The data sources used in this study are land cover data and driving factor data. The land cover data used in the study was based on the classification results of Landsat images in 1987, 1993, 1998, 2003, 2007, 2013 and 2017, with a spatial resolution of 30 m [65]. The 2017 dataset was classified according to the support vector machine method with a high-precision manual correction; the overall accuracy was 98.8%. The classified 2017 image was then used as a reference for the visual interpretation of the remaining images [27]. The land cover types were categorized into 10 classes: mangrove forests (MF), tidal sandflats (TS), aquaculture ponds (AP), water (WT), cultivated land (CL), wetlands (WL), bare land (BL), other forests (OF), suitable land for mangrove (SLM) and building land (BDL).

The driving factor data were mainly selected from 12 indices related to mangrove distribution, including two terrain indices, two vegetation indices, seven location indices and one correlation index (Table 1).

Table 1. The driving factor data used in this study.

Type	Factors	Unit
Terrain	Elevation	m
	Slope	degree
Vegetation	EVI	-
	EVI change trends	-
Location	Distance to major road	m
	Distance to minor road	m
	Distance to sea	m
	Distance to river	m
	Distance to aquaculture ponds	m
	Distance to building land	m
	Distance to suitable land for mangrove	m
Correlation	Spatial autocorrelation factor	m

For the terrain indices, the elevation was derived from the 30 m resolution SRTM 1 Arc-Second Global (SRTMGL1) dataset. The slope was then calculated from the elevation gradient.

For the vegetation indices, the EVI was calculated from Landsat images. This index solves the saturation problem under high vegetation coverage [66], and is more suitable for mangrove change studies [33]. In this study, the Landsat TM/ETM+/OLI images with a spatial resolution of 30 m were acquired each year for Huiwen from 1999 to 2003 (for simulation) and for all the study areas from 2013 to 2017 (for prediction). Due to the insignificant phenology effect of mangrove forests, the impacts of cloud cover and tide level were the main parameters considered when screening the images. In this study, the image with the lowest tide level among images, with cloud cover less than 10%, was selected as a representative image for that year. The EVI change trends were calculated by the Theil-Sen median trend analysis using the EVI images of the last five years (Equation 1). It has been proven that this method can well reflect the trends of long time series data, and has been widely used in vegetation studies [33,34,67,68].

$$S = \text{median}\left(\frac{\text{EVI}_j - \text{EVI}_i}{j - i}\right) \tag{1}$$

where EVI_i and EVI_j are the EVI values of years i and j. $S > 0$ indicates an increasing trend of the series; all other values indicate a decreasing trend.

The location indices were calculated by the inverse distance weighting method according to the different information. The road information was obtained from OpenStreetMap (OSM). In this study, motorway, trunk, primary, secondary, tertiary and highway links were merged as major roads, while minor roads constituted the remaining residential, pedestrian, cycleways and so forth. The information of sea, river, aquaculture ponds, building land and suitable land for mangrove were obtained from land cover data.

For the correlation index, the spatial autocorrelation factor was selected as follows

$$\text{autocov}_{(\text{i, k})} = \sum_{\text{j}\neq\text{i}} \text{w}_{\text{ij}}\text{y}_{\text{j}} / \sum_{\text{j}\neq\text{i}} \text{w}_{\text{ij}} \tag{2}$$

where k is the land cover type of pixel i. y_j is the state in which land cover type k exists in pixel j (1 means exists, 0 means does not exist). w_{ij} is the inverse distance weight between pixel i and j, which can be expressed as:

$$\text{w}_{\text{ij}} = \begin{cases} 1/\text{D}_{\text{ij}} & ,\text{D}_{\text{ij}} < d \\ 0 & ,\text{D}_{\text{ij}} \geq d \end{cases} \tag{3}$$

where D_{ij} is the Euclidean distance between pixels i and j. d is the threshold distance. In this study, d = 500 m is taken according to the pixel size (30 m) of this study.

3. Methods

A spatio-temporal simulation method for mangrove forests was proposed (Figure 2). According to the characteristics of mangrove forests, the simulation accuracy of different models such as logistic regression, support vector regression (SVR) and random forest, in terms of spatial characteristics, was compared, and the applicability of driving factors such as vegetation and correlation indices in mangrove simulation were explored. Then, according to the characteristics of the different protected areas, three development scenarios of NGS, EDS and MPS were set. The CLUE-S model was used to predict the spatio-temporal distribution of mangrove forests from 2022 to 2037 under different scenarios. Finally, based on the prediction results of spatio-temporal distribution of mangrove forests, the future change trends of mangrove forests from 2017 to 2037 were analyzed.

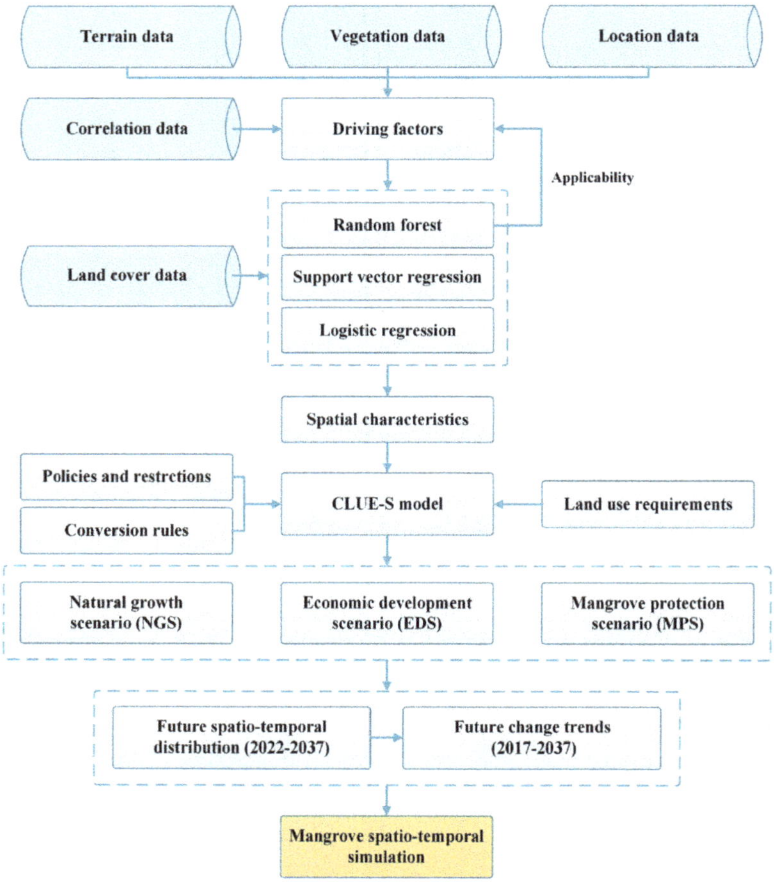

Figure 2. Flowchart of the spatio-temporal simulation of mangrove forests.

3.1. CLUE-S Model

The CLUE-S model was chosen to simulate mangrove forests in protected areas. It was developed by researchers at Wageningen University based on the CLUE model. Considering the competing mechanisms of land-use changes, this model can effectively link the land-use change processes and environmental driving factors, and is suitable for mesoscale and small-scale land-use simulation studies. The CLUE-S model consists of four input modules (spatial policies and restrictions, land-use type conversion rules, land-use requirements and spatial characteristics) and a spatial allocation module [46].

The spatial policies and restrictions module is used to restrict land-use layout. It refers to areas where land-use changes are not allowed to occur, usually including policy restrictions (nature reserves, national parks and so forth) and spatial restrictions (rivers, oceans and so forth). In this study, rivers and estuaries are usually not encroached upon by other land-use types. Therefore, these areas were set as restricted areas to ensure the connectivity of water bodies.

The land-use type conversion rules module is used to quantify the possibility and intensity of the land-use type conversion, mainly containing conversion sequence and conversion elasticity. Conversion sequence is a parameter that indicates whether conversion can occur between various land-use types. It is represented by a matrix in the CLUE-S model, and the value is 0 (not transferable) or 1 (transferable). The conversion sequence of this study was adjusted according to different scenarios, as described in Section 3.3. Table S2 shows the conversion sequence under different scenarios. Conversion elasticity is a parameter that quantifies the reversibility of land-use changes, with a value range of 0 (easy to transfer) to 1 (not easy to transfer). According to different scenarios, the conversion elasticity of this study was adjusted for different land-use types separately, as described in Section 3.3. Table S3 shows the conversion elasticity under different scenarios.

The land-use requirements module is used to calculate future area demands, which is a key parameter affecting the iterations of the CLUE-S model. It can be estimated by a statistical model [52,53,69–72], a system dynamics model [73,74], a professional model [75,76] and so forth. In this study, linear regression and scenario analysis were used to predict the future area demands for 2022–2037, as described in Section 3.3.

The spatial characteristics module is used to calculate the spatial suitability probability of various land-use types at specific locations. It is key for the CLUE-S model to assess the potential for land-use changes. Spatial suitability probability is generally calculated from land-use data and driving factors by different models. In this study, in addition to the traditional logistic regression, two commonly used machine learning models that support vector regression and random forest methods were added to compare the simulation accuracy in terms of spatial characteristics. Support vector regression uses different kernel functions to nonlinearly map the original low-dimensional feature to the high-dimensional feature space, and then it constructs a linear decision function to solve the nonlinear problem [77]. This method is well suited for solving high-dimensional and nonlinear regression problems. Random forest is based on multiple decision trees. Parts of the features on the tree node are randomly selected for training, and then the optimal feature is selected to divide the node. The final regression result is the average of all the results of the decision tree [78]. This method reduces the generalization error of a single decision tree, and greatly improves the regression accuracy. To reduce the problem of uneven sample selection, both training and validation samples in this study were generated by a stratified random sampling method, where the number of training samples constituted 2% of the total pixels.

The space allocation module is used to calculate the final allocation result. It is based on the results of the four input modules and the land-use map of the starting year. According to the total occurrence probability of various land-use types in each pixel, the space is allocated by several iterations until the land-use requirements are satisfied [46]. The CLUE-S model was implemented through the lulcc-package in R software [79].

3.2. Model Validation Indices

The area under curve (AUC) was chosen to evaluate the simulation accuracy of the different models in terms of spatial characteristics. It represents the area under the receiver operating characteristic (ROC) curve, and is commonly used as a performance evaluation index for the classification or fitting algorithms [80]. The value of AUC ranges from 0.5 to

1.0; the closer to 1 the value is, the better performance of the model. An AUC above 0.7 is generally considered as good, and above 0.9 is excellent [51].

Four types of indices were selected to assess the accuracy of the simulation results: the overall accuracy (OA), the Kappa coefficient, the three-dimensional approach and the Figure of Merit (FoM). The OA is the ratio of the number of correctly classified pixels to the total number of pixels, and is often used in the accuracy assessment of classification algorithms. It is also widely used for consistency checking between simulation and observation maps of land-use change models [52,53,72]. The Kappa coefficient represents the proportion of error reduction in the evaluated classification, compared with a completely random classification. It is an index designed for evaluating the accuracy of classification results [81], and is further developed into four different forms that can quantify either quantity error or location error [82]; KQuantity believes that the results have the ability to consider location precision and is applicable to area change evaluation; KLocation believes that the results have the ability to consider quantity precision and is applicable to location change evaluation; and KStandard and KNo consider location and quantity precision together. KNo believes that the results do not have the ability to consider quantity and location precision, which is more objective. KStandard, KNo and KLocation were selected in this study to evaluate the simulation results in terms of location changes. The three-dimensional approach is based on the three-map comparison method, which divides the simulation results into two agreement and three disagreement components [83]. The agreement components are Correct rejections (persistence simulated correctly) and Hits (change simulated correctly). The disagreement components are Misses (change simulated as persistence), Wrong hits (change simulated as change to wrong category) and False alarms (persistence simulated as change). This approach uses a three-dimensional, spatio-temporal comparison method to evaluate results, which is a good supplement to the Kappa coefficient [84]. The FoM is the intersection of the observed change and simulated change divided by the union of the observed change and simulated change [85]. It is often used to compare the consistency of the simulation and observation maps, which is expressed as follows

$$\mathrm{FoM} = \frac{\mathrm{b}}{\mathrm{a} + \mathrm{b} + \mathrm{c} + \mathrm{d}} \tag{4}$$

where a is Misses, b is Hits, c is Wrong hits and d is False alarms.

To explore the contribution of different driving factors to spatial characteristic simulation, the applicability of different driving factors was analyzed using IncNodePurity (Increase in node purity) in random forest. The principle of this index is to calculate the sum of the squares of the residuals. It represents the heterogeneous impact of each feature on the observations of decision tree nodes, reflecting the feature importance in the context of decreasing accuracy. The larger the IncNodePurity, the greater the importance of the driving factor in spatial characteristic simulation.

3.3. Scenario Setting

Three development scenarios, NGS, EDS and MPS, were set up to analyze the impact of different development directions on the future spatio-temporal distribution of mangrove forests.

The NGS is designed to maintain current development trends. Based on the land cover changes from 2003 to 2017, the conversion sequence (Table S2) and the conversion elasticity (Table S3) were determined. The area demands for 2022–2037 were predicted based on the area of 2003, 2007, 2013 and 2017 by linear regression. The results were appropriately adjusted to ensure the rationality of the changes (Table S4).

The EDS is designed to prioritize economic development. Under this scenario, the area of building land and aquaculture ponds will increase and the area of mangrove forests will decrease. The conversion sequence of aquaculture ponds and building land was restricted. The two types were only allowed to be transferred into economic construction land (aquaculture ponds, cultivated land and building land) (Table S2). The conversion elasticity was adjusted to 0.9 (the highest value) for aquaculture ponds and to 0.35 (the lowest value) for mangrove forests (Table S3). The area demands for 2022–2037 were predicted based on the area changes during the economic development phase (rapid growth of building land and aquaculture ponds) from 1987 to 2017. The results were appropriately adjusted to ensure that the growth rate of building land and aquaculture ponds under this scenario was higher than the NGS (Table S4).

The MPS is designed to prioritize the protection of mangrove forests. Under this scenario, the area of mangrove forests will increase and the area of building land and aquaculture ponds will decrease. The conversion sequence of mangrove forests was restricted. Mangrove forests were not allowed to be transferred into aquaculture ponds, cultivated land and building land (Table S2). The conversion elasticity was adjusted to 0.9 (the highest value) for mangrove forests and to 0.35 (the lowest value) for aquaculture ponds and building land (Table S3). The area demands for 2022–2037 were predicted based on the area changes during the mangrove protection phase (rapid growth of mangrove forests) from 1987 to 2017. The results were appropriately adjusted to ensure that the growth rate of mangrove forests under this scenario was higher than the NGS (Table S4).

4. Results

4.1. Simulation Accuracy of Spatial Characteristic

Based on the observation map and driving factors of Huiwen in 2003, the AUC values of various land cover types in different models were calculated (Table 2). Autologistic, AutoSVR and AutoRF are improved models of Logistic, SVR and RF with the spatial autocorrelation factor. AutoRF had the highest AUC values, with all types of land cover types exceeding 0.95. AutoSVR also achieved excellent accuracy, with all AUC values above 0.9. Autologistic was slightly worse, with AUC values below 0.9 for suitable land for mangrove and cultivated land, but greater than 0.7, indicating that the model also maintained good accuracy. The AUC values of Autologistic, AutoSVR and AutoRF were all higher than those of the unimproved model, indicating that the spatial autocorrelation factor can indeed improve the simulation accuracy. The AUC values of the above six models were greater than 0.7 for all types of land cover types, especially for mangrove forests (above 0.95), showing that the 12 driving factors selected in this study could well simulate the changes in mangrove forests. It is of interest that the AUC values of RF were higher than those of the improved Autologistic and AutoSVR, proving that the random forest had significant advantages in spatial characteristic simulation.

Table 2. AUC values of various land cover types in different models.

Model	AP	WT	CL	WL	BL	MF	OF	SLM	BDL
Logistic	1.000	0.981	0.808	0.910	0.779	0.948	0.931	0.708	1.000
SVR	0.978	0.986	0.932	0.889	0.913	0.959	0.932	0.965	0.946
RF	1.000	0.996	0.968	0.905	0.919	0.993	0.969	1.000	1.000
Autologistic	1.000	0.986	0.808	0.973	0.903	0.956	0.936	0.703	1.000
AutoSVR	0.981	0.991	0.930	0.967	0.952	0.960	0.950	0.994	0.981
AutoRF	1.000	0.999	0.976	0.979	0.958	0.995	0.978	1.000	1.000

4.2. Simulation Results and Accuracy Assessment

Based on the observation map and spatial characteristics in 2003, the spatial distribution in 2007, 2013 and 2017 were simulated using the CLUE-S model. To assess the accuracy of the simulation results, the OA and Kappa coefficients were calculated for the simulation results of different models based on the observation map of 2007, 2013 and 2017 (Table 3). The simulation accuracy decreased with the increase of simulation time. The highest accuracy was in 2007, the second highest was in 2013 and the lowest was in 2017, all maintaining good levels (OA above 75%). For different models, similar to the results of AUC values, RF is more advantageous than SVR and Logistic and the model with spatial autocorrelation is more ideal than the model without it. AutoRF had the highest simulation accuracy, with an OA of 77.94% in 2017, indicating that the simulation results of this model are in high agreement with the observation. KStandard, Kno and Klocation were 0.7638, 0.7835 and 0.8293, respectively, indicating that the model had high simulation accuracy for location changes.

Table 3. OA and Kappa coefficients of different models in 2007, 2013 and 2017.

Year	Model	OA	KStandard	Kno	Klocation
2007	Logistic	91.28%	0.8919	0.9019	0.8925
	SVR	90.63%	0.8839	0.8946	0.8848
	RF	91.30%	0.8922	0.9021	0.8928
	Autologistic	91.38%	0.8931	0.9030	0.8936
	AutoSVR	91.48%	0.8944	0.9041	0.8951
	AutoRF	92.00%	0.9008	0.9100	0.9014
2013	Logistic	82.14%	0.7808	0.7991	0.7814
	SVR	82.07%	0.7799	0.7983	0.7809
	RF	83.33%	0.7954	0.8124	0.7958
	Autologistic	82.46%	0.7847	0.8027	0.7854
	AutoSVR	83.75%	0.7882	0.8059	0.7886
	AutoRF	83.76%	0.8007	0.8173	0.8012
2017	Logistic	76.38%	0.7118	0.7343	0.7123
	SVR	76.57%	0.7140	0.7364	0.7147
	RF	77.61%	0.7268	0.7481	0.7274
	Autologistic	76.69%	0.7155	0.7378	0.7160
	AutoSVR	77.31%	0.7231	0.7447	0.7236
	AutoRF	77.94%	0.7638	0.7835	0.8293

The three-dimensional approach and FoM were also used to further analyze the simulation results of different models in 2017 (Table 4). The results of FoM differed from the previous results. The SVR is more advantageous than RF and Logistic, and the model without spatial autocorrelation is more ideal than a model with it. This is because AutoRF is more conservative in change strategy compared to other models. It ignored some pixels that should have been changed (Misses and Hits) but reduced some false pixels (False alarms + Wrong hits). This led to a lower FoM for AutoRF, but the correct predicted pixels (Correct rejections + Hits) were higher than other models, indicating that AutoRF has some advantages in balancing the overall simulation accuracy.

Table 4. Three-dimensional approach and FoM of different models in 2017.

Model	Misses	Hits	Wrong Hits	False Alarms	Correct Rejections	FoM
Logistic	0.1515	0.0103	0.0247	0.0600	0.7536	0.0417
SVR	0.1393	0.0207	0.0264	0.0686	0.7449	0.0813
RF	0.1430	0.0139	0.0295	0.0514	0.7622	0.0586
Autologistic	0.1475	0.0115	0.0274	0.0582	0.7554	0.0469
AutoSVR	0.1408	0.0190	0.0266	0.0595	0.7541	0.0772
AutoRF	0.1459	0.0123	0.0283	0.0464	0.7672	0.0526

Figure 3 compares the simulation maps and observation maps for AutoRF in 2007, 2013 and 2017. The similarity between the simulation maps and the observation maps was high and both reflected the real spatial distribution of various land cover types. Mangrove forests in simulation maps were all distributed near the sea, which was consistent with the actual growth of mangrove forests. There were also some deviations in simulation maps. The simulation maps were more fragmented compared to the observation maps, which may be related to the high simulation resolution (30 m). The new bridges (Zone A1), new roads (Zone A2) and aquaculture ponds governance (Zone A3), which were greatly affected by policy, were not well simulated. In general, however, the spatial agreement between the simulation maps and observation maps was good.

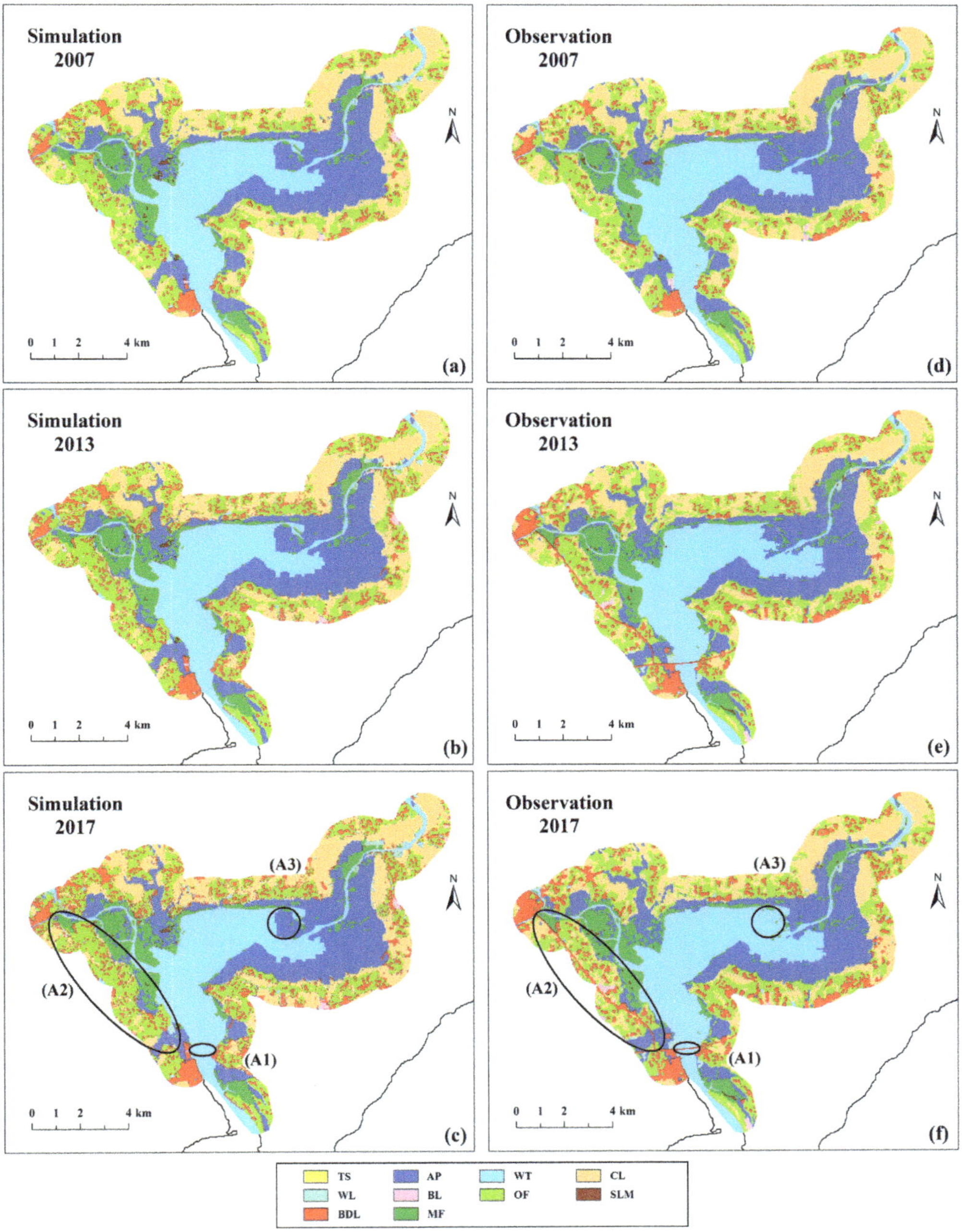

Figure 3. Comparison between simulation maps (**a–c**) and observation maps (**d–f**) for AutoRF.

4.3. Applicability of Driving Factors

The IncNodePurity was used to evaluate the applicability of driving factors in the mangrove simulation (Figure 4). The results of AutoRF and RF were basically the same. EVI had the greatest importance, indicating that vegetation indices will significantly improve the accuracy of the mangrove simulation. The distance to suitable land for mangrove was the second greatest, suggesting that the changes of mangrove forests are closely related to the distribution of suitable land for mangrove. The importance of the two location indices, distance to river and distance to sea, were also greater, as mangrove forests mostly grow on mudflats between land and sea. The greater importance of the distance to aquaculture ponds and building land indicated that the presence of those two land cover types exerts a strong influence on mangrove growth, which is consistent with the results of previous studies [27]. The spatial autocorrelation factor was equally important, showing that this factor is also essential in machine learning models. The distance to major road was also an important influencing factor, indicating that mangrove growth is closely related to the frequency of human activities. The four indices of EVI change trends, distance to minor road, slope and elevation were weakly related. EVI change trends may better reflect mangrove trends within time spans, and have limited ability to predict future trends. The distance to minor road reflects the impact of residential, pedestrian, cycleway and other trails. The frequency of human activity on these roads is weaker than on major roads, so the impact on mangrove forests was weaker. Since the study area is small and belongs to the coastal area, the two terrain indices, slope and elevation, had little variation, so they were the least important.

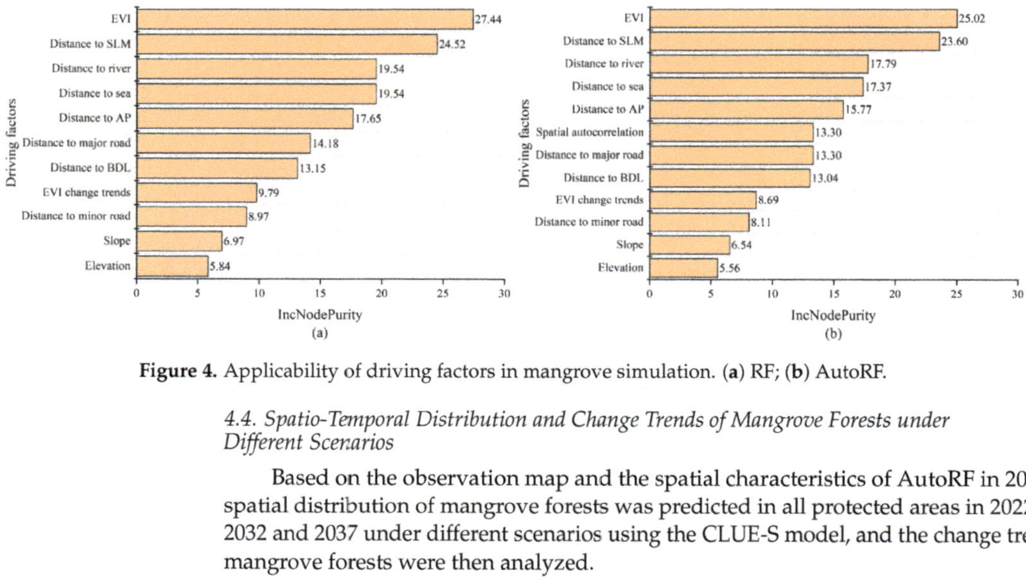

Figure 4. Applicability of driving factors in mangrove simulation. (**a**) RF; (**b**) AutoRF.

4.4. *Spatio-Temporal Distribution and Change Trends of Mangrove Forests under Different Scenarios*

Based on the observation map and the spatial characteristics of AutoRF in 2017, the spatial distribution of mangrove forests was predicted in all protected areas in 2022, 2027, 2032 and 2037 under different scenarios using the CLUE-S model, and the change trends of mangrove forests were then analyzed.

4.4.1. Spatio-Temporal Distribution of Mangrove Forests

Figure S1 shows the observation maps for 2017 and the prediction maps for 2037 under different scenarios. The results of Huiwen were shown in Figure 5. Mangrove forests in Huiwen were mainly distributed in the junction areas between land and sea along the coast, interlacing with aquaculture ponds. It can be seen that the CLUE-S model used in this study can predict the future spatio-temporal distribution well. The simulation maps in 2037 varied under different scenarios, but the spatial distribution of various land cover types was similar to the observation map in 2017. Under NGS, mangrove forests had few changes as a whole, showing a slow expansion trend. The expansion areas were mainly concentrated in the southern (Zone A1) and central (Zone A2) coastal areas, which were transferred from aquaculture ponds and other forests. Aquaculture ponds showed a contraction trend, especially in the eastern coastal region (Zone A3), which provided a more suitable growth environment for surrounding mangrove forests. Under EDS, mangrove forests showed a significant contraction trend. This is because EDS was designed to ensure economic development. The economic construction land will be aggressively expanded, and the growth areas of mangrove forests will inevitably be occupied. Only three patches of mangrove forests remained in the region, in the west (Zone B1), northeast (Zone B2) and south (Zone B3), noting that the area has been significantly reduced compared to 2017. Mangrove forests in all other areas were severely contracted, and most of them have been developed for building land and aquaculture ponds (Zones B4, B5 and B6). The remaining scattered mangrove forests were surrounded by other land cover types. The quality of the mangrove habitats was worrisome. Under MPS, mangrove forests showed a significant expansion trend. This is because MPS focuses on the protection of mangrove forests. The economic construction land that destroys mangrove habitats will drastically contract, and mangrove forests will inevitably expand. Similar to NGS, the southern (Zone C1) and central (Zone C2) coastal areas were the priority areas for mangrove growth. The two patches of mangrove forests in the west (Zone C3) and northeast (Zone C4) also expanded further outward, which was associated with the contraction of surrounding aquaculture ponds. Aquaculture ponds showed a significant contraction trend, and mangrove forests have begun to dominate the eastern coastal areas (Zone C5).

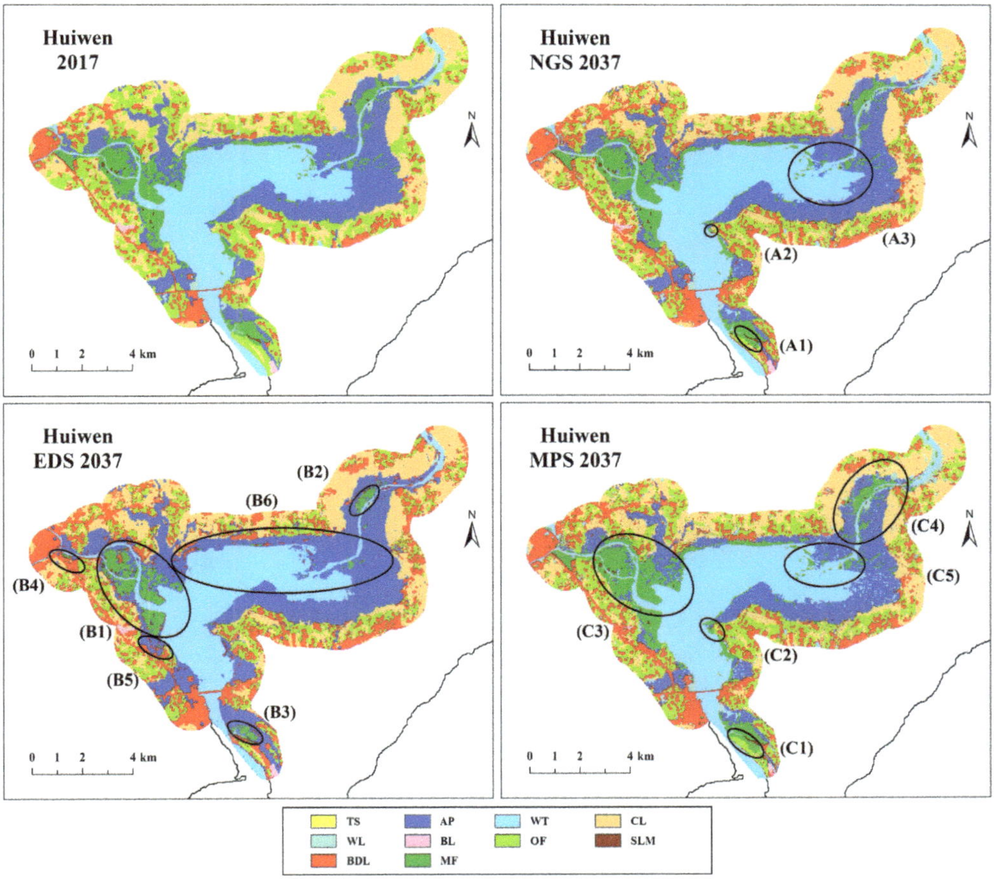

Figure 5. Comparison between the observation maps for 2017 and the simulation maps for 2037 in Huiwen under different scenarios.

4.4.2. Temporal Change Trends of Mangrove Forests

To explore the temporal change trends of mangrove forests under different scenarios, the area changes of mangrove forests and three related land cover types (aquaculture ponds, building land and other forests) were calculated in all protected areas from 2017 to 2037 (Figure 6). By 2037, the mangrove area increased slowly under NGS, decreased significantly under EDS and increased significantly under MPS with 4460, 2704 and 5456 ha respectively. The reasons for different changes of mangrove forests under different scenarios were as follows. Under EDS, the area of aquaculture ponds and building land increased significantly. The decreased area of other forests cannot provide enough space for economic construction land; mangrove forests will be seriously encroached upon. Thus, the mangrove area will decrease significantly. Under NGS, aquaculture ponds and building land still maintained their growth trends, but slowed down significantly compared to EDS. Thus, the impact of economic development on mangrove habitats was greatly reduced and the mangrove area will increase slowly. Under MPS, although the area of building land still increased slightly, the area of aquaculture ponds was reduced significantly, providing enough space for mangrove growth. Thus, the mangrove area will increase significantly.

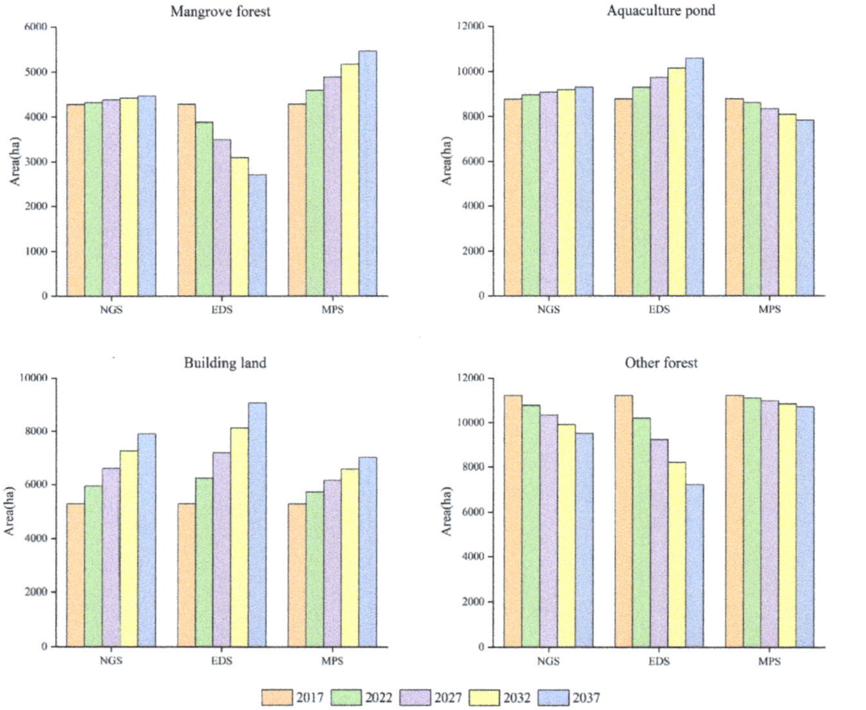

Figure 6. Area changes of mangrove forests and three related land cover types (aquaculture ponds, building land and other forests) from 2017 to 2037.

4.4.3. Spatial Change Trends of Mangrove Forests

To further analyze the causes of mangrove changes, the conversion between mangrove forests and other land cover types was analyzed from 2017 to 2037 under different scenarios (Figure 7). Under EGS, the two largest land cover types transformed from mangrove forests were aquaculture ponds (916 ha) and building land (557 ha), much higher than other land cover types, indicating that mangrove contraction is indeed closely related to the expansion of economic construction land. Cultivated land also had a larger area (100 ha) transformed from mangrove forests, showing that some mangrove forests were encroached upon for crop production. Under MPS, aquaculture ponds, cultivated land and other forests were the three largest types transformed into mangrove forests with 422, 326 and 322 ha, respectively. The reason for the larger area transformed from aquaculture ponds and cultivated land is due to the strict implementation of the policy of converting fishponds to wetlands and converting cultivated land to wetlands. Other forests can transform into mangrove forests because there is a transformation route: other forests—aquaculture ponds/suitable land for mangrove—mangrove forests. Under NGS, the changes of mangrove forests were between EGS and MPS. Other forests had the largest area (139 ha) transformed into mangrove forests. Water, wetlands, aquaculture ponds and cultivated land also had a larger area (about 50 ha per type). Building land (102 ha) and aquaculture ponds (85 ha) were the two largest land cover types transformed from mangrove forests.

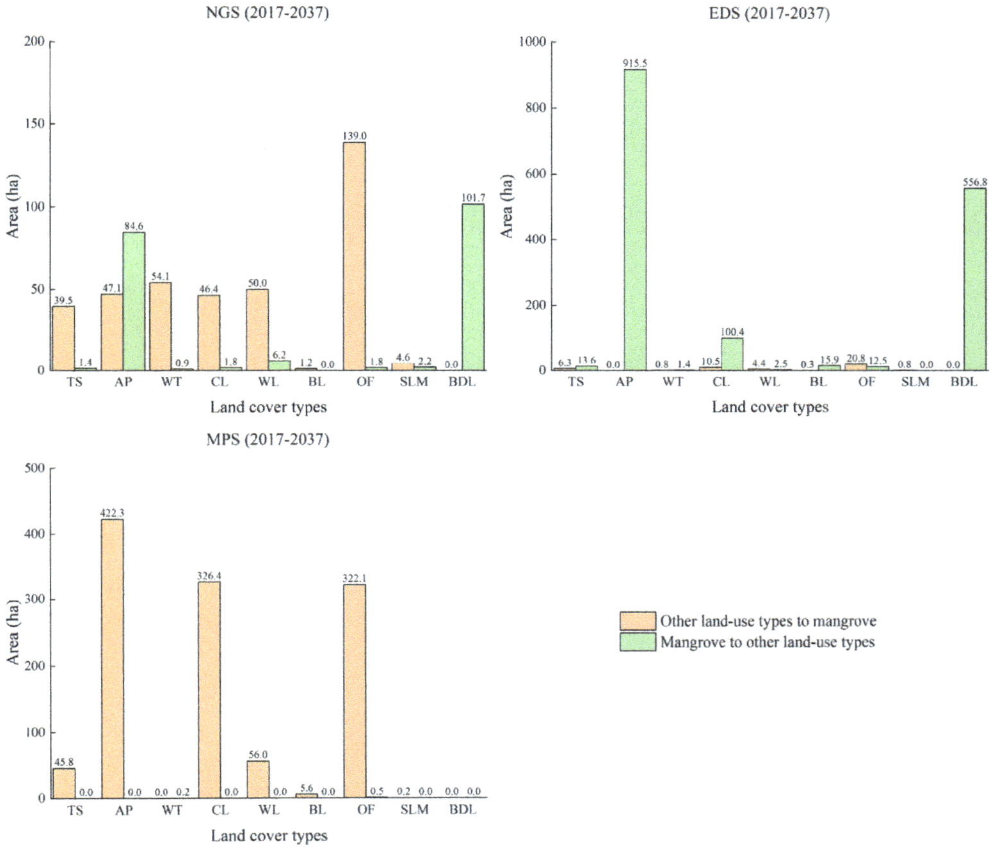

Figure 7. Conversion between mangrove forests and other land cover types from 2017 to 2037 under different scenarios.

Figure 8 shows the land cover changes of mangrove forests from 2017 to 2037 under NGS. The mangrove area of all protected areas generally increased. Only the coastal of Puqian (Zones B1 and B2), the central of Guanan (Zones C1 and C2) and the western of Dongzhaigang (Zone E1) decreased significantly. Most of the mangrove growth areas were around existing mangrove forests (Zones A1, D1, D2, F1, F2, G1, H1 and I1) or along the riverbank (Zone I2).

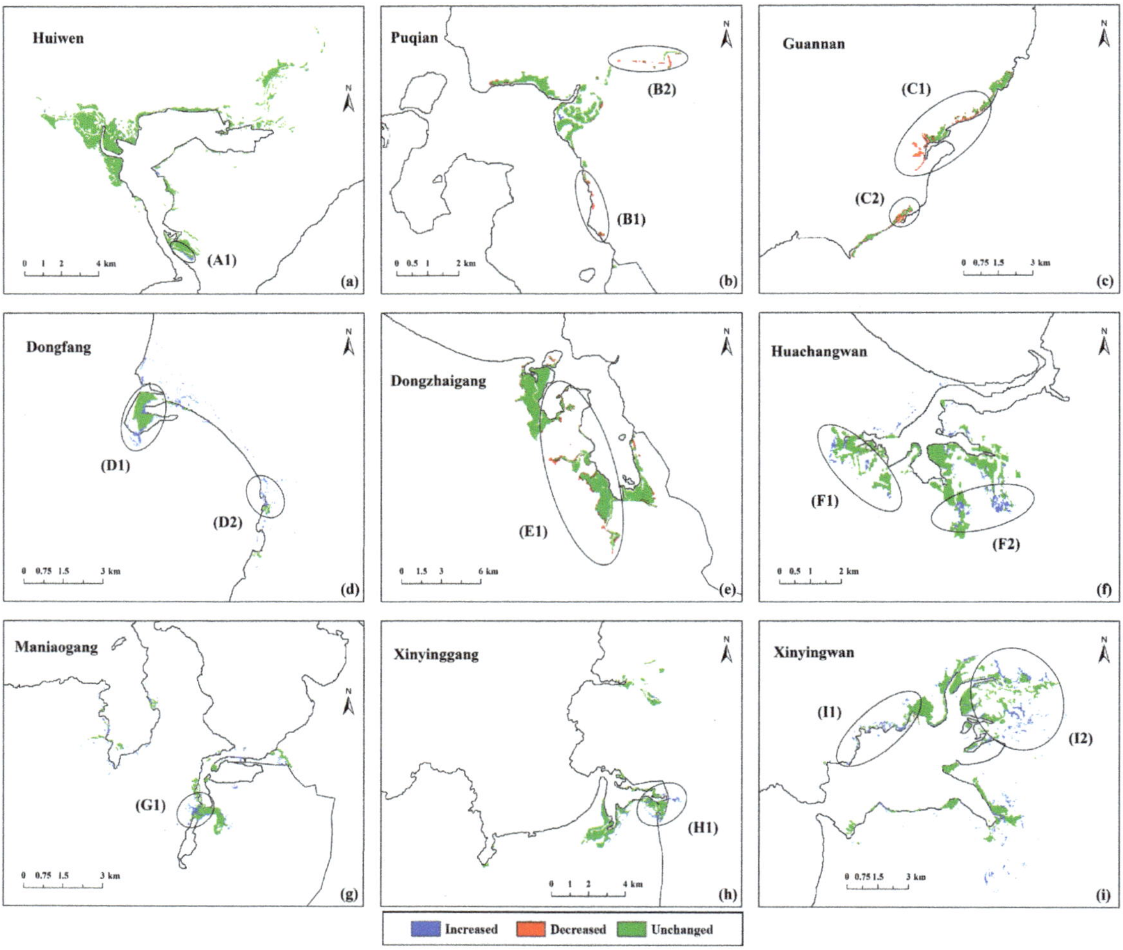

Figure 8. Land cover changes of mangrove forests (**a**–**i**) from 2017 to 2037 under NGS.

Figure 9 shows the land cover changes of mangrove forests from 2017 to 2037 under EDS. The mangrove area of all protected areas showed a significant decrease. The main types transformed from mangrove forests are aquaculture ponds (Zones A1, A2, A3, B1, B2, C1, D1, F1 and G1), building land (Zones A4, B2, C2, D1, E1, H1 and H2) and cultivated land (Zones F2, F3 and I1). Most of the contraction areas are comprised of marginal mangrove forests, as mangrove forests here are vulnerable to external disturbances. Their health conditions were worse compared to the interior mangrove forests. Therefore, these areas are more easily transformed into economic construction land, and will be the priority encroachment area during the economic development phase. The fragmented and scattered mangrove forests (Zones A1, B1, B2 and C2) contracted more than the intact and contiguous mangrove forests. Since these mangrove forests are mostly surrounded by other land cover types, their ability to resist stress is weak. With the decline of the habitat quality, the growth areas of fragmented and scattered mangrove forests will be rapidly encroached upon.

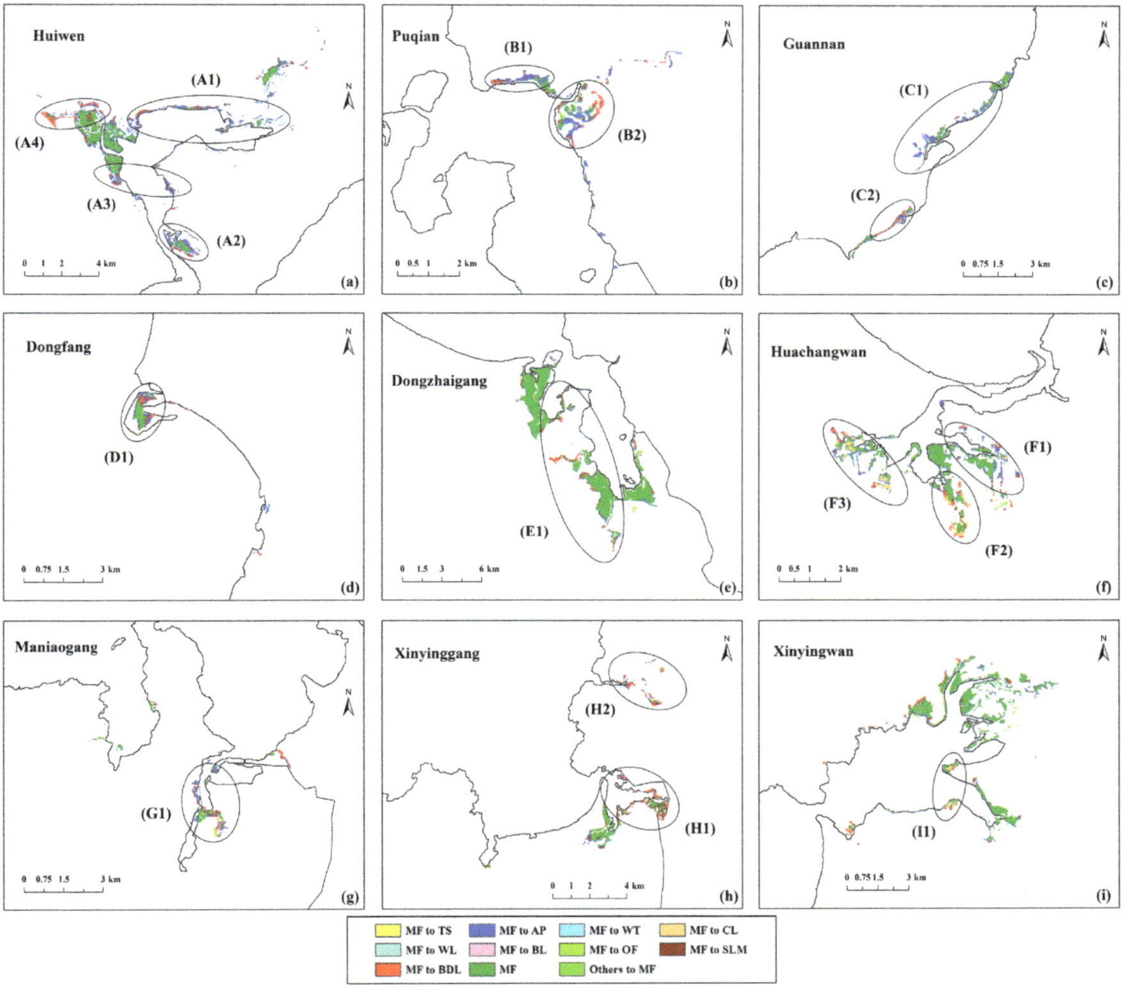

Figure 9. Land cover changes of mangrove forests (**a**–**i**) from 2017 to 2037 under EDS.

Figure 10 shows the land cover changes of mangrove forests under MPS from 2017 to 2037. The mangrove area of the most protected areas showed a significant increase. The main types transformed into mangrove forests were aquaculture ponds (Zones B1, D1, D2, F1, G1, G2 and H2), cultivated land (Zones B1, D2, E2, F1 and H1) and other forests (Zones A1, D3, E1, F2, I1 and I2). Most of these expansion areas were located around existing mangrove forests, as the habitat quality there is better. Therefore, after the governance of economic construction land, these areas are more easily transformed into mangrove forests, and will be the priority restoration area during the mangrove protection phase. Other expansion areas were along the riverbank (Zones E1 and I2). This is because these areas are close to water and away from other land cover types, which are conducive to mangrove growth, especially for mangrove seedlings [33]. Only the mangrove area in Guannan did not increase significantly under this scenario, as the ecological management of this region has always been poor. Aquaculture ponds have been threatening surrounding mangrove forests [65].

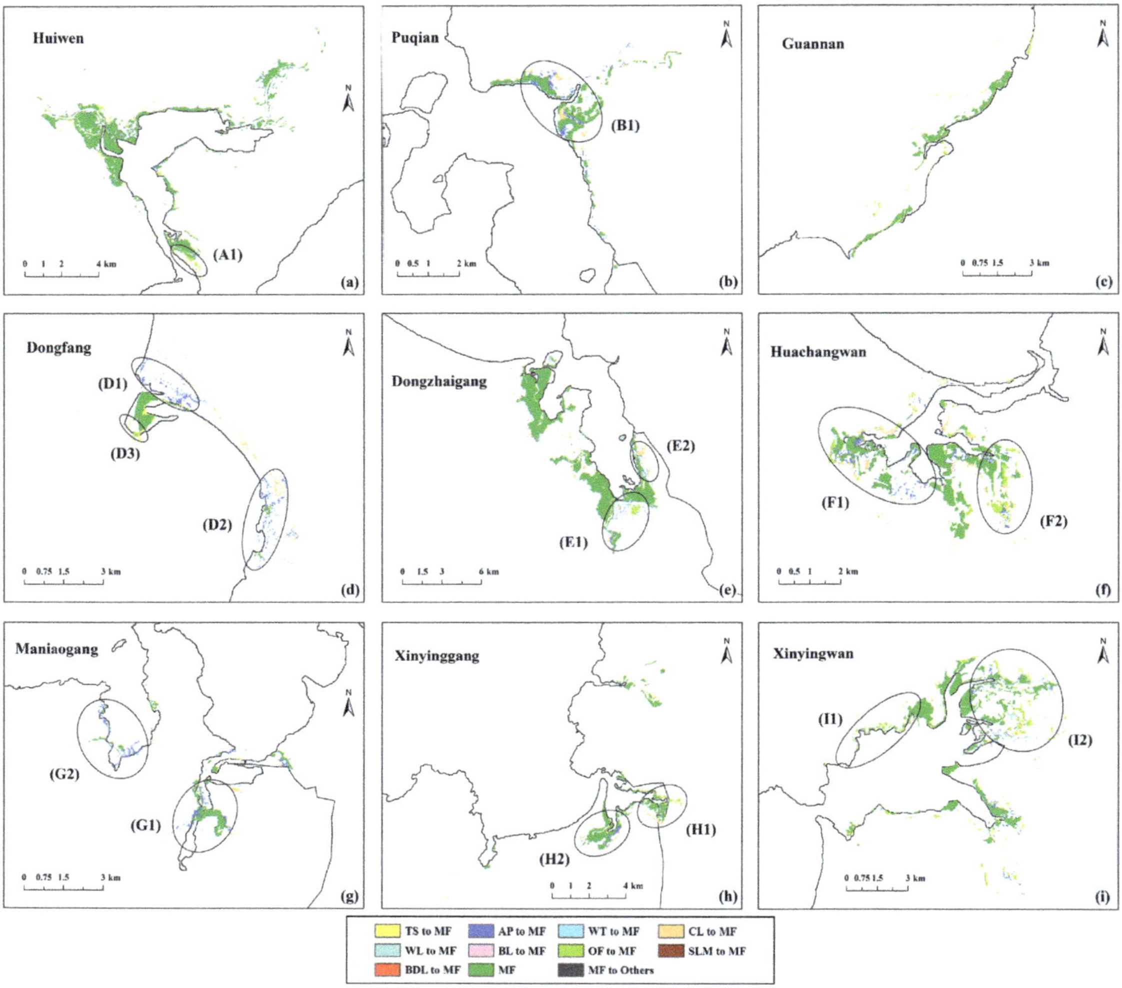

Figure 10. Land cover changes of mangrove forests (**a**–**i**) from 2017 to 2037 under MPS.

5. Discussion

5.1. Comparison of the Spatio-Temporal Simulation Methods of Mangrove Forests

In this study, the simulation accuracy of logistic regression, support vector regression and random forest were first compared in terms of spatial characteristics, and the simulation potential of different models was evaluated. The results showed that: (1) RF is more advantageous than SVR and Logistic; (2) the model with spatial autocorrelation is more ideal than the model without it; and (3) AutoRF performs the best in spatial characteristic simulation. AutoRF was then combined with the CLUE-S model to simulate the spatio-temporal distribution of mangrove forests. The results showed that the simulation maps of the model are in good spatial agreement with the observation maps, and the OA (77.94%) is better compared to the other models. This indicated that AutoRF proposed in this study has significant advantages for mangrove simulation. Driving factors reflecting the characteristics of mangrove forests were also explored in this study, and the applicability of mangrove simulation was assessed. The results showed that the vegetation index EVI can significantly improve the accuracy of mangrove simulation; location indices such as distance to suitable land for mangrove, river, sea, aquaculture ponds, major road and building land are also very important constraints; and the spatial autocorrelation factor is also indispensable in machine learning models. This indicated that the driving factors selected in this study play an important role in mangrove simulation. Overall, the spatio-temporal simulation method proposed in this study can effectively simulate the relationship between mangrove change processes and environmental driving factors, accurately predict the future spatio-temporal distribution of mangrove forests, and provide a theoretical basis and decision support for mangrove protection, restoration and utilization.

5.2. Future Changes of Mangrove Forests in Hainan Island

The future changes of mangrove forests in Hainan Island from 2022 to 2037 vary under different scenarios. By 2037, the mangrove area increased slowly under NGS, decreased significantly under EDS and increased significantly under MPS occupying 4460, 2704 and 5456 ha, respectively. The changes in area are associated with changes in aquaculture ponds, building land, cultivated land and other forests. The contraction of mangrove forests is closely related to the expansion of aquaculture ponds, building land and cultivated land. The marginal mangrove forests are more prone to contraction than the interior mangrove forests and the fragmented mangrove forests contracted more than the intact mangrove forests. Mangrove forests in these areas should be protected as a priority in the future. The expansion of mangrove forests is attributed to the contraction of aquaculture ponds, cultivated land and other forests. The areas around existing mangrove forests and on both sides of the riverbank are typical areas for mangrove expansion; these areas are preferable for artificial planting. Comparing different development scenarios, the MPS should be the most suitable development direction in the future. The reason is that this scenario maintains a steady growth of economic construction land and significant growth of mangrove forests, which reasonably balances the relationship between economic development and mangrove protection.

5.3. Limitations and Future Perspective of the Study

The potential of land cover changes was assessed in this study through spatial suitability probability, considering only the relationship between land cover types and driving factors from a mono-temporal perspective. Although this model may be more stable for long-term simulation with non-stationary change patterns, it is difficult to capture and distinguish different change processes, as the spatial configuration of past changes is not considered [54]. Therefore, other forests can directly transform into mangrove forests in this study. This is because the transformation route of other forests-aquaculture ponds/suitable land for mangrove-mangrove forests was simplified to other forests-mangrove forests. Another similar phenomenon was that only a small area of suitable land for mangrove was transformed into mangrove forests. This is because the transformation route of aquaculture ponds-suitable land for mangrove-mangrove forests was simplified to aquaculture ponds-mangrove forests. Another limitation of this method is the difficulty in predicting changes affected by policy such as new bridges, new roads and aquaculture ponds governance (Zones A1, A2 and A3 in Figure 3). Meanwhile, the future area demands under different scenarios were predicted by linear regression and scenario analysis. Although these methods are simple and effective [53], they cannot reflect the complex processes in land-use changes. This study also found that the simulated maps were more fragmented compared to the observation maps, which may be related to the high simulation resolution (30 m).

Therefore, models that consider the relationship between past changes and driving factors, such as DINAMICA and the Land Change Modeler (LCM), should be combined in future studies to better capture the actual process of land cover changes [54]. Future planning data should be added to the model to reduce the uncertainty of simulation results caused by policy. At the same time, system dynamics models or artificial neural networks should also be used to estimate area demands. Finally, it is also the goal of future efforts to explore the simulation accuracy of models and the applicability of driving factors under different resolutions, especially for high resolutions (integrated with Sentinel-1/2), and select the optimal resolution applicable to mangrove spatio-temporal simulation.

6. Conclusions

Mangrove forests are important wetland ecosystems that grow in the intertidal zone between land and sea. They provide important social, ecological and economic services to coastal areas. In recent years, mangrove protection has reached a broad international consensus and effective mangrove monitoring methods are essential. A spatio-temporal simulation method for mangrove forests was proposed in this study. This method compared the simulation accuracy of different models in terms of spatial characteristics, evaluated the applicability of driving factors in mangrove simulation and predicted the future spatio-temporal distribution and change trends of mangrove forests under different scenarios.

The simulation results of different models reveal that for spatial characteristic simulation: (1) RF is more advantageous than SVR and Logistic; (2) the model with spatial autocorrelation is more ideal than the model without it; and (3) AutoRF performs the best. For different driving factors, knowledge of vegetation index EVI, various location indices and the spatial autocorrelation factor can significantly improve the accuracy of mangrove simulation. The prediction results of Hainan Island showed that the future spatio-temporal distribution of mangrove forests varies under different scenarios. The mangrove area increased slowly under NGS, decreased significantly under EDS and increased significantly under MPS. The contraction of mangrove forests is closely related to the expansion of aquaculture ponds, building land and cultivated land. Mangrove contraction is more severe in marginal or fragmented areas. The expansion of mangrove forests is due to the contraction of aquaculture ponds, cultivated land and other forests. The areas around existing mangrove forests and on both sides of the riverbank are typical areas of mangrove expansion. The MPS should be the most suitable development direction for the future, as it can reasonably balance economic development with mangrove protection. This will provide a theoretical basis and decision support for mangrove protection, restoration and utilization.

Supplementary Materials: The following are available online at https://www.mdpi.com/article/10.3390/rs13204059/s1, Figure S1: Comparison between the observation maps for 2017 and the simulation maps for 2037 under different scenarios, Table S1: The details of the selected reserves in this study, Table S2: The conversion sequence under different scenarios, Table S3: The conversion elasticity under different scenarios, and Table S4: The annual area changes under different scenarios (pixel/y).

Author Contributions: Conceptualization, J.L.; methodology, B.Z. and J.L.; software, B.Z.; validation, B.Z.; formal analysis, B.Z., J.L. and G.S.; investigation, B.Z. and J.L.; resources, J.L.; data curation, B.Z.; writing—original draft preparation, B.Z., J.L. and G.S.; writing—review and editing, B.Z., J.L. and G.S.; visualization, B.Z.; supervision, J.L.; project administration, J.L.; and funding acquisition, J.L. All authors have read and agreed to the published version of the manuscript.

Funding: This research was funded by Hainan Provincial Department of Science and Technology (Grant No. ZDKJ2019006) and Strategic Priority Research Program of the Chinese Academy of Sciences (Grant No. XDA19030302).

Data Availability Statement: The data are not publicly available due to the confidentiality of the research projects.

Acknowledgments: The authors would like to express thanks to the anonymous reviewers for their voluntary work and the constructive comments to improve this manuscript.

Conflicts of Interest: The authors declare no conflict of interest.

References

1. Giri, C. Recent Advancement in Mangrove Forests Mapping and Monitoring of the World Using Earth Observation Satellite Data. *Remote Sens.* **2021**, *13*, 563. [CrossRef]
2. Mumby, P.J.; Edwards, A.J.; Arias-Gonzalez, J.E.; Lindeman, K.C.; Blackwell, P.G.; Gall, A.; Gorczynska, M.I.; Harborne, A.R.; Pescod, C.L.; Renken, H.; et al. Mangroves enhance the biomass of coral reef fish communities in the Caribbean. *Nature* **2004**, *427*, 533–536. [CrossRef] [PubMed]
3. Barbier, E.B. Natural barriers to natural disasters: Replanting mangroves after the tsunami. *Front. Ecol. Environ.* **2006**, *4*, 124–131. [CrossRef]
4. Walters, B.B.; Ronnback, P.; Kovacs, J.M.; Crona, B.; Hussain, S.A.; Badola, R.; Primavera, J.H.; Barbier, E.; Dahdouh-Guebas, F. Ethnobiology, socio-economics and management of mangrove forests: A review. *Aquat. Bot.* **2008**, *89*, 220–236. [CrossRef]
5. Barbier, E.B.; Hacker, S.D.; Kennedy, C.; Koch, E.W.; Stier, A.C.; Silliman, B.R. The value of estuarine and coastal ecosystem services. *Ecol. Monogr.* **2011**, *81*, 169–193. [CrossRef]
6. Donato, D.C.; Kauffman, J.B.; Murdiyarso, D.; Kurnianto, S.; Stidham, M.; Kanninen, M. Mangroves among the most carbon-rich forests in the tropics. *Nat. Geosci.* **2011**, *4*, 293–297. [CrossRef]
7. Vo, Q.T.; Kuenzer, C.; Vo, Q.M.; Moder, F.; Oppelt, N. Review of valuation methods for mangrove ecosystem services. *Ecol. Indic.* **2012**, *23*, 431–446. [CrossRef]
8. Goldberg, L.; Lagomasino, D.; Thomas, N.; Fatoyinbo, T. Global declines in human-driven mangrove loss. *Glob. Chang. Biol.* **2020**, *26*, 5844–5855. [CrossRef]
9. Yirga, A.; Addisu Legesse, S.; Mekuriaw, A. Carbon Stock and Mitigation Potentials of Zeghie Natural Forest for Climate Change Disaster Reduction, Blue Nile Basin, Ethiopia. *Earth Syst. Environ.* **2020**, *4*, 27–41. [CrossRef]
10. Valipour, M.; Bateni, S.M.; Jun, C. Global Surface Temperature: A New Insight. *Climate* **2021**, *9*, 81. [CrossRef]
11. Richards, D.R.; Friess, D.A. Rates and drivers of mangrove deforestation in Southeast Asia, 2000-2012. *Proc. Natl. Acad. Sci. USA* **2016**, *113*, 344–349. [CrossRef]
12. Friess, D.A.; Thompson, B.S.; Brown, B.; Amir, A.A.; Cameron, C.; Koldewey, H.J.; Sasmito, S.D.; Sidik, F. Policy challenges and approaches for the conservation of mangrove forests in Southeast Asia. *Conserv. Biol.* **2016**, *30*, 933–949. [CrossRef]
13. Thomas, N.; Lucas, R.; Bunting, P.; Hardy, A.; Rosenqvist, A.; Simard, M. Distribution and drivers of global mangrove forest change, 1996–2010. *PLoS ONE* **2017**, *12*, e0179302. [CrossRef]
14. Murakami, H.; Delworth, T.L.; Cooke, W.F.; Zhao, M.; Xiang, B.Q.; Hsu, P.C. Detected climatic change in global distribution of tropical cyclones. *Proc. Natl. Acad. Sci. USA* **2020**, *117*, 10706–10714. [CrossRef] [PubMed]
15. Alongi, D.M. Present state and future of the world's mangrove forests. *Environ. Conserv.* **2002**, *29*, 331–349. [CrossRef]
16. Mayaux, P.; Holmgren, P.; Achard, F.; Eva, H.; Stibig, H.; Branthomme, A. Tropical forest cover change in the 1990s and options for future monitoring. *Philos. Trans. R. Soc. B-Biol. Sci.* **2005**, *360*, 373–384. [CrossRef]
17. Hamilton, S.E.; Casey, D. Creation of a high spatio-temporal resolution global database of continuous mangrove forest cover for the 21st century (CGMFC-21). *Glob. Ecol. Biogeogr.* **2016**, *25*, 729–738. [CrossRef]
18. Conchedda, G.; Durieux, L.; Mayaux, P. An object-based method for mapping and change analysis in mangrove ecosystems. *ISPRS J. Photogramm. Remote Sens.* **2008**, *63*, 578–589. [CrossRef]
19. Giri, C.; Ochieng, E.; Tieszen, L.L.; Zhu, Z.; Singh, A.; Loveland, T.; Masek, J.; Duke, N. Status and distribution of mangrove forests of the world using earth observation satellite data. *Glob. Ecol. Biogeogr.* **2011**, *20*, 154–159. [CrossRef]
20. Nascimento, W.R.; Souza, P.W.M.; Proisy, C.; Lucas, R.M.; Rosenqvist, A. Mapping changes in the largest continuous Amazonian mangrove belt using object-based classification of multisensor satellite imagery. *Estuar. Coast. Shelf Sci.* **2013**, *117*, 83–93. [CrossRef]
21. Jia, M.M. *Remote Sensing Analysis of China' Mangrove Forests Dynamic During 1973 to 2013*; University of Chinese Academy of Sciences: Beijing, China, 2014.
22. Chen, B.Q.; Xiao, X.M.; Li, X.P.; Pan, L.H.; Doughty, R.; Ma, J.; Dong, J.W.; Qin, Y.W.; Zhao, B.; Wu, Z.X.; et al. A mangrove forest map of China in 2015: Analysis of time series Landsat 7/8 and Sentinel-1A imagery in Google Earth Engine cloud computing platform. *ISPRS J. Photogramm. Remote Sens.* **2017**, *131*, 104–120. [CrossRef]
23. Bunting, P.; Rosenqvist, A.; Lucas, R.M.; Rebelo, L.M.; Hilarides, L.; Thomas, N.; Hardy, A.; Itoh, T.; Shimada, M.; Finlayson, C.M. The Global Mangrove WatchA New 2010 Global Baseline of Mangrove Extent. *Remote Sens.* **2018**, *10*, 1669. [CrossRef]
24. Suyadi; Gao, J.; Lundquist, C.J.; Schwendenmann, L. Characterizing landscape patterns in changing mangrove ecosystems at high, latitudes using spatial metrics. *Estuar. Coast. Shelf Sci.* **2018**, *215*, 1–10. [CrossRef]
25. Jia, M.M.; Wang, Z.M.; Zhang, Y.Z.; Mao, D.H.; Wang, C. Monitoring loss and recovery of mangrove forests during 42 years: The achievements of mangrove conservation in China. *Int. J. Appl. Earth Obs. Geoinf.* **2018**, *73*, 535–545.
26. Buitre, M.J.C.; Zhang, H.S.; Lin, H. The Mangrove Forests Change and Impacts from Tropical Cyclones in the Philippines Using Time Series Satellite Imagery. *Remote Sens.* **2019**, *11*, 688. [CrossRef]
27. Liao, J.J.; Zhen, J.N.; Zhang, L.; Metternicht, G. Understanding Dynamics of Mangrove Forest on Protected Areas of Hainan Island, China: 30 Years of Evidence from Remote Sensing. *Sustainability* **2019**, *11*, 5356. [CrossRef]
28. Chamberlain, D.; Phinn, S.; Possingham, H. Remote Sensing of Mangroves and Estuarine Communities in Central Queensland, Australia. *Remote Sens.* **2020**, *12*, 197. [CrossRef]
29. Guo, Y.J.; Liao, J.J.; Shen, G.Z. Mapping Large-Scale Mangroves along the Maritime Silk Road from 1990 to 2015 Using a Novel Deep Learning Model and Landsat Data. *Remote Sens.* **2021**, *13*, 245. [CrossRef]

30. Anwar, M.S.; Takewaka, S. Analyses on phenological and morphological variations of mangrove forests along the southwest coast of Bangladesh. *J. Coast. Conserv.* **2014**, *18*, 339–357. [CrossRef]
31. Berlanga-Robles, C.A.; Ruiz-Luna, A.; Villanueva, M.R.N. Seasonal trend analysis (STA) of MODIS vegetation index time series for the mangrove canopy of the Teacapan-Agua Brava lagoon system, Mexico. *Giscience Remote Sens.* **2019**, *56*, 338–361.
32. Le, H.T.; Tran, T.V.; Gyeltshen, S.; Nguyen, C.P.T.; Tran, D.X.; Luu, T.H.; Duong, M.B. Characterizing Spatiotemporal Patterns of Mangrove Forests in Can Gio Biosphere Reserve Using Sentinel-2 Imagery. *Appl. Sci. Basel* **2020**, *10*, 4058. [CrossRef]
33. Zhu, B.; Liao, J.; Shen, G. Combining time series and land cover data for analyzing spatio-temporal changes in mangrove forests: A case study of Qinglangang Nature Reserve, Hainan, China. *Ecol. Indic.* **2021**, *131*, 108135. [CrossRef]
34. Jiang, W.G.; Yuan, L.H.; Wang, W.J.; Cao, R.; Zhang, Y.F.; Shen, W.M. Spatio-temporal analysis of vegetation variation in the Yellow River Basin. *Ecol. Indic.* **2015**, *51*, 117–126. [CrossRef]
35. Stephenne, N.; Lambin, E.F. A dynamic simulation model of land-use changes in Sudano-sahelian countries of Africa (SALU). *Agric. Ecosyst Env.* **2001**, *85*, 145–161. [CrossRef]
36. Delgado-Matas, C.; Pukkala, T. Optimisation of the traditional land-use system in the Angolan highlands using linear programming. *Int J. Sustain. Dev. World Ecol.* **2014**, *21*, 138–148. [CrossRef]
37. Taromi, R.; DuRoss, M.; Chen, B.T.; Faghri, A.; Li, M.X.; DeLiberty, T. A multiobjective land development optimization model: The case of New Castle County, Delaware. *Transp. Plan. Technol.* **2015**, *38*, 277–304. [CrossRef]
38. Saysel, A.K.; Barlas, Y.; Yenigun, O. Environmental sustainability in an agricultural development project: A system dynamics approach. *J. Environ. Manag.* **2002**, *64*, 247–260. [CrossRef]
39. Liu, X.P.; Ou, J.P.; Li, X.; Ai, B. Combining system dynamics and hybrid particle swarm optimization for land use allocation. *Ecol. Model.* **2013**, *257*, 11–24. [CrossRef]
40. Chen, C.F.; Son, N.T.; Chang, N.B.; Chen, C.R.; Chang, L.Y.; Valdez, M.; Centeno, G.; Thompson, C.A.; Aceituno, J.L. Multi-Decadal Mangrove Forest Change Detection and Prediction in Honduras, Central America, with Landsat Imagery and a Markov Chain Model. *Remote Sens.* **2013**, *5*, 6408–6426. [CrossRef]
41. Wu, D.Q.; Liu, J.; Zhang, G.S.; Ding, W.J.; Wang, W.; Wang, R.Q. Incorporating spatial autocorrelation into cellular automata model: An application to the dynamics of Chinese tamarisk (*Tamarix chinensis* Lour.). *Ecol. Model.* **2009**, *220*, 3490–3498. [CrossRef]
42. Basse, R.M.; Omrani, H.; Charif, O.; Gerber, P.; Bodis, K. Land use changes modelling using advanced methods: Cellular automata and artificial neural networks. The spatial and explicit representation of land cover dynamics at the cross-border region scale. *Appl. Geogr.* **2014**, *53*, 160–171. [CrossRef]
43. Castella, J.C.; Verburg, P.H. Combination of process-oriented and pattern-oriented models of land-use change in a mountain area of Vietnam. *Ecol. Model.* **2007**, *202*, 410–420. [CrossRef]
44. Liu, X.P.; Liang, X.; Li, X.; Xu, X.C.; Ou, J.P.; Chen, Y.M.; Li, S.Y.; Wang, S.J.; Pei, F.S. A future land use simulation model (FLUS) for simulating multiple land use scenarios by coupling human and natural effects. *Landsc. Urban. Plan.* **2017**, *168*, 94–116. [CrossRef]
45. Roetter, R.P.; Hoanh, C.T.; Laborte, A.G.; Van Keulen, H.; Van Ittersum, M.K.; Dreiser, C.; Van Diepen, C.A.; De Ridder, N.; Van Laar, H.H. Integration of Systems Network (SysNet) tools for regional land use scenario analysis in Asia. *Environ. Model. Softw.* **2005**, *20*, 291–307. [CrossRef]
46. Verburg, P.H.; Soepboer, W.; Veldkamp, A.; Limpiada, R.; Espaldon, V.; Mastura, S.S.A. Modeling the spatial dynamics of regional land use: The CLUE-S model. *Environ. Manag.* **2002**, *30*, 391–405. [CrossRef]
47. Mukhopadhyay, A.; Mondal, P.; Barik, J.; Chowdhury, S.M.; Ghosh, T.; Hazra, S. Changes in mangrove species assemblages and future prediction of the Bangladesh Sundarbans using Markov chain model and cellular automata. *Environ. Sci. Process. Impacts* **2015**, *17*, 1111–1117. [CrossRef]
48. Bozkaya, A.G.; Balcik, F.B.; Goksel, C.; Esbah, H. Forecasting land-cover growth using remotely sensed data: A case study of the Igneada protection area in Turkey. *Environ. Monit. Assess.* **2015**, *187*, 59. [CrossRef]
49. DasGupta, R.; Hashimoto, S.; Okuro, T.; Basu, M. Scenario-based land change modelling in the Indian Sundarban delta: An exploratory analysis of plausible alternative regional futures. *Sustain. Sci.* **2019**, *14*, 221–240. [CrossRef]
50. Tajbakhsh, A.; Karimi, A.; Zhang, A.L. Modeling land cover change dynamic using a hybrid model approach in Qeshm Island, Southern Iran. *Environ. Monit. Assess.* **2020**, *192*, 303. [CrossRef]
51. Lin, Y.-P.; Chu, H.-J.; Wu, C.-F.; Verburg, P.H. Predictive ability of logistic regression, auto-logistic regression and neural network models in empirical land-use change modeling—A case study. *Int. J. Geogr. Inf. Sci.* **2011**, *25*, 65–87. [CrossRef]
52. Jiang, W.G.; Chen, Z.; Lei, X.; He, B.; Jia, K.; Zhang, Y.F. Simulation of urban agglomeration ecosystem spatial distributions under different scenarios: A case study of the Changsha-Zhuzhou-Xiangtan urban agglomeration. *Ecol. Eng.* **2016**, *88*, 112–121. [CrossRef]
53. Peng, K.F.; Jiang, W.G.; Deng, Y.; Liu, Y.H.; Wu, Z.F.; Chen, Z. Simulating wetland changes under different scenarios based on integrating the random forest and CLUE-S models: A case study of Wuhan Urban Agglomeration. *Ecol. Indic.* **2020**, *117*, 106671. [CrossRef]
54. Mas, J.F.; Kolb, M.; Paegelow, M.; Olmedo, M.T.C.; Houet, T. Inductive pattern-based land use/cover change models: A comparison of four software packages. *Environ. Model. Softw.* **2014**, *51*, 94–111. [CrossRef]
55. Tang, F.; Fu, M.C.; Wang, L.; Zhang, P.T. Land-use change in Changli County, China: Predicting its spatio-temporal evolution in habitat quality. *Ecol. Indic.* **2020**, *117*, 106719. [CrossRef]

56. Mei, Z.X.; Wu, H.; Li, S.Y. Simulating land-use changes by incorporating spatial autocorrelation and self-organization in CLUE-S modeling: A case study in Zengcheng District, Guangzhou, China. *Front. Earth Sci.* **2018**, *12*, 299–310. [CrossRef]
57. Jiang, W.G.; Chen, Z.; Lei, X.; Jia, K.; Wu, Y.F. Simulating urban land use change by incorporating an autologistic regression model into a CLUE-S model. *J. Geogr. Sci.* **2015**, *25*, 836–850. [CrossRef]
58. Li, Z.; Jiang, W.G.; Wang, W.J.; Lei, X.; Deng, Y. Exploring spatial-temporal change and gravity center movement of construction land in the Chang-Zhu-Tan urban agglomeration. *J. Geogr. Sci.* **2019**, *29*, 1363–1380. [CrossRef]
59. Peng, K.F.; Jiang, W.G.; Ling, Z.Y.; Hou, P.; Deng, Y.W. Evaluating the potential impacts of land use changes on ecosystem service value under multiple scenarios in support of SDG reporting: A case study of the Wuhan urban agglomeration. *J. Clean. Prod.* **2021**, *307*, 127321. [CrossRef]
60. Piao, S.L.; Fang, J.Y. Dynamic vegetation cover change over the last 18 years in China. *Quat. Sci.* **2001**, *21*, 294–302.
61. Besag, J.E. Nearest-Neighbour Systems and the Auto-Logistic Model for Binary Data. *J. R. Stat. Soc. Ser. B Stat. Methodol.* **1972**, *34*, 75–83. [CrossRef]
62. Wang, W.Q.; Wang, M. *The Mangroves of China*; Science Press: Beijing, China, 2007.
63. Xin, X.; Song, X.Q.; Lei, J.R.; Fang, Z.S.; Meng, Q.W. Mangrove Plants Resources and Its Conservation Strategies on Hainan. *J. Trop. Biol.* **2016**, *7*, 477–483.
64. Chen, H.X.; Chen, E.Y. Distribution of Mangrove in Hainan Island at Present. *J. Trop Oceanogr.* **1985**, *1985*, 74–79.
65. Zhen, J.N. *Monitoring and Dynamic Analysis of Mangrove Forests in Hainan Island using Multi-Temporal Remote Sensing Images*; University of Chinese Academy of Sciences: Beijing, China, 2019.
66. Liu, H.Q.; Huete, A. A feedback based modification of the NDVI to minimize canopy background and atmospheric noise. *ITGRS* **1995**, *33*, 457–465. [CrossRef]
67. Cao, R.; Jiang, W.G.; Yuan, L.H.; Wang, W.J.; Lv, Z.L.; Chen, Z. Inter-annual variations in vegetation and their response to climatic factors in the upper catchments of the Yellow River from 2000 to 2010. *J. Geogr. Sci.* **2014**, *24*, 963–979. [CrossRef]
68. Lv, J.X.; Jiang, W.G.; Wang, W.J.; Wu, Z.F.; Liu, Y.H.; Wang, X.Y.; Li, Z. Wetland Loss Identification and Evaluation Based on Landscape and Remote Sensing Indices in Xiong'an New Area. *Remote Sens.* **2019**, *11*, 2834. [CrossRef]
69. Chen, Y.; Li, X.; Su, W.; Li, Y. Simulating the optimal land-use pattern in the farming-pastoral transitional zone of Northern China. *Comput. Environ. Urban. Syst.* **2008**, *32*, 407–414. [CrossRef]
70. Zhang, L.P.; Zhang, S.W.; Huang, Y.J.; Cao, M.; Huang, Y.F.; Zhang, H.Y. Exploring an Ecologically Sustainable Scheme for Landscape Restoration of Abandoned Mine Land: Scenario-Based Simulation Integrated Linear Programming and CLUE-S Model. *Int. J. Environ. Res. Public Health* **2016**, *13*, 354. [CrossRef]
71. Zhao, X.; Li, S.; Pu, J.; Miao, P.; Wang, Q.; Tan, K. Optimization of the National Land Space Based on the Coordination of Urban-Agricultural-Ecological Functions in the Karst Areas of Southwest China. *Sustainability* **2019**, *11*, 6752. [CrossRef]
72. Zhang, Z.; Hu, B.; Jiang, W.; Qiu, H. Identification and scenario prediction of degree of wetland damage in Guangxi based on the CA-Markov model. *Ecol. Indic.* **2021**, *127*, 107764. [CrossRef]
73. Wu, M.; Ren, X.; Che, Y.; Yang, K. A Coupled SD and CLUE-S Model for Exploring the Impact of Land Use Change on Ecosystem Service Value: A Case Study in Baoshan District, Shanghai, China. *Environ. Manag.* **2015**, *56*, 402–419. [CrossRef]
74. He, X.; Mai, X.; Shen, G. Delineation of Urban Growth Boundaries with SD and CLUE-s Models under Multi-Scenarios in Chengdu Metropolitan Area. *Sustainability* **2019**, *11*, 5919. [CrossRef]
75. Lourdes Lima, M.; Romanelli, A.; Massone, H.E. Assessing groundwater pollution hazard changes under different socio-economic and environmental scenarios in an agricultural watershed. *Sci. Total Environ.* **2015**, *530*, 333–346. [CrossRef]
76. Adhikari, R.K.; Mohanasundaram, S.; Shrestha, S. Impacts of land-use changes on the groundwater recharge in the Ho Chi Minh city, Vietnam. *Environ. Res.* **2020**, *185*, 109440. [CrossRef]
77. Cortes, C.; Vapnik, V. Support-vector networks. *MLear* **1995**, *20*, 273–297. [CrossRef]
78. Breiman, L. Random forests. *MLear* **2001**, *45*, 5–32.
79. Moulds, S.; Buytaert, W.; Mijic, A. An open and extensible framework for spatially explicit land use change modelling: The lulcc R package. *Geosci. Model. Dev.* **2015**, *8*, 3215–3229. [CrossRef]
80. Bradley, A.P. The use of the area under the roc curve in the evaluation of machine learning algorithms. *Pattern Recognit.* **1997**, *30*, 1145–1159. [CrossRef]
81. Cohen, J.A. Coefficient of Agreement for Nominal Scales. *Educ. Psychol. Meas.* **1960**, *20*, 37–46. [CrossRef]
82. Pontius, R.G. Quantification error versus location error in comparison of categorical maps. *Photogramm. Eng. Remote Sens.* **2000**, *66*, 1011–1016.
83. Pontius, R.G., Jr.; Peethambaram, S.; Castella, J.-C. Comparison of Three Maps at Multiple Resolutions: A Case Study of Land Change Simulation in Cho Don District, Vietnam. *Ann. Assoc. Am. Geogr.* **2011**, *101*, 45–62. [CrossRef]
84. Varga, O.G.; Pontius, R.G., Jr.; Szabo, Z.; Szabo, S. Effects of Category Aggregation on Land Change Simulation Based on Corine Land Cover Data. *Remote Sens.* **2020**, *12*, 1314. [CrossRef]
85. Pontius, R.G., Jr.; Boersma, W.; Castella, J.-C.; Clarke, K.; de Nijs, T.; Dietzel, C.; Duan, Z.; Fotsing, E.; Goldstein, N.; Kok, K.; et al. Comparing the input, output, and validation maps for several models of land change. *Ann. Reg. Sci.* **2008**, *42*, 11–37. [CrossRef]

remote sensing

MDPI

Article

Spectral Reflectance-Based Mangrove Species Mapping from WorldView-2 Imagery of Karimunjawa and Kemujan Island, Central Java Province, Indonesia

Arie Dwika Rahmandhana [1], Muhammad Kamal [2,*] and Pramaditya Wicaksono [2]

1 Master of Science Study Program in Remote Sensing, Faculty of Geography, Universitas Gadjah Mada, Yogyakarta 55281, Indonesia; ariedwika@mail.ugm.ac.id

2 Department of Geographic Information Science, Faculty of Geography, Universitas Gadjah Mada, Yogyakarta 55281, Indonesia; prama.wicaksono@ugm.ac.id

* Correspondence: m.kamal@ugm.ac.id

Abstract: Mangrove mapping at the species level enables the creation of a detailed inventory of mangrove forest biodiversity and supports coastal ecosystem management. The Karimunjawa National Park in Central Java Province is one of Indonesia's mangrove habitats with high biodiversity, namely, 44 species representing 25 true mangroves and 19 mangrove associates. This study aims to (1) classify and group mangrove species by their spectral reflectance characteristics, (2) map mangrove species by applying their spectral reflectance to WorldView-2 satellite imagery with the spectral angle mapper (SAM), spectral information divergence (SID), and spectral feature fitting (SFF) algorithms, and (3) assess the accuracy of the produced mangrove species mapping of the Karimunjawa and Kemujan Islands. The collected field data included (1) mangrove species identification, (2) coordinate locations of targeted mangrove species, and (3) the spectral reflectance of mangrove species measured with a field spectrometer. Dendrogram analysis was conducted with the Ward linkage method to classify mangrove species based on the distance between the closest clusters of spectral reflectance patterns. The dendrogram showed that the 24 mangrove species found in the field could be grouped into four levels. They consisted of two, four, and five species groups for Levels 1 to 3, respectively, and individual species for Level 4. The mapping results indicated that the SID algorithm had the highest overall accuracy (OA) at 49.72%, 22.60%, and 15.20% for Levels 1 to 3, respectively, while SFF produced the most accurate results for individual species mapping (Level 4) with an OA of 5.08%. The results suggest that the greater the number of classes to be mapped, the lower the mapping accuracy. The results can be used to model the spatial distribution of mangrove species or the composition of mangrove forests and update databases related to coastal management.

Keywords: mangrove species; spectrometer; spectral reflectance; WorldView-2; dendrogram

Citation: Rahmandhana, A.D.; Kamal, M.; Wicaksono, P. Spectral Reflectance-Based Mangrove Species Mapping from WorldView-2 Imagery of Karimunjawa and Kemujan Island, Central Java Province, Indonesia. *Remote Sens.* **2022**, *14*, 183. https://doi.org/10.3390/rs14010183

Academic Editors: Brigitte Leblon and Chandra Giri

Received: 15 November 2021
Accepted: 30 December 2021
Published: 1 January 2022

1. Introduction

Indonesia is a global ecological hotspot, judging from the extent and rich biodiversity of its mangrove ecosystem. Bunting et al. [1] mapped and reported the latest data on the world's mangrove area based on a combined analysis of ALOS PALSAR radar images and optical Landsat 5 TM and Landsat 7 ETM+ images taken between 2009 and 2011. The estimated global mangrove area is approximately 137,600 km^2, and Indonesia has the largest mangrove forest, with a total area of 26,890 km^2. It accounts for 19.5% of the mangroves worldwide and 50.4% of those in Asia. Indonesia is also estimated to contain 43 of around 75 true mangrove species in the world [2,3] or about 57% of all mangrove species worldwide. The characteristics of mangrove forests generally differ from those of mainland forests. For instance, their habitats are not climate-dependent but are shaped by tides, the extent of seawater-inundated soils, elevation, and the presence of canopy structures. They especially thrive on low-lying land without canopy structures [4]. Illegal logging and the creation of ponds are currently degrading mangrove forests. For example, Ilman et al. [5] reported that the estimated decline in mangrove area in Indonesia due to land clearing for ponds and logging between 1800 and 2012 was 913,000 hectares. Put another way, the average rate of decrease in mangrove area is 4300 hectares per year. The conversion of mangrove forests for aquaculture, tourism, and agricultural purposes has disrupted ecosystem stability and reduced physical and biological mangrove functions, affecting the existence of vulnerable mangrove species that are rare or limited.

The Karimunjawa Islands in Jepara Regency, Central Java, Indonesia, are a mangrove ecosystem with high species diversity. This ecosystem is relatively undisturbed and well-preserved because most of its area lies within the Karimunjawa National Park. Karimunjawa and Kemujan Island (two members of the Karimunjawa archipelago) have 3.964 km^2 of mangrove forests under the management and protection of the national park. A total of 25 true mangroves species have been recorded in this archipelago [4]. The inventory of mangrove distribution in the Karimunjawa National Park in 2002 found 44 mangrove species, including both true mangroves and their associates. Due to the widespread land conversion and logging in the mangrove area and its surroundings, mangrove species must be regularly mapped to maintain reliable information about the biodiversity of coastal ecosystems.

Cruising, a necessity for mapping activities, can be more challenging in mangrove areas than terrestrial forests. An alternative to cruising is needed to map and monitor the distribution of mangrove species efficiently and thoroughly. Remote sensing technology is an effective substitute in this context because it can minimize fieldwork and reduce the requisite time, cost, and effort of mapping. It is particularly cost- and time-efficient compared with field sampling, plant identification, and vegetation classification [6]. Furthermore, using remote sensing products, mainly satellite imagery, allows past conditions of the observed area to be recorded, enabling multi-temporal analyses. Satellite imagery also provides a synoptic overview of a large expanse, which is efficient for large-extent studies of the Earth's surface.

Mangroves differ from non-mangrove vegetation. They can be identified from specific tones and colors in remote sensing images and their association with coastal areas [7,8]. Healthy mangroves with green chlorophyll have low spectral reflectance in the blue and red bands, high in the green band, and significantly high in the near-infrared band; the reflectance increases with decreased water content in leaves [7]. Visually, true-color composite images for mangroves show a darker green color than vegetation in general because of the absorption of mangrove substrates in the near-infrared band. In satellite imagery, spectral reflectance allows tropical mangroves to be distinguished at the species level on a laboratory scale [9]. The accurate mapping of mangrove species requires images with high spatial and spectral resolution and the spectral reflectance pattern of each mangrove species to be measured in the field.

The spectral reflectance values are measured in the field using a hand-held spectrometer with high spectral detail that accommodates mangrove species mapping. They are used as the basis for a spectral library of mangrove species that comprises endmembers for mapping with a pixel-based classification algorithm. Classification is necessary to create visual results. Furthermore, the grouping of mangrove species based on spectral characteristics entails a cluster analysis to determine the optimal number of classes to be mapped. This study explores the use of three classification algorithms, spectral angle mapper (SAM), spectral information divergence (SID), and spectral feature fitting (SFF), to map the mangrove distribution on Karimunjawa and Kemujan Island. Specifically, the objectives of this study were to (1) classify and group mangrove species by their spectral reflectance characteristics, (2) map their spatial distribution with field-measured spectral reflectance and by applying the SAM, SFF, and SID classification algorithms to WorldView-2 images, and (3) assess the classification accuracy of each algorithm.

2. Materials and Methods

2.1. Study Site

The research area included Karimunjawa and Kemujan Island of Indonesia (5°49′33″–5°48′23″ S, 110°24′34″–110°28′37″ E). Their mangrove ecosystems comprise highly diverse mangrove species and are conservation areas in the Karimunjawa National Park, making them pristine and well-preserved. These two islands are located in the Karimunjawa District of Jepara Regency, Central Jawa Province, about 85 km north of Java Island (Figure 1). Several environmental factors that influence the growth of mangroves at the research site are the coastal physiography (topography), tides (length, duration, and range), waves and currents, climate (light, rainfall, temperature, and wind), salinity, dissolved oxygen, soil, and nutrients [10]. Based on the Schmidt and Ferguson climate classification, Karimunjawa and Kemujan Island have a C climate type with average rainfall of 3000 mm/year and air temperatures of 30–31 °C. Topographically, the Karimunjawa National Park area is located 0–506 m above sea level, spanning from flat coastal areas to undulating lowlands [11]. Hilly regions stretch from the east (highest peak) to the west and from the middle to the south. Karimunjawa experiences mixed tides with a prevailing diurnal pattern, i.e., one high and low tide but sometimes two high and low tides each day. The lowest low water level (LLWL) of the island waters is 20 cm, and the highest high-water level (HHWL) is 138 cm, making the maximum tidal range—the difference between the HHWL and LLWL—118 cm; this is classified as micro-tidal [12]. Karimunjawa and Kemujan Island have dark gray grumusol developed from quartz sandstone, micaceous sandstone, and quartz siltstone, and a few coral fragments included in mixed substrates (sand substrates and mixed gravels), with clay deposits along the coast. The soil physical properties on the two islands are fairly similar [11].

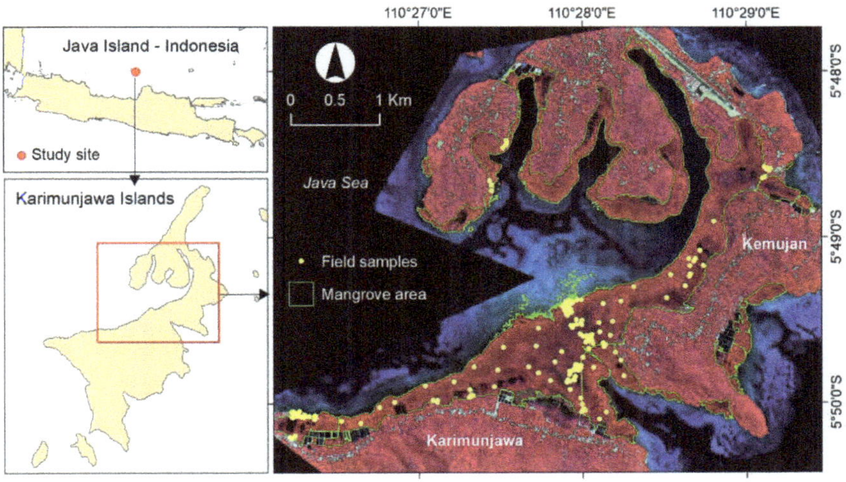

Figure 1. Study area map—The Karimunjawa and Kemujan Islands displayed on a Landsat 8 OLI color composite image with RGB 546 band combination.

The Karimunjawa archipelago has five types of ecosystems: coral reefs, mangroves, seagrass beds, lowland tropical rainforests, and coastal forests. Mangrove ecosystems can be found on the coasts of Karimunjawa, Kemujan, Cemara Kecil, Cemara Besar, Krakal Kecil, Krakal Besar, Merican, Menyawakan, and Sintok Island, over a total area of 3.964 km^2. There are 25 species (13 families) of true mangroves and nine species (seven families) of mangrove associates within the national park area, and five species (five families) of mangrove associates outside the national park [4]. Most of the mangroves on Karimunjawa and Kemujan Island are under the Karimunjawa National Park's jurisdiction, and the community manages the rest. In general, the park consists of eight zones: core, wilderness, marine protection, marine tourism use, historical-cultural-religious use, rehabilitation, marine use, and traditional fisheries. The research sites are located in the wilderness and marine use zones. The mangrove area on Kemujan Island is used for trekking tours through the forest and bird watching towers.

2.2. Image Data and Processing

The primary data source in this study was the WorldView-2 (WV-2) image covering parts of Karimunjawa and Kemujan Island. It was acquired on 27 June 2020, with a 2 m spatial resolution (multispectral sensor) and eight multispectral bands as described in Table 1 [13]. The image was selected to match the acquisition time to the fieldwork that was conducted at the end of the rainy season and ensure a cloud-free image. In this study, image pre-processing was required to extract the spectral signatures of the targeted objects. The first part of image pre-processing was a radiometric correction to convert the image pixel value from digital numbers into at-sensor reflectance with Updike and Comp's [14] procedure. The second part involved producing pixel values from the at-surface reflectance with fast line-of-sight atmospheric analysis of hypercubes (FLAASH) [15], an atmospheric correction model for the WV-2 image. The atmospheric visibility parameter was estimated from the moderate-resolution imaging spectroradiometer (MODIS) aerosol product [16].

Table 1. Band characteristics of WorldView-2 imagery [13].

Spectral Band	Wavelength	Spatial Resolution
Panchromatic	450–800 nm	0.46 m
Multispectral–Coastal	400–450 nm	1.84 m
Multispectral–Blue	450–510 nm	1.84 m
Multispectral–Green	510–580 nm	1.84 m
Multispectral–Yellow	585–625 nm	1.84 m
Multispectral–Red	630–690 nm	1.84 m
Multispectral–Red Edge	705–745 nm	1.84 m
Multispectral–Near-IR1	770–895 nm	1.84 m
Multispectral–Near-IR2	860–1040 nm	1.84 m

Differences in the radiometric correction results at each level are visible in the object's spectral reflectance curve (Z profile), presented in Figure 2. The spectral reflectance values in bands 1 (coastal) and 2 (blue) at the DN level (Figure 2a) decreased after correction at the surface reflectance level (Figure 2c), meaning that the atmospheric effects in the image had been corrected. The highest spectral reflectance values were in the red edge, near-IR1, and near-IR2 bands; in other words, the vegetation objects (mangroves) showed high reflectance values in these three bands. The spectral reflectance of mangroves forms a pattern similar to vegetation in general: low in the visible bands and high in the infrared bands (near-IR1 and near-IR2). This means that the radiometric and atmospheric corrections had been successfully applied to the WV-2 image.

The WV-2 image was visually interpreted to distinguish between mangrove and non-mangrove objects. True-color (RGB; 532) and false-color (near-IR1, red, blue; 752) composite images were used in this process, and mangrove and non-mangrove objects were identified based on their colors, tones, textures, and associations. The false-color composite image (752) depicts any vegetation features in red. However, it does not distinguish between mangroves and non-mangroves in areas formerly used as fish ponds. The true-color image composition was added to assist in this discrimination process. Mangrove objects can also be identified from their associations with surrounding objects, i.e., mangroves are located in coastal areas or border the sea. This mangrove object delineation output was then inputted for image masking to separate mangrove objects from others.

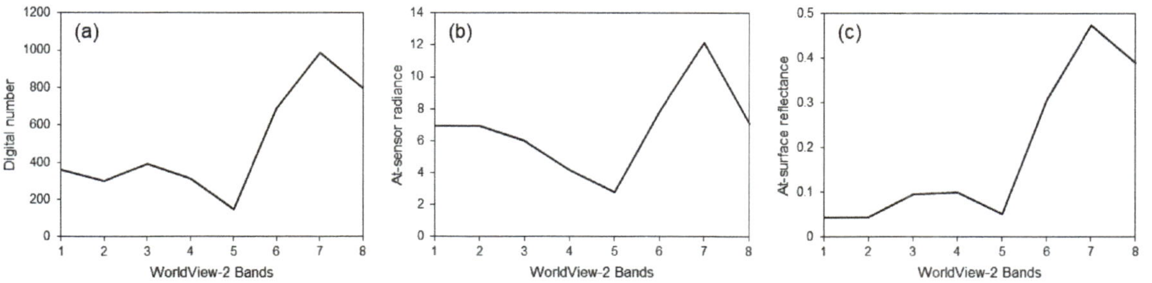

Figure 2. Spectral reflectance curves of mangrove objects at three correction levels: (**a**) digital number, (**b**) at-sensor radiance, and (**c**) at-surface reflectance.

2.3. Field Data Collection

The spectral reflectance values of the targeted mangrove species were collected directly from the sampling points during 5–10 March 2021. The number and locations of field samples were selected purposively based on the Pixel Purity Index (PPI) values [17] and the sites' accessibility. The PPI, in this case, was used to locate the purest mangrove pixels in the WV-2 image for spectral reflectance collection. The PPI image was produced from the minimum noise fraction (MNF) algorithm to remove the noise in the data and reduce the computational requirements for further processing. This research used 10,000 iterations because, according to Plaza and Chang [18], PPI produces pure pixels after 10,000 to 100,000 repetitions. These samples were then plotted onto a map that was later used to read the spectral reflectance of mangrove species in the field.

The field data collection resulted in 201 point samples covering 24 targeted mangrove species (Table 2). The distribution of the field samples is presented in Figure 1. The field samples were divided into two major groups: modeling samples for developing the spectral library (24 samples, Figure 3a) and validation models for the accuracy assessment of the resulting map (177 samples, Figure 3b). The modeling samples were spectral reflectance mangrove species collected in the areas with high species diversity. This research only focused on the true mangrove species, and researchers were assisted by park managers or rangers familiar with the targeted species' location to collect these samples. The validation samples were mangrove species identified in the field. The validation sample locations were selected from "white pixels" (i.e., pure pixels) derived from the PPI calculation. In addition to PPI, the validation samples were also determined with aerial photos to identify a mix of two species (*Avicennia marina* and *Ceriops tagal*) found in the field.

Table 2. Mangrove species identified in the field survey.

No	Mangrove Species	No	Mangrove Species	No	Mangrove Species
1	*Acanthus ebracteatus*	9	*Ceriops tagal*	17	*Rhizophora mucronata*
2	*Acanthus ilicifolius*	10	*Excoecaria agallocha*	18	*Rhizophora stylosa*
3	*Acrostichum aureum*	11	*Heritiera littoralis*	19	*Scyphiphora hydrophyllacea*
4	*Aegiceras corniculatum*	12	*Lumnitzera littorea*	20	*Sonneratia alba*
5	*Avicennia marina*	13	*Lumnitzera racemosa*	21	*Sonneratia caseolaris*
6	*Bruguiera cylindrica*	14	*Pemphis acidula*	22	*Sonneratia ovata*
7	*Bruguiera gymnorrhiza*	15	*Rhizophora apiculata*	23	*Xylocarpus granatum*
8	*Bruguiera sexangula*	16	*Rhizophora lamarckii*	24	*Xylocarpus moluccensis*

The field spectral reflectance values of mangrove species were measured with a JAZ EL-350 portable spectrometer from Ocean Optics (https://oceanoptics.com/, accessed on 14 November 2021). First, the white and dark references were measured. A white reference produces a standard "white object" spectral reflectance reading to calculate the object's spectral sample, while a dark reference produces a reading of an object with perfect absorption to create a "black body" reference [19]. Second, the spectral reflectance values of mangrove species were measured at the leaf level (Figure 3a), based on the criterion that healthy leaves appeared entirely green or contained chlorophyll.

Figure 3. Field data collection: (**a**) spectral reflectance measurement of the targeted mangrove species and (**b**) field validation sample collection.

Some important aspects to consider in collecting spectral reflectance data for mangrove species in the field are (1) the field of view (FOV) of the spectrometer sensor, (2) the distance between the spectrometer and the targeted object, (3) the angle and direction of measurement, and (4) the light conditions at the time of observation [19]. Kamal et al. [20] collected the spectral reflectance of *Rhizophora stylosa* on Karimunjawa Island at 2 cm, 50 cm, 1 m, 2 m, and 5 m distances with ten readings at each distance. According to this study, the spectral reflectance curves recorded at close range to the leaf (i.e., 2 cm) and from the furthest distance (i.e., 5 m) showed the lowest curve variation between readings. Therefore, this study used a distance of 2 cm for leaf spectral measurements to ensure that only the mangrove leaf was read (Figure 3a) and ensure highly consistent spectrometer results across readings [20,21]. The spectrometer's measurement angle was set at 45° to the nadir and facing the sun to avoid shadows on the target object. In addition, the number of spectrometer reading repetitions affects the degree of confidence in the reading results. The leaf measurements were repeated six times to ensure the consistency of the readings in similar natural lighting conditions in the field.

The field-measured spectral reflectance profiles were stored in a file in .jaz format. A .jaz file contains reference spectrum data (white reflectance), dark reflectance, and the targeted object spectral reflectance data, which are used to construct the object's spectral reflectance curve. The six replicates used to measure each mangrove species object avoided errors such as saturation in the spectral reflectance measurement, which increased the spectrometer readings' precision. The field measurements produced data on objects' light intensity at a particular wavelength; thus, the reflectance values needed to be calculated. These values were normalized and calculated with the formula described in Kamal et al. [19] and Wicaksono et al. [22]. The normalized and mean values were then used as input to build a spectral library.

2.4. Mangrove Species Clustering Analysis

White and dark reference readings were collected for each measurement of mangrove species samples using the spectrometer to normalize the spectral reflectance of each mangrove species and create a standard range of spectral reflectance values. This allows the derived spectral reflectance curves of different mangrove species to be compared directly. Mangrove species' spectral reflectance can be analyzed effectively at wavelengths of 350–900 nm with a JAZ EL-350 spectrometer [20]. This wavelength range was selected because, based on the specifications of the spectrometer used, noise is likely to occur below 350 nm and above 900 nm. The mangrove species spectra compiled in the spectral library were then resampled according to the center wavelength of the WV-2 image bands and used to develop an optical dendrogram. An optical dendrogram determines how components (i.e., mangrove species) are grouped spectrally. It is also used to identify similarities and cluster distances between species or groups. The optical dendrogram for this study was created in the IBM SPSS Statistics 24 program based on Wicaksono et al.'s [22] work, which used seagrass as the research object and the Ward linkage method. The derived dendrogram was then used as the basis for the mangrove species classification scheme in remote sensing-based mapping.

The dendrogram was built using the Ward linkage method as described by Wicaksono et al. [22]. It analyzes clusters hierarchically by determining the distance between two clusters expressed as an increase in the "error sum of squares" (ESS). The Ward linkage method selects grouping steps sequentially to minimize the ESS at each step. The dendrogram was not developed from the field-measured spectral reflectance but rather the resampled data; therefore, eight wavelengths were selected to represent the eight WV-2 image bands. The resampled spectral reflectance values were inputted in the dendrogram for further use in pixel-based mapping. This dendrogram identifies similarities and cluster distances between mangrove species, producing several large groups according to the combination of cluster distances.

2.5. Pixel-Based Classification and Accuracy Assessment

The field-collected spectral samples were extracted and divided into three types: white reference, dark reference, and the object's spectral reference. This extraction was performed to produce the spectral reflectance curves of mangrove species for a spectral library, i.e., a collection of references containing the spectral reflectance values of various objects [19,20]. A normalization process was conducted to obtain the appropriate spectral reflectance curves allowing for easy understanding and high representativeness. The extracted spectral reflectance samples of mangrove species were then collected in one container to be inputted into a spectral library in the ENVI 5.3 program. The compiled spectral library of mangrove species from field measurements was spectrally resampled to align with the spectral resolution of the WV-2 image as the basis of mangrove species mapping.

The mangrove species were mapped in ENVI 5.3 using the spectral library from the resampled spectra as the classification input or endmember. For this purpose, three pixel-based classification algorithms were applied and evaluated for mapping mangrove species, namely the spectral angle mapper (SAM), spectral information divergence (SID), and spectral feature fitting (SFF). The SAM algorithm can classify objects based on their spectral reflectance values by considering the illumination angle reflected by the object. SAM determines the similarity between two spectral reflectance objects by calculating the "spectral angle" created between them and treating them as vectors in a space with dimensionality equal to the number (n) of image bands used [23,24]. Because it only factors in the direction of the spectrum, the spectrum length and other factors are considered equal. It determines the similarity of the unknown spectrum t to the reference spectrum r using Equation (1), below:

$$\alpha = cos^{-1}(t \cdot r||t||\ ||r||) \quad (1)$$

The SID algorithm can determine the target pixel based on differences in the object's spectral reflectance information. It calculates the information difference between the target pixel (r_1) and the target reference (r_2) from the sum of the relative difference between r_1 and r_2 ($D(r_1||r_2)$), and the relative difference between r_2 and r_1 ($D(r_2||r_1)$). This algorithm is presented in Equation (2) below [25]:

$$SID(r_1, r_2) = (D(r_1||r_2)) + (D(r_2||r_1)) \quad (2)$$

Finally, the SFF algorithm compares the spectral image with the endmember reference. The reference spectra were scaled to match those of the image after the continuum was removed from both datasets (reference and object spectra) [26].

The spectral reflectance of each mangrove species was measured in the field simultaneously with the validation samples. The validation samples are different datasets from those used to map the mangrove species. They contain mangrove species data and their coordinate locations in the field. However, both modeling and validation samples were determined with the same purposive sampling technique because of high object heterogeneity in the field and low accessibility. Both sets of samples were also selected with the same criteria, i.e., the PPI calculation results.

The accuracy assessment was conducted with a confusion matrix to measure the extent to which the mangrove species classification results from field-collected and post-field-processed data were similar. In addition to percent accuracy, it also evaluated each algorithm's misclassification by observing the presence or absence of a logical error in the species classification. This accuracy assessment resulted in user's accuracy (UA), producer's accuracy (PA), and overall accuracy (OA) values, following Congalton and Green's procedures [27].

3. Results

3.1. Spectral Reflectance of Mangrove Species

The spectral library for each species was developed by displaying the spectral reflectance curve of the replicated samples with their averages and observing the corresponding curves of all species. The normalized spectral reflectance of 24 mangrove species (10 of primary data and 14 of secondary data) were combined into one spectral reflectance curve, as shown in Figure 4a. The field-measured spectral reflectance has a very high spectral resolution (1586 bands) while that from the WV-2 image has a low resolution (eight bands). Therefore, spectral resampling was conducted to match the spectral resolution derived from the spectrometer to the WV-2 image bands. The center wavelengths of the WV-2 image obtained from the "Radiometric Use of WorldView-2 Imagery" guidelines by DigitalGlobe [13] were used as the targeted spectra in the spectral resampling process. The center wavelength for each WV-2 band is as follows: 427 nm (coastal), 478.3 nm (blue), 545.8 nm (green, 607.7 nm (yellow), 658.8 nm (red), 724.1 nm (red-edge), 832.9 nm (NIR1), and 949.3 nm (NIR2). Besides degrading spectral resolution, spectral resampling also simplified the spectral reflectance of objects measured in the field. In general, both field-collected and resampled data had the same pattern: low in the coastal, blue, and red bands, slightly high in the green band, high in the red-edge band, and very high in the near-IR1 and near-IR2 bands (Figure 4a,b).

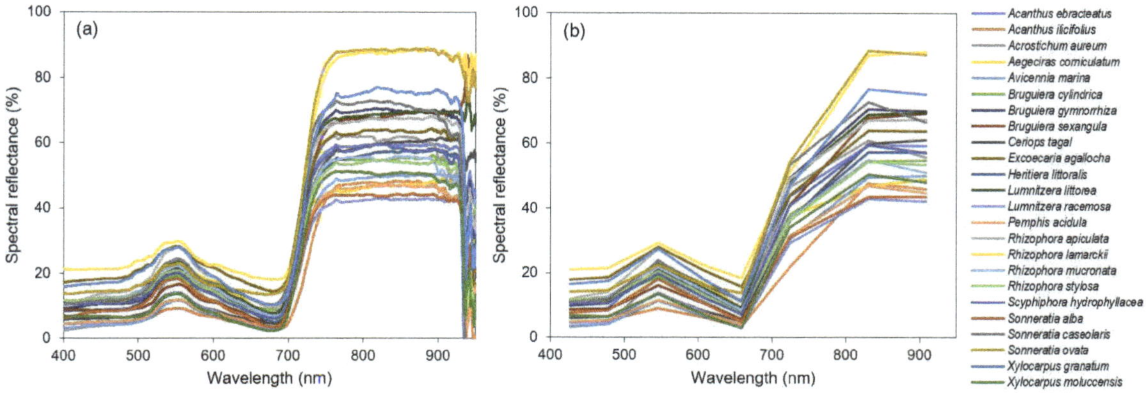

Figure 4. Spectral reflectance curves of the field-measured data (**a**) and the resampling results based on WorldView-2 bands (**b**) for the 24 mangrove species found in the study site.

3.2. Clustering Analysis of Mangrove Species

The clustering of mangrove species was conducted based on the resampled mangrove spectra. Indirectly, this clustering also grouped several morphological features of plants, especially their leaves. Klančnik and Gaberščik [28] explained that every leaf shape has structural characteristics with distinctive optical properties. The redundancy analysis showed that the leaves' morphological and biochemical characteristics had particular relationships with leaf spectral reflectance. In general, the most dominant plant part that can be identified by satellite imagery is the leaf canopy. In this study, the spectral reflectance approach used as input in the clustering analysis indirectly captured some of the morphological characteristics of the mangroves, especially the leaves.

Based on the resulting dendrogram (Figure 5), there were four possible classification schemes. The cluster distance for the most distinguishable class was 25, while the cluster distance for the least distinguishable class was 0. The dendrogram showed four levels with different numbers of species groups. Level 1 (two groups), Level 2 (four groups), and Level 3 (five groups) can be used as references in pixel-based mangrove species mapping because they have a relatively large distance between clusters (Table 3). Meanwhile, the Level-4 classification scheme cannot be used for this purpose because the distances between clusters are too small. The results of this species clustering are used to design scenarios for mangrove species mapping using the WV-2 image.

Table 3. Mangrove species grouping of Levels 1–4 resulting from the cluster analysis.

Level	Group	Mangrove Species
Level 1	A	*A. ebracteatus, A. ilicifolius, A. aureum, A. corniculatum, A. marina, B. cylindrica, B. sexangular, C. tagal, H. littoralis, L. racemosa, P. acidula, R. mucronata, R. stylosa, S. hydrophyllacea, S. alba, X. moluccensis*
	B	*B. gymnorrhiza, E. agallocha, L. littorea, R. apiculata, R. lamarckii, S. caseolaris, S. ovata, X. granatum*
Level 2	A	*A. ebracteatus, A. aureum, A. corniculatum, B. cylindrica, B. sexangular, C. tagal, H. littoralis, R. mucronata, R. stylosa, S. hydrophyllacea*
	B	*A. ilicifolius, A. marina, L. racemosa, P. acidula, S. alba, X. moluccensis*
	C	*R. lamarckii, S. ovata*
	D	*B. gymnorrhiza, E. agallocha, L. littorea, R. apiculata, S. caseolaris, X. granatum*
Level 3	A	*A. corniculatum, B. cylindrica, R. mucronata, R. stylosa*
	B	*A. ebracteatus, A. aureum, B. sexangula, C. tagal, H. littoralis, S. hydrophyllacea*
	C	*A. ilicifolius, A. marina, L. racemosa, P. acidula, S. alba, X. moluccensis*
	D	*R. lamarckii, S. ovata*
	E	*B. gymnorrhiza, E. agallocha, L. littorea, R. apiculata, S. caseolaris, X. granatum*
Level 4		Individual mangrove species

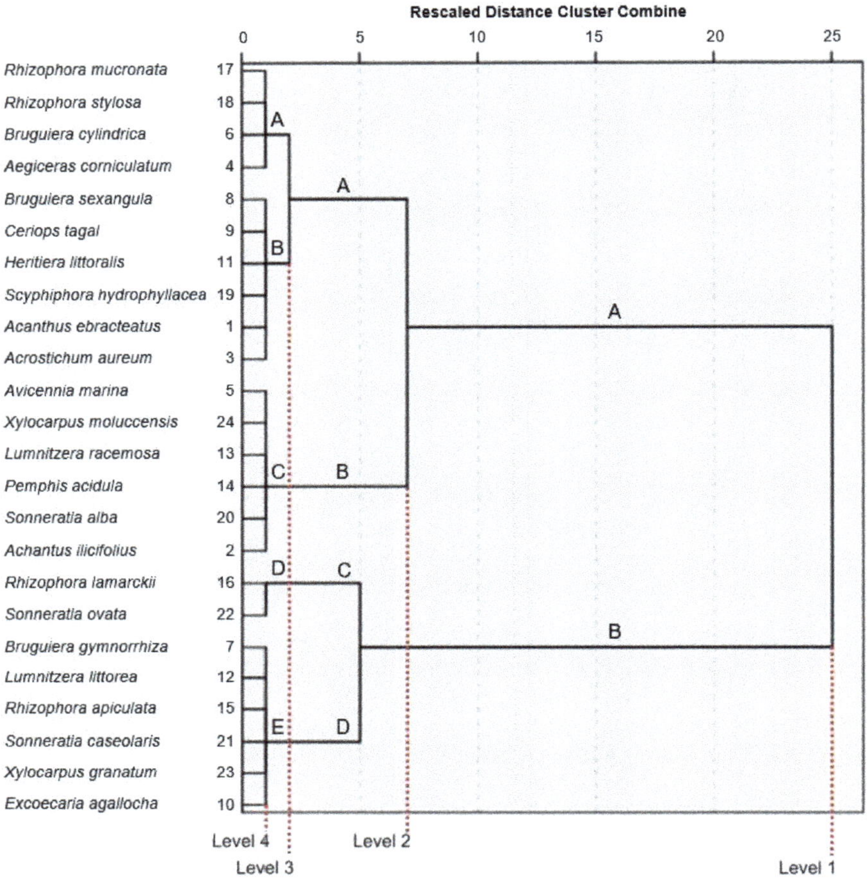

Figure 5. Dendrogram of mangrove species obtained with the Ward linkage method showing four levels of species grouping.

3.3. Pixel-Based Classification

The classification mapping was based on the dendrogram analysis results, which produced four levels of mangrove species groups. First, the spectral reflectance curves for each group of mangrove species at each level of the dendrogram were averaged. Then, these results were inputted into the three algorithms of mangrove species mapping. The SAM- and SID-based classifications used default thresholds of 0.1 and 0.05, respectively. Because there was no default threshold for the SFF, the threshold value was set at 10. For each dendrogram level, the threshold values of the classification algorithms were made equal to reduce user intervention and maximize software performance. The mapping results of using these three classification algorithms are presented in Figure 6. Overall, the SAM-based mangrove species classification results at Levels 1, 2, and 3 could not classify all mangrove areas, as was evident from the extensive gray areas in the mangrove delineation results. Some of the mangrove areas were unclassified; this could be due to an inappropriate threshold value for the classification. The front/distal formation bordering the seawater was generally dominated by *Rhizophora* groups and some *Bruguiera* individuals. The SAM algorithm found *Pemphis acidula* in the distal formation, but this species was not entirely classified in the Level-3 mapping. Meanwhile, the Level-1 mapping with the SID and SFF algorithms classified the distal formation as Group B with eight species, three of which were *Bruguiera gymnorrhiza*, *Rhizophora lamarckii*, and *Rhizophora apiculata*. According to field reports, *Rhizophora* groups and some *Bruguiera* individuals generally dominated the distal formation bordering the sea, meaning that *Rhizophora apiculata* is highly likely to have been among the species classified.

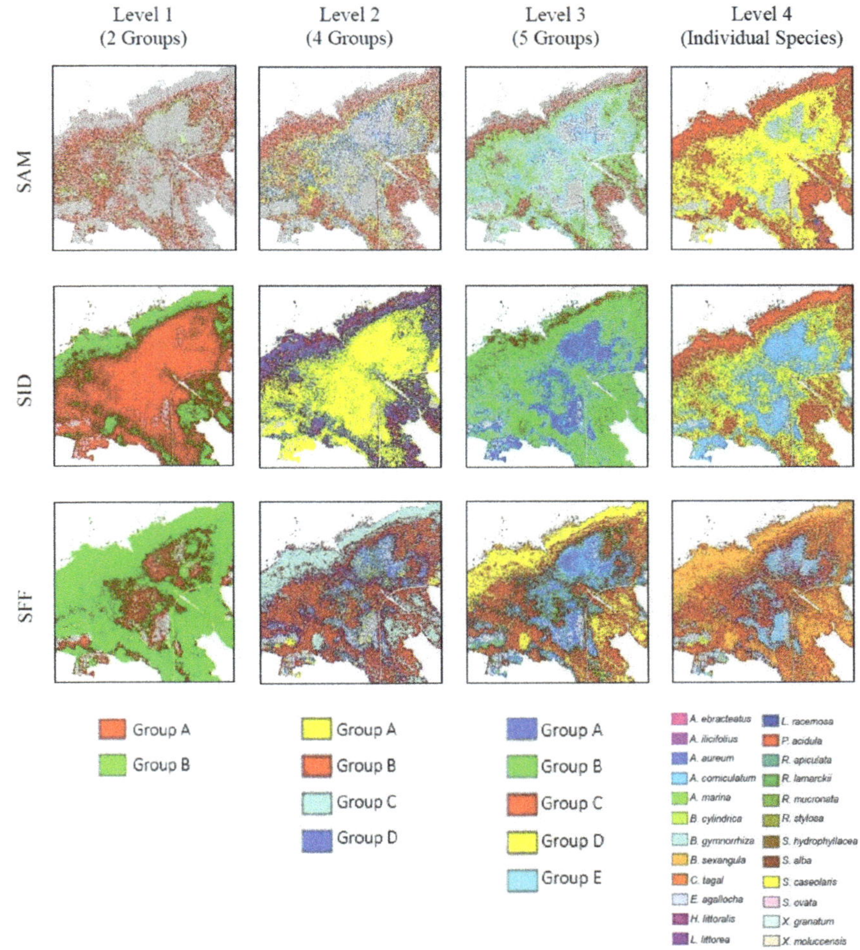

Figure 6. Mangrove species mapping results using the SAM, SID, and SFF algorithms at group Levels 1–4.

The Level-2 SID-based mapping showed Group A as the dominant species constituent. Group A consisted of several understory genera, such as *Achantus*, *Acrosticum*, *Aegiceras*, and *Scyphiphora* and some trees, such as *Bruguiera sexangula*, *Bruguiera cylindrica*, *Ceriops tagal*, *Rhizophora mucronata*, and *Rhizophora stylosa*, located in the medial and proximal formations. Meanwhile, the distal formation was dominated by Groups B and D, creating a mixed pattern. In contrast, the Level-2 SFF-based classification showed that the distal formation was dominated by Group C, consisting of *Rhizophora lamarckii* and *Sporomusa ovata*. However, these results were inaccurate as these two species cannot grow large and are rarely present in tidal areas but instead reside in the medial and the proximal areas. Meanwhile, the SID- and SFF-based classifications at Level 3 also showed significant differences in species composition; the former showed Group B predominating and the latter, codominant Groups A, C, and D.

The individual species mapping results (Level 4) differed from the other three classifications. The SAM and SID algorithms classified *Pemphis acidula* as the dominant species in the distal formation, although it is rarely found in tidal areas such as this distal formation, typically growing under the canopy of the proximal formation. This causes a visual classification error and subsequently affects the accuracy of the classification mapping using both algorithms. Meanwhile, SFF classified *Bruguiera sexangula* as the dominant species in the distal formation. It is commonly found in tidal areas and is associated with *Rhizophora* but not as a major species with broad distribution. The SFF-based classification suggested the presence of *Bruguiera sexangular* along the tidal area, another imprecise result that affects the mapping accuracy value.

4. Discussion

4.1. Mangrove Species Clusters

Mangrove species clustering can adopt plant morphological approaches commonly used in compiling taxonomies. For example, leaf shape, canopy shape, stem characteristics, and plant habitat characteristics such as salinity level, type of substrate, and length of tidal inundation can be used. However, the spectral reflectance curve of the mangrove species object on the WV-2 image was integrated with the field-measured curve for this study. The result of the spectral reflectance resampling was then inputted into the clustering in dendrograms for pixel-based mapping classification.

The results of the dendrogram analysis showed four levels with varying numbers of species groups (Figure 5 and Table 3). Level 1 (2 groups), Level 2 (4 groups), and Level 3 (5 groups) could be used as references for the pixel-based mapping of mangrove species because the clusters were quite far apart. In contrast, Level 4 could not be used for the same purpose because the clusters were too close together. The Level 1 scheme had a cluster distance of 25, resulting in two large clusters in which Bruguiera, Lumnitzera, Rhizophora, Sonneratia, and Xylocarpus were divided into two groups. In the Level 2 scheme with a cluster distance of 5, the mangrove species were divided into four groups. The species Rhizophora lamarckii and Sonneratia ovata were clustered into one group for the similarity of their spectral reflectance; physiologically, the two species also have more wax coating on the leaf surfaces compared with other species. The Level 3 scheme had a cluster distance of two, which divided the mangrove species into five groups where group A, at a rescaled distance of 5, was split into two new groups while the other three groups remained. Unlike in Levels 1–3, the Level 4 scheme grouped mangroves by single species, without forming new groups.

4.2. Accuracy Assessment of the Resulting Maps

The mapping accuracy was assessed with a confusion matrix to calculate the overall accuracy (OA), producer's accuracy (PA), and user's accuracy (UA). The validation samples were point data obtained directly from the field survey in 2021 (primary data) and secondary data with the attributes of mangrove species. The primary data were acquired using the average point method at 2 m $\times$ 2 m pure pixel points that had been created. Additionally, the samples were incidentally collected when encountering certain minor species that had gone unidentified during the interpretation of the aerial photographs. To assess the accuracy of the mapping results, a total of 177 sample points were analyzed using the "extract values to point" tool in ArcGIS on the model developed previously to compile the matrix.

In this step, the mangrove species maps generated using the three classification algorithms were evaluated for their OA, as presented in Tables 4 and 5. In general, the accuracy of the three classification algorithms at all four levels was low. The SID-based classifications at Levels 1 (with two species groups), 2 (four), and 3 (five) had the highest OA at 49.72%, 22.60%, and 15.25%, respectively. Meanwhile, the SFF-based classification for the Level-4 mapping of 24 species had the highest OA at 5.08%. The results demonstrated that the greater the number of classes to be mapped, the smaller the accuracy value. This finding corresponds to Andréfouët et al. [29] and Kamal et al. [30], who explained that the percentage of classification accuracy expressed in OA decreased with the increase in the number of classes. In addition, the high heterogeneity of objects with natural properties, such as mangroves and mixed species in one pixel, affected the OA. The characteristics of mapped objects also affect the accuracy because if more pixels are mixed, the objects will be more difficult to classify.

Table 4. Accuracy of mangrove species mapping with SAM, SID, and SFF-based classifications at Levels 1, 2, and 3.

Classification Algorithms	Group	Level 1 (%) PA	UA	OA	Group	Level 2 (%) PA	UA	OA	Group	Level 3 (%) PA	UA	OA
SAM	A	8.40	47.62	7.34	A	5.88	50.00	5.08	A	3.33	25.00	5.65
	B	5.17	37.50		B	5.88	15.00		B	13.16	33.33	
					C	0.00	0.00		C	5.88	21.43	
					D	3.85	16.67		D	0.00	0.00	
									E	1.92	8.33	
SID	A	58.82	66.67	49.72	A	36.76	31.25	22.60	A	13.33	5.97	15.25
	B	31.03	39.13		B	3.92	7.14		B	42.11	27.59	
					C	0.00	0.00		C	1.96	5.26	
					D	25.00	30.23		D	0.00	0.00	
									E	11.54	35.29	
SFF	A	21.85	70.27	38.42	A	10.29	31.82	10.73	A	6.67	4.65	7.91
	B	72.41	43.30		B	5.88	7.50		B	5.26	50.00	
					C	50.00	0.00		C	5.88	7.89	
					D	11.54	20.69		D	50.00	6.25	
									E	7.69	23.53	

Table 5. Accuracy of mangrove species mapping with SAM, SID, and SFF-based classifications at Level 4.

Class	SAM PA (%)	SAM UA (%)	SAM OA (%)	SID PA (%)	SID UA (%)	SID OA (%)	SFF PA (%)	SFF UA (%)	SFF OA (%)
A	0	0	0.56	0	0	0	0	0	5.08
B	0	0		0	0		0	0	
C	0	0		0	0		0	0	
D	0	0		0	0		0	0	
E	0	0		0	0		0	0	
F	0	0		0	0		0	0	
G	0	0		0	0		0	0	
H	0	0		0	0		33.33	2.27	
I	0	0		0	0		0	0	
J	0	0		0	0		6.25	9.09	
K	0	0		0	0		0	0	
L	0	0		0	0		46.15	37.50	
M	0	0		0	0		0	0	
N	0	0		0	0		0	0	
O	0	0		0	0		0	0	
P	0	0		0	0		0	0	
Q	0	0		0	0		0	0	
R	0	0		0	0		0	0	
S	0	0		0	0		0	0	
T	0	0		0	0		0	0	
U	33.33	2.44		0	0		33.33	16.67	
V	0	0		0	0		0	0	
W	0	0		0	0		0	0	
X	0	0		0	0		0	0	

The mapping accuracy assessment also calculated UA and PA for each classification algorithm and each dendrogram level to determine which errors reduced the accuracy. In general, the UA of all classification algorithms at each level was higher than the PA, as not all of the validation samples matched the classification model; many fell into a different class (unclassified or class "0"). With many samples considered unclassified, the PA decreased because, in contrast to UA, the number of correctly classified samples as the divisor increased. At Level 1 classifications with the SAM, SID, and SFF algorithms, there were, respectively, 148, 26, and 43 unclassified points among 177 samples. Furthermore, the SAM, SID, and SFF-based classifications led to 137, 24, and 38 unclassified points at Level 2; 132, 15, and 27 unclassified points at Level 3; and 63, 5, and 13 unclassified points at Level 4. The SAM-based classification had the highest number of unclassified points at all levels, resulting in low PA, UA, and OA.

The SAM algorithm demonstrated the lowest accuracy at all levels because the default threshold of 0.1 was used, leaving the mangrove area partially unclassified. The accuracy points extracted from the SAM model did not overlap because the mangrove area was not classified. The number of unclassified points significantly affected the accuracy of the SAM-based mapping results. Not all accuracy points that overlap with the model show accurate or desirable results. At Levels 1, 2, and 3, the points that did not overlap in the class that should have been the majority were placed in the last class, namely classes B, D, and E, respectively. At Level 4, the inaccurate points were mostly classified as *Pemphis acidula* and *Sonneratia caseolaris*. These issues significantly decreased the OA of SAM-based classification.

Mapping single species with the SID- and SFF-based classifications resulted in overlapping classes with more accuracy points than using SAM. The accuracy points in the SID-based classification overlapped more with *Acrostichum aureum*, *Aegiceras corniculatum*, *Pemphis acidula*, and *Sonneratia caseolaris* than with other species. Meanwhile, in the SFF-based classification, the accuracy points overlapped more with *Aegiceras corniculatum*, *Bruguiera sexangula*, and *Pemphis acidula*. However, according to the classification results, there were only nine overlapped accuracy points with the correct values when using the SFF and none when using the SID. In pixel-based mapping, increasing the classes to be mapped will reduce the accuracy value. In contrast, Kamal et al. [30] mapped *Rhizophora stylosa* and produced an OA of 52%. Hirano et al. [31] conducted a similar study mapping *Rhizophora mangle* that resulted in an OA of 40%. Mixed pixels are one of the problems that arise in remote sensing images because one pixel in the image can consist of two or more types of objects. Likewise, mixed pixels were common in the WV-2 satellite imagery used in this study because of the high heterogeneity and species diversity of the area. Hyperspectral field data used as the input in multispectral images also affects the mapping accuracy. Ideally, this type of study should be based on hyperspectral image data so the spectral resolution would be similar.

4.3. Classification Performance Evaluation

The three classification algorithms have different characteristics, each with particular advantages and disadvantages. They have dissimilar capacities in recognizing objects even from the same type of data source, i.e., the field-measured spectral reflectance of mangrove species. In the SAM algorithm, the similarity between the two spectra is determined by calculating the "spectral angle" between the image spectrum and the object spectrum and then assuming a vector in the same dimensional space with the same number of bands [32]. Its disadvantages include insensitivity to other known factors (besides angle) and using the same treatment for all illumination because SAM only uses spectrum "direction" and not spectrum "length." SAM only considers the spectral reflectance pattern of the object, not the difference in the intensity of the object's spectral reflectance [33]. The SAM-based classification uses vector directions to distinguish between features' spectral reflectance properties. Features with smaller spectral angles are categorized into the same class. SAM classifies pure classes and leaves the rest unclassified, thus failing to classify entire plant species with pure pixels in the mangrove area [32,34]. In addition, the factors considered in determining the threshold value affect the classification results, especially at Levels 1, 2, and 3.

In contrast to SAM, the SID algorithm considers each pixel a random variable, using a spectral histogram to obtain a mapped probability. Muhammad and Mirza [34] suggest that problems arising from the SAM classification algorithm can be minimized by using SID. SID was applied to the same endmember to classify unclassified species and impure pixels in SAM-based classification results. The SID algorithm uses the size of the divergence to match pixels to spectral references. A smaller divergence means that the pixel is more likely to be similar to the spectral reference. According to Nidamanuri and Zbell [35], SID-based classification can measure the spectral variability of a single mixed pixel and determine similar spectra. However, according to Shanmugam and Srinivasaperumal [36], SID's weakness is that it is more effective on mixed pixel targets.

The SFF classification algorithm produces a separate scale image and root mean square (RMS) image or a combination of both. SFF is an absorption feature-based algorithm that matches the image spectrum (pixels) with the object spectrum (reference) [37]. In this study, SFF classified areas of high species diversity better than SAM and SID in single species mapping (Level 4). However, according to Muhammad and Mirza [34], this algorithm is the most time-consuming to use and the resulting classification still needs improvements. For the input, SFF requires a reference spectrum of the image or spectral library. The second requirement is removing the continuum from the image and spectral reference (object) before analysis [37], for instance, by resampling that degrades the spectral resolution of the data to that of the image.

5. Conclusions

This study identified 24 mangrove species on Karimunjawa and Kemujan Island. Based on their spectral reflectance characteristics, there were four dendrogram levels: Level 1 (consisting of two groups), 2 (four groups), 3 (five groups), and 4 (single species). The SAM-based classifications at Levels 1, 2, and 3 did not entirely classify the mangroves. The SAM and SID algorithms successfully mapped *Pemphis acidula* massively in the distal formation at Level 4. Using SID, Group B was found to prevail at Level 1 while Group A was dominant at Level 2. At the same level, the SFF algorithm classified Group C as dominant in the distal formation (*Rhizophora lamarckii* and *Sonneratia ovata*). Meanwhile, the SID- and SFF-based classifications at Level 3 showed Group B prevailing in the former and codominant Groups A, C, and D in the latter. The SFF algorithm classified *Bruguiera sexangula* in the distal formation. The best accuracy for mapping mangrove species distribution was obtained by applying the SID-based classification at Levels 1, 2, and 3, with overall accuracies of 49.72%, 22.60%, and 15.20%, respectively. Meanwhile, the best single-species mapping accuracy (Level 4) was obtained with SFF-based classification, with an overall accuracy of 5.08%. In conclusion, the three classification algorithms offered low mapping accuracy due to the high heterogeneity of species in the field, which resulted in many mixed pixels and limited access to obtain evenly distributed accuracy points. The greater the number of classes to be mapped, the smaller the accuracy. A predefined threshold value that is less than optimal is also a source of low accuracy. Future research can focus on assessing whether the number of mangrove species affects the accuracy of mapping results. This can be achieved by replicating the mapping method in a mangrove environment with low species variation.

Author Contributions: Conceptualization and methodology, A.D.R., M.K. and P.W.; validation and formal analysis, A.D.R.; writing—original draft preparation, A.D.R.; writing—review and editing, M.K and P.W.; visualization, A.D.R.; supervision, M.K. and P.W.; funding acquisition, M.K. All authors have read and agreed to the published version of the manuscript.

Funding: This research was funded by the 2021 Penelitian Terapan Unggulan Perguruan Tinggi Grant scheme provided by the Ministry of Education, Culture, Research and Technology of the Republic of Indonesia (contract number 1738/UN1/DITLIT/DIT-LIT/PT/2021).

Institutional Review Board Statement: Not applicable.

Informed Consent Statement: Not applicable.

Data Availability Statement: The data presented in this study are available from the first author on request.

Acknowledgments: The authors would like to thank (1) the Department of Geographic Information Science at the Faculty of Geography, Universitas Gadjah Mada, for providing research facilities and equipment, (2) the Karimunjawa National Park management for granting fieldwork permits, and (3) S.M. Ridha and M.S. Usni for their invaluable assistance during the fieldwork.

Conflicts of Interest: The authors declare no conflict of interest. The funders had no role in the design of the study; the collection, analysis, or interpretation of data; the writing of the manuscript; or the decision to publish the results.

References

1. Bunting, P.; Rosenqvist, A.; Lucas, R.M.; Rebelo, L.-M.; Hilarides, L.; Thomas, N.; Hardy, A.; Itoh, T.; Shimada, M.; Finlayson, C.M. The Global Mangrove Watch—A New 2010 Global Baseline of Mangrove Extent. *Remote Sens.* **2018**, *10*, 1669. [CrossRef]
2. Kusmana, C. Distribution and Current Status of Mangrove Forests in Indonesia. In *Mangrove Ecosystems of Asia: Status, Challenges and Management Strategies*; Faridah-Hanum, I., Latiff, A., Hakeem, K.R., Ozturk, M., Eds.; Springer: New York, NY, USA, 2014; pp. 37–60. [CrossRef]
3. Spalding, M.; Kainuma, M.; Collins, L. *World Atlas of Mangroves*; Earthscan: London, UK, 2010.
4. Karimunjawa National Park. *Jenis Mangrove Taman Nasional Karimunjawa [Mangrove Species in the Karimunjawa National Park]*; Karimunjawa National Park: Semarang, Indonesia, 2012. (In Indonesian)
5. Ilman, M.; Dargusch, P.; Dart, P. A historical analysis of the drivers of loss and degradation of Indonesia's mangroves. *Land Use Policy* **2016**, *54*, 448–459. [CrossRef]
6. Lewis, D.; Phinn, S.; Arroyo, L. Cost-effectiveness of seven approaches to map vegetation communities—A case study from Northern Australia's tropical savannas. *Remote Sens.* **2013**, *5*, 377–414. [CrossRef]
7. Kuenzer, C.; Bluemel, A.; Gebhardt, S.; Quoc, T.V.; Dech, S. Remote sensing of mangrove ecosystems: A review. *Remote Sens.* **2011**, *3*, 878–928. [CrossRef]
8. Lucas, R.; Lule, A.V.; Rodríguez, M.T.; Kamal, M.; Thomas, N.; Asbridge, E.; Kuenzer, C. Spatial Ecology of Mangrove Forests: A Remote Sensing Perspective. In *Mangrove Ecosystems: A Global Biogeographic Perspective: Structure, Function, and Services*; Rivera-Monroy, V.H., Lee, S.Y., Kristensen, E., Twilley, R.R., Eds.; Springer: Cham, Switzerland, 2017; pp. 87–112.
9. Vaiphasa, C.; Ongsomwang, S.; Vaiphasa, T.; Skidmore, A.K. Tropical mangrove species discrimination using hyperspectral data: A laboratory study. *Estuar. Coast Shelf Sci.* **2005**, *65*, 371–379. [CrossRef]
10. Alwidakdo, A.; Azham, Z.; Kamarubayana, L. Studi Pertumbuhan Mangrove Pada Kegiatan Rehabilitasi Hutan Mangrove. *Agrifor Jurnal Ilmu Pertanian dan Kehutanan* **2014**, *13*, 11–18. (In Indonesian)
11. Department of Forestry. Buku Informasi 50 Taman Nasional di Indonesia. In *Book of Information on 50 National Parks in Indonesia*; Directorate General of Forest Protection and Nature Conservation: Jakarta, Indonesia, 2007. (In Indonesian)
12. Musa, M.; Handoyo, G.; Setyono, H. Peramalan Pasang di Perairan Pulau Karimunjawa, Kabupaten Jepara, Menggunakan Program "Worldtides". *J. Oceanogr.* **2013**, *3*, 1–7. (In Indonesian)
13. Data Sheet WorldView-2. Available online: https://dg-cms-uploads-production.s3.amazonaws.com/uploads/document/file/98/WorldView2-DS-WV2-rev2.pdf (accessed on 1 October 2016).
14. Updike, T.; Comp, C. Radiometric Use of WorldView-2 Imagery. Available online: https://www.digitalglobe.com/sites/default/files/Radiometric_Use_of_WorldView-2_Imagery%20%281%29.pdf (accessed on 1 August 2013).
15. Matthew, M.W.; Adler-Golden, S.M.; Berk, A.; Richtsmeier, S.C.; Levine, R.Y.; Bernstein, L.S.; Acharya, P.K.; Anderson, G.P.; Felde, G.W.; Hoke, M.P.; et al. Status of Atmospheric Correction Using a MODTRAN4-Based Algorithm. In Proceedings of the SPIE 4049 Algorithms for Multispectral, Hyperspectral, and Ultraspectral Imagery VI, Orlando, FL, USA, 23 August 2000.
16. Level 1 and Atmosphere Archive and Distribution System (LAADS) DAAC. Available online: https://earthdata.nasa.gov/eosdis/daacs/laads (accessed on 15 December 2020).
17. Boardman, J.W. Automated Spectral Unmixing of AVIRIS Data using Convex Geometry Concepts. In Proceedings of the Summaries of the 4th Annual JPL Air-Borne Geosciences Workshop, Pasadena, CA, USA, 25 October 1993; pp. 11–14.
18. Plaza, A.J.; Chang, C.I. *High Performance Computing in Remote Sensing*; Chapman and Hall/CRC: Boca Raton, FL, USA, 2008.
19. Kamal, M.; Arjasakusuma, S.; Adi, N.S. *JAZ EL-350 VIS NIR Portable Spectrometer Operational Guideline*; Remote Sensing Laboratory, Faculty of Geography Universitas Gadjah Mada: Yogyakarta, Indonesia, 2012.

20. Kamal, M.; Ningam, M.U.L.; Alqorina, F. The Effect of Field Spectral Reflectance Measurement Distance to the Spectral Reflectance of *Rhizophora stylosa*. In Proceedings of the IOP Conference Series: Earth and Environmental Science, Yogyakarta, Indonesia, 27–28 September 2017; Volume 98.
21. Kamal, M.; Ningam, M.U.L.; Alqorina, F.; Wicaksono, P.; Murti, S.H. Combining field and image spectral reflectance for mangrove species identification and mapping using WorldView-2 image. In Proceedings of the SPIE Remote Sensing 10790, Earth Resources and Environmental Remote Sensing/GIS Applications IX, 107901P, Berlin, Germany, 9 October 2018.
22. Wicaksono, P.; Fauzan, M.A.; Kumara, I.S.W.; Yogyantoro, R.N.; Lazuardi, W.; Zhafarina, Z. Analysis of reflectance spectra of tropical seagrass species and their value for mapping using multispectral satellite images. *Int. J. Remote Sens.* **2019**, *40*, 8955–8978. [CrossRef]
23. Kruse, F.A.; Lefkoff, A.B.; Boardman, J.W.; Heidebrecht, K.B.; Shapiro, A.T.; Barloon, P.J.; Goetz, A.F.H. The Spectral Image Processing System (SIPS)—Interactive Visualization and Analysis of Imaging Spectrometer Data. *Remote Sens. Environ.* **1993**, *44*, 145–163. [CrossRef]
24. Borengasser, M.; Hungate, W.S.; Watkins, R. *Hyperspectral Remote Sensing: Principles and Applications*; Taylor & Francis in Remote Sensing Applications, CRC Press: New York, NY, USA, 2008.
25. Chang, C.I. An Information-Theoretic Approach to Spectral Variability, Similarity, and Discrimination for Hyperspectral Image Analysis. *IEEE Tras. Inf. Theory* **2000**, *46*, 1927–1932. [CrossRef]
26. Clark, R.N.; Swayze, G.A.; Gallagher, A.; Gorelick, N.; Kruse, F.A. Mapping with imaging spectrometer data using the complete band shape least-squares algorithm simultaneously fit to multiple spectral features from multiple materials. In *Proceedings of the 3rd Airborne Visible/Infrared Imaging Spectrometer (AVIRIS) Workshop*; JPL Publication 91-28; Jet Propulsion Laboratory: Pasadena, CA, USA, 20–21 May 1991; pp. 2–3.
27. Congalton, R.G.; Green, K. *Assessing the Accuracy of Remotely Sensed Data: Principles and Practices*, 2nd ed.; CRC Press: Boca Raton, FL, USA, 2009.
28. Klančnik, K.; Gaberščik, A. Leaf spectral signatures differ in plant species colonizing habitats along a hydrological gradient. *J. Plant Ecol.* **2016**, *9*, 442–450. [CrossRef]
29. Andréfouët, S.; Kramer, P.; Torres-Pulliza, D.; Joyce, K.E.; Hochberg, E.J.; Garza-Pérez, R.; Muller-Karger, F.E. Multi-site evaluation of IKONOS data for classification of tropical coral reef environments. *Remote Sens. Environ.* **2003**, *88*, 128–143. [CrossRef]
30. Kamal, M.; Phinn, S.; Johansen, K. Object-Based Approach for Multi-Scale Mangrove Composition Mapping Using Multi-Resolution Image Datasets. *Remote Sens.* **2015**, *7*, 4753–4783. [CrossRef]
31. Hirano, A.; Madden, M.; Welch, R. Hyperspectral image data for mapping wetland vegetation. *Wetland* **2003**, *23*, 436–448. [CrossRef]
32. Rashmi, S.; Addamani, S.; Venkat, S. Spectral Angle Mapper Algorithm for Seagrass and Other Benthic Habitats in Bolinao, Pangasinan using Worldview-2 Satellite Image. In Proceedings of the Geoscience and Remote Sensing Symposium (IGARSS): IEEE International, Melbourne, Australia, 21–26 July 2013; pp. 1579–1582. [CrossRef]
33. Rahmandhana, A.D. Pemetaan Distribusi Jenis Mangrove Melalui Integrasi Citra WorldView-2 dan Pengukuran Spektrometer Lapangan di Pulau Karimunjawa dan Kemujan, Kabupaten Jepara [Mapping the Distribution of Mangrove Species through the Integration of WorldView-2 Image and Field Spectrometer Measurements in Karimunjawa and Kemujan Island, Jepara Regency]. Master's Thesis, Universitas Gadjah Mada, Yogyakarta, Indonesia, 14 September 2021. (In Indonesian).
34. Muhammad, S.; Mirza, M.W. Hyperspectral Mapping Methods for Differentiating Mangrove Species along Karachi Coast. *Int. J. Environ. Ecol. Eng.* **2013**, *7*, 963–965. [CrossRef]
35. Nidamanuri, R.R.; Zbell, B. A method for selecting optimal spectral resolution and comparison metric for material mapping by spectral library search. *Prog. Phys. Geog.* **2010**, *34*, 47–58. [CrossRef]
36. Shanmugam, S.; Srinivasaperumal, P. Spectral Matching Approaches in Hyperspectral Image Processing. *Int. J. Remote Sens.* **2014**, *35*, 8217–8251. [CrossRef]
37. Hyperspectral Analytics in ENVI Target Detection and Spectral Mapping Methods. Available online: http://www.spectroexpo.com/wp-content/uploads/2021/03/Hyperspectral_Whitepaper.pdf (accessed on 10 December 2020).

remote sensing

MDPI

Article

Global Mangrove Watch: Updated 2010 Mangrove Forest Extent (v2.5)

Pete Bunting [1,*], Ake Rosenqvist [2], Lammert Hilarides [3], Richard M. Lucas [1] and Nathan Thomas [4,5]

1 Department Geography and Earth Sciences, Aberystwyth University, Aberystwyth SY23 3DB, UK; richard.lucas@aber.ac.uk
2 solo Earth Observation (soloEO), Tokyo 104-0054, Japan; ake.rosenqvist@soloEO.com
3 Wetlands International, 6700 AL Wageningen, The Netherlands; Lammert.Hilarides@wetlands.org
4 Earth System Science Interdisciplinary Research Center (ESSIC), University of Maryland, College Park, MD 20740, USA; nathan.m.thomas@nasa.gov
5 Biospheric Sciences, NASA Goddard Space Flight Center, Greenbelt, MD 20771, USA
* Correspondence: pfb@aber.ac.uk; Tel.: +44-1970-622615

Abstract: This study presents an updated global mangrove forest baseline for 2010: Global Mangrove Watch (GMW) v2.5. The previous GMW maps (v2.0) of the mangrove extent are currently considered the most comprehensive available global products, however areas were identified as missing or poorly mapped. Therefore, this study has updated the 2010 baseline map to increase the mapping quality and completeness of the mangrove extent. This revision resulted in an additional 2660 km^2 of mangroves being mapped yielding a revised global mangrove extent for 2010 of some 140,260 km^2. The overall map accuracy was estimated to be 95.1% with a 95th confidence interval of 93.8–96.5%, as assessed using 50,750 reference points located across 60 globally distributed sites. Of these 60 validation sites, 26 were located in areas that were remapped to produce the v2.5 map and the overall accuracy for these was found to have increased from 82.6% (95th confidence interval: 80.1–84.9) for the v2.0 map to 95.0% (95th confidence interval: 93.7–96.4) for the v2.5 map. Overall, the improved GMW v2.5 map provides a more robust product to support the conservation and sustainable use of mangroves globally.

Keywords: mangroves; extent; mapping; sentinel-2; global mangrove watch

Citation: Bunting, P.; Rosenqvist, A.; Hilarides, L.; Lucas, R.M.; Thomas, T. Global Mangrove Watch: Updated 2010 Mangrove Forest Extent (v2.5). *Remote Sens.* **2022**, *14*, 1034. https://doi.org/10.3390/rs14041034

Academic Editor: Chandra Giri

Received: 19 January 2022
Accepted: 17 February 2022
Published: 21 February 2022

Publisher's Note: MDPI stays neutral with regard to jurisdictional claims in published maps and institutional affiliations.

1. Introduction

At the United Nations Framework Convention on Climate Change (UNFCCC) Conference of the Parties 26 (COP26) in 2021, an international agreement was made to end deforestation by 2030. To ensure adherence to this, accurate global scale maps of forested ecosystems will be critical. One such ecosystem is mangrove forests, which have witnessed an elevated rate of loss compared to terrestrial forests over the past decades [1] with regional losses exceeding 3%, driven by anthropogenic disturbances [1–3] such as conversion to aquaculture [4] or agriculture [5], urban expansion [6], oil palm plantations [7], and climate change [8]. Mangrove forests support a large number of ecosystem services [9], such as carbon storage and sequestration [10], coastal protection [11], food production [9], and tourism [12]. The ecosystem services of tidal mangroves and marshes were estimated to be worth 193,843 USD per hectare per year for 2007, equating to 25 trillion USD annually [13]. Accurate baseline maps of extent are therefore essential for a local and global ecosystem service accounting as well as verifying COP26 goals. Indeed, the ambitious goals set by the Global Mangrove Alliance (GMA), to restore 20% of mangrove forests by 2030, require accurate baselines upon which their efforts can be built. Furthermore, baseline maps are the keystone for mapping environmental descriptors that characterise this ecosystem, such as biomass [14], understanding the drivers of land cover change [2], and locating primary regions for potential restoration.

The Bunting et al. [15] Global Mangrove Watch (GMW) version 2.0 extent maps have emerged as the primary global dataset for characterising mangrove extent. There are a number of initiatives (e.g., GMA, GMW Portal; https://globalmangrovewatch.org; accessed 8 January 2022) that are aiming to preserve and restore mangroves and wider international objectives such as the UN Sustainable Development Goals (SDGs), for which the existing Global Mangrove Watch (GMW) version 2.0 layers [15,16] are a key dataset and are already used for reporting against. Currently, the GMW v2.0 [16] is the most up-to-date mangrove extent at the highest spatial resolution available. However, all mangrove datasets (i.e., [15–18]) published to date have areas that are missing (i.e., not mapped) or where mapping quality is poor. These limitations are evident globally and are caused by, for example, sensor specific characteristics (e.g., Landsat 7 Enhanced Thematic Mapper (ETM+) scan-line error), limited data availability, excessive cloud cover, or a combination of the above. These limitations degrade the performance of the map to meet the needs of the COP26 and GMA global initiatives by the year 2030.

For the GMW version 2.0, Bunting et al. [15] used two random forest classifiers to classify mangrove extent for the year 2010 from a combination of ALOS PALSAR and Landsat sensor data. As demonstrated by [15,19,20], the L-band radar data used in GMW v2.0 are sensitive to mapping mangrove change, while providing limited capability to classify the mangrove extent. However, optical remotely sensed data, particularly those with a Shortwave Infrared (SWIR) waveband, are well suited to the mapping of mangrove forest extent [21]. More recently, a number of studies (e.g., [21–25]) have made use of Sentinel-2 imagery and have demonstrated typical classification accuracies >90% for mangrove extent using ensemble machine learning classification approaches (e.g., random forests) through the Google Earth Engine platform. However, these studies have typically been undertaken over small spatial extents or for a few countries (e.g., [25]), single countries (e.g., [24]), or particular areas of interest at sub-national scales (e.g., [23]). Alternative approaches to mangrove mapping that have focused on mapping through time have also been proposed, such as [26,27], which have used the COntinuous monitoring of Land Disturbance (COLD) [28] method to provide individual site level time-series maps of mangrove extent. However, these time-series approaches are computationally intensive and therefore difficult to apply at a global scale. Nevertheless, they have demonstrated the ability of advanced machine learning and intensive computational processing for delivering maps at the quality required for international reporting.

The aim of this work was to produce an update to the 2010 Global Mangrove Watch version 2.0 [15] suitable for fully supporting the needs of ambitious global level targets relating to mangrove forest preservation and restoration. Specific regions were identified as missing or of poor quality within the GMW v2.0 product and a new method was, therefore, proposed for achieving vastly improved mapping, combining higher-resolution data at higher imaging cadence with advanced machine learning models. These results were combined with the GMW v2.0 map to create the most complete map of mangrove extent currently available and will form the basis of a subsequent study, updating the estimates of mangrove change.

2. Materials and Methods

The analysis was undertaken on the SuperComputing Wales (SCW) High Performance Computing (HPC) infrastructure using the Remote Sensing and GIS Library (RSGISLib) of tools [29], the KEA image format [30] and the pbprocesstools [31] workload management library to manage the workflow of tasks on the HPC.

2.1. Areas to Be Mapped

Through user feedback on the GMW v2.0 maps, 204 regions were identified to be either missing or poorly mapped. As detailed by Bunting et al. [15], the choice of 2010 for the baseline map was driven by the use of ALOS PALSAR data, coverage of which was most complete for the year 2010 [15]. However, this resulted in Landsat-5 TM and Landsat-7

ETM+ temporal composites affected by ETM+ scan-line error artefacts, particularly in areas of high cloud cover (e.g., Niger Delta). These artefacts were, in some cases, present with the GMW v2.0 maps. Areas identified that required (re-)mapping as part of this study are shown in Figure 1.

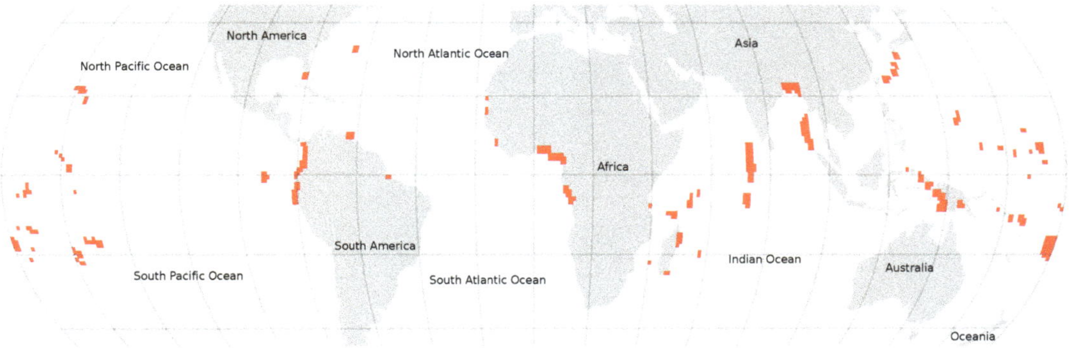

Figure 1. Map of regions identified as in need of updating as part of the Global Mangrove Watch (GMW) v2.5 analysis.

2.2. *Mangrove Habitat Mask*

Bunting et al. [15] developed a mangrove habitat mask which was used to limit the classification of mangroves to those where mangroves can be expected to be present. For example, mangroves must generally be close to water and at or close to mean sea level. However, this mask was found to have been too tight in a number of regions (e.g., Florida, USA) and therefore caused under-classification of mangroves. Ahead of the mapping, the habitat mask was therefore revised, primarily through manually digitising regions to be added, but also intersecting the maps of [17,18] to ensure that all regions mapped in those products were fully located within the GMW habitat mask.

2.3. *Mangrove Mapping*

Bunting et al. [15] used the most appropriate and available data (i.e., Landsat TM and ETM+ and ALOS PALSAR) for 2010 in the original mapping (v2.0). Therefore, to improve the mapping, an approach that used alternative datasets was required. In this case, Sentinel-2 imagery was used to map the areas outlined in Figure 1. For the analysis, 100 global mangrove/non-mangrove XGBoost classifiers [32] were trained and applied to each Sentinel-2 acquisition. To produce a single unified classification, results from all acquisitions and classifiers were merged to create a probability for each pixel to be mangroves. This probability surface was then thresholded to produce the binary map, which was subject to a manual Quality Assurance (QA) process to produce the Sentinel-2-derived maps. These were then combined with the v2.0 2010 baseline and a change detection using the 2010 ALOS PALSAR data was applied to create a revised GMW v2.5 2010 baseline. The processing stages are outlined in Figure 2.

2.3.1. Sentinel-2 Processing

The Sentinel-2 imagery was downloaded from the Google Cloud public dataset [33] and, for the purpose of this work, processed to a 20-m pixel resolution (i.e., the 10-m resolution Sentinel-2 image bands were resampled to 20 m using averaging) orthorectified standardised surface reflectance product. For the 383 Sentinel-2 granules identified as intersecting with the regions of interest (Figure 1), individual acquisitions were selected for download based on cloud cover. Initially, the 10 acquisitions with the lowest cloud cover (maximum cloud cover of 20%) were identified from the entire Sentinel-2 archive (2015–2020). An iterative process was then followed where, for each granule, the scenes were downloaded and processed to produce a cloud mask. The combined cloud masks were checked to estimate whether data were available for all mangrove regions within the scene, as defined by the mangrove habitat mask. If more acquisitions were required, the thresholds for the number of acquisitions and maximum cloud cover were increased. These were increased to a maximum of 100 scenes and a maximum cloud threshold of 75%. In total, 11,262 Sentinel-2 acquisitions were downloaded and used for this analysis.

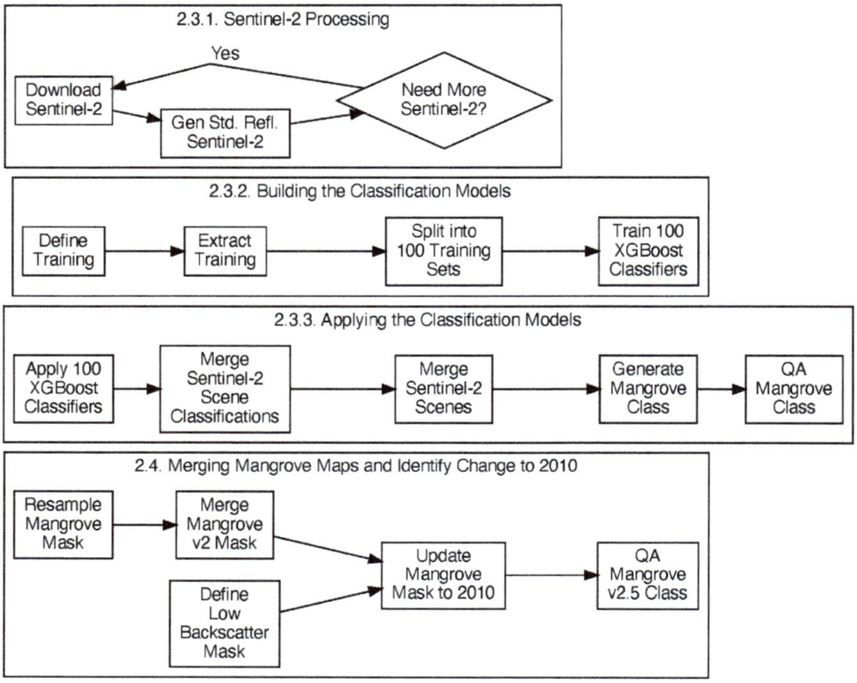

Figure 2. Flowchart of the methods used to generate the GMW v2.5 map for 2010.

To generate a standardised reflectance product for the classifications, the ARCSI software [34] was employed, as successfully demonstrated in past studies [15,26,27,35]. The ARCSI software uses the 6S model [36] through the Py6S module [37] parameterised using the image header information and an aerosol optical depth estimated from a dark object subtraction [15]. Using the method of Shepherd and Dymond [38], the resulting images were normalised for a sensor view angle and local topography producing a standardised reflectance product. A tropical atmosphere and maritime aerosol profile was used for all scenes.

Cloud masking was undertaken using the product of two approaches. The FMask [39,40] algorithm was applied using the Python-FMask implementation [41]. The s2cloudless [42] LightGBM classifier [43] was also applied to each scene. The resulting s2cloudless classification was further refined using a morphological closing (with a 5×5 circular operator) followed by a morphological dilation (with a 7×7 circular operator). Finally, any cloud objects were removed if they were less than 10 pixels in size. The final cloud mask was defined as the intersection of the two masks. The cloud shadow mask was derived using the approach implemented within FMask, as described in Zhu et al. [39].

Finally, a 'clear sky' mask was derived for each acquisition, defining the areas of the scene to be used for further analysis. The 'clear-sky' mask aims to identify the larger continuous parts of the image, removing small areas between clouds. The first step buffered the cloud and cloud shadows by 30 km, clumping the remaining non-cloud regions. The non-cloud clumps with an area greater than 3000 pixels were then selected and grown to the 10-km contour of the cloud and cloud shadow pixels. An example of the 'clear sky' mask is shown in Figure 3c.

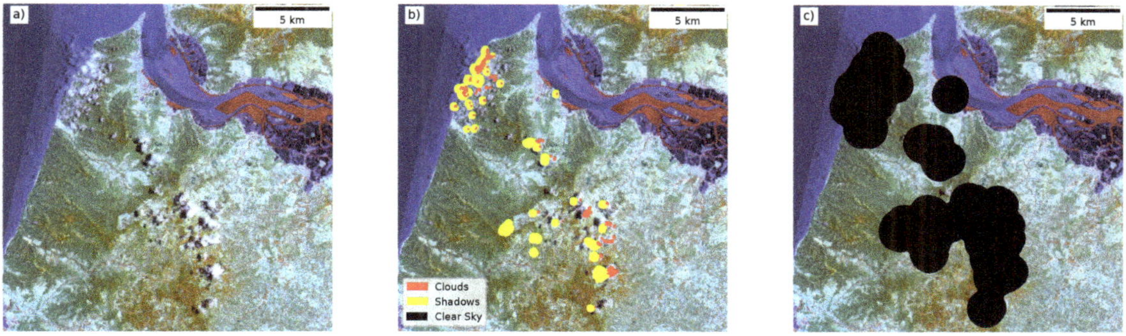

Figure 3. An example of the clear sky mask for part of a Sentinel-2 scene, where (**a**) is the original scene (false colour: near infrared (NIR), shortwave infrared band 1 (SWIR-1), and red bands) and (**b**) is the resulting cloud and cloud shadow mask, which can been seen to have missed some clouds, and (**c**) is the clear sky mask which has masked the regions around the cloud and cloud shadows.

2.3.2. Building the Classification Models

Classification of mangrove extent was undertaken on a scene-by-scene basis rather than through the creation of image composites (i.e., merging multiple scenes using a metric such as the greenest pixel). Image composites, whilst relevant for visualisation, often have artefacts due to prevailing environmental conditions (e.g., wet or dry season or, in the case of mangroves, tidal regimes) at the time of the acquisitions or processing errors (e.g., missed cloud or cloud shadows). These artefacts can then impact the classification result. An alternative is to classify each of the scenes independently and then merge those results to create a single map.

To derive training data for the classification, 10,284 samples were created from the existing GMW v2.0 map. These were manually checked against the Sentinel-2 imagery. For regions not already within the GMW v2.0 product training regions, these were manually defined. Non-mangrove regions were defined as regions outside of the GMW habitat product, with points sampled randomly within this region and through manual selection of regions giving a total of 52,555 sample points for training.

The resulting samples were then intersected with all 11,262 Sentinel-2 acquisitions, with each scene masked to the relevant valid clear sky area. This resulted in 4,421,644 mangrove and 9,830,388 non-mangrove pixel values to train the classifier. Given the volume of sample data available, it was decided to split the training data into 100 sets, each with 400,000 samples (200,000 for mangroves and 200,000 for non-mangroves). Those samples were then split into 3 sets for training (100,000 for each class), testing (50,000 for each class), and validating (50,000 for each class) the model.

The XGBoost [32] binary classification algorithm was used for the analysis given its ability to use large training datasets and allow transfer learning (i.e., further training of an existing model). This method has been shown by John et al. [35] to provide good results for the classification of land cover from Earth observation data. To optimise the hyper-parameters of the XGBoost model, a subset of 20% was selected from the training (20,000 per class) and validation (10,000 per class) samples. Bayesian optimisation was used to identify the optimal hyper-parameters for each of the 100 classifiers. The range of values for the parameters optimised is given in Table 1. Following identification of the hyper-parameters, each of the 100 models was trained using the full dataset (i.e., 400,000 samples). The testing accuracies of the models (using the 50,000 samples per class) were between 97–99%.

2.3.3. Applying the Classification Models

To apply the 100 global XGBoost classifiers to the individual Sentinel-2 acquisitions, the models were first further trained using the local training data from the Sentinel-2 acquisition, which was limited to 25,000 samples. This allowed the global classifier to be locally optimised for the individual acquisitions. The classifiers were then applied to all the acquisitions, with this creating 112,620 classifications. To avoid incorrect classification of mangroves in areas where they would not be located (e.g., in mountainous areas), the classification was only applied within an updated version of the mangrove habitat layer of Bunting et al. [15].

Table 1. The range of hyper-parameter values from the 100 models.

Parameter	Min.	Low. Quartile	Median	Up. Quartile	Max.	Mode
eta	0.0835	0.236	0.299	0.299	0.489	0.299
gamma	0	0	0	4	4	0
max_delta_step	0	6	7	7	10	7
max_depth	8	13	13	16	20	13
min_child_weight	1	10	10	10	10	10
num_boost_round	68	100	100	100	100	100
subsample	0.752	0.818	0.839	1.0	1.0	0.818

The individual classifications were then merged in two steps to create a mangrove probability for each pixel. The first step merged the 100 classifications applied to a scene to create a single probability output image for the scene. The probability was calculated as the number of times each pixel had been classified as mangroves (i.e., a value of 1 meant that all 100 classifiers classified the pixel as mangroves, while a value of 0.1 meant that only 10 classifiers classified the pixel as mangroves). The second step calculated the mean probability from all the acquisitions for each pixel, providing a single probability surface for all the areas mapped.

To derive the final binary mask of mangrove extent, a global threshold was applied to the probability surface. The threshold was identified through a sensitivity analysis using the mangrove samples based on the 0.1 increments (from 0.2 to 0.8). A mangrove mask was generated for each threshold where the mask with the best agreement with the mangrove samples used to train and test the XGBoost classifiers selected. A threshold of 0.5 provided the greatest correspondence and was therefore applied to all the regions updated using the Sentinel-2 imagery. For studies focused on specific regions, a further local optimisation could be undertaken by selecting a local threshold. However, for this study, a global threshold was applied as defining local regions would be difficult and could result in boundary artefacts within the resulting maps.

Finally, a visual assessment of the mangrove extent was undertaken where polygons identifying regions as incorrectly classified as mangroves were digitised with reference to the Sentinel-2 imagery and high-resolution Google Earth, Mapbox Satellite, and Bing maps imagery. The areas were then removed from the mangrove extent mask.

2.4. Merging Mangrove Maps and Identify Change to 2010

In addition to the new map produced from the Sentinel-2 analysis, two other mangrove maps were used to resolve issues for particular areas. For the Sundarbans, in India and Bangladesh, the mapping of Awty-Carroll et al. [26] for the year 2010 was added to the Sentinel-2 maps to be merged with the GMW v2.0 products. The Sundarbans were significantly affected by stripping from the Landsat ETM+ data within GMW v2.0. Additionally, mangrove maps for the French overseas territories, where there was found to be a high prevalence of cloud cover that reduces the availability of useable Sentinel-2 data, generated by the French National Mangrove Observation Network [44], were used to improve the new maps.

Following generation of the revised maps, these were merged with the existing GMW v2.0 baseline for 2010 to create the updated 2010 GMW v2.5 baseline. However, the updated areas had been mapped with data acquired over the period from 2015 to 2020 and a change detection was therefore required to backcast the map for 2010. 2010 ALOS PALSAR data were used for this and therefore the new mapping was resampled (nearest neighbour) onto the same 0.000222 degrees (∼25 m) pixel grid of the GMW v2.0 and ALOS PALSAR data layers.

As demonstrated by Thomas et al. [3,19,20], mangroves produce a high backscatter response in the L-band SAR data while the majority of non-mangrove surfaces (e.g., water bodies and mudflats) have a low L-band backscatter. As a result, there is a change trajectory between mangroves and non-mangroves, which was used by Thomas et al. [20] as the basis for a methodology for mapping mangrove change. This was applied globally to produce the GMW v2.0 change layers [16].

For implementation, a low backscatter mask was created for 2010 and used to remove mangroves that were within the new map but not present in 2010. The mask was defined using a combination of the ALOS PALSAR 2010 layer and the Landsat-based Pekel et al. [45] water occurrence layer generated for the period 1984–2020. The analysis was undertaken on a 1 × 1 degree grid, where the water occurrence layer was used to define areas that could be considered as 'permanent' waterbodies, defined as a water occurrence between >90 and <100. However, if no pixels were identified, then the threshold was lowered to >70. For the pixels associated with 'permanent' waterbodies, the 99th percentile of the SAR backscatter was calculated for both the Horizontal-Horizontal (HH) and Horizontal-Vertical (HV) polarisations. The thresholds for classifying the water extent were then calculated for both polarisations as:

$$\text{SAR threshold} = \text{99th percentile} - (0.15 \times \text{99th percentile}). \tag{1}$$

If no 'permanent' waterbody pixels were identified, then the SAR thresholds where defined as −14 dB in the HH and −17 dB in the HV polarisations. To produce the low backscatter mask, the SAR backscatter was thresholded with values below those calculated above were used and the water occurrence layer had a value > 5.

The low backscatter mask was then used to mask all the tiles, including areas which have not been remapped, updating the mangrove mask and aligning it with the ALOS PALSAR data for 2010. Finally, a Quality Assurance (QA) process was undertaken where the product was visually assessed against a variety of image sources, including high-resolution Google Earth, Mapbox Satellite and Bing Maps imagery, the Sentinel-2 data, 2010 ALOS PALSAR, and 2010 Landsat imagery data. Polygons were manually drawn for regions which should be removed from the map (i.e., not mangroves but areas that had been mapped as mangroves) or added to the map (i.e., mangroves but areas that had not been mapped as such). These QA edits were then rasterised and applied to the map producing the final GMW v2.5 layer.

2.5. Accuracy Assessment

To assess the accuracy of the new v2.5 layer, 26 sites (Figure 4) where new mapping had occurred were selected, representing a range of different mangrove settings, types, and extents. Additionally, a further 34 sites (Figure 4) were distributed globally for assessing the overall product accuracy. For each site, an area of 0.2 × 0.2 degrees was defined and 1000 random stratified points were defined for each class (mangroves and non-mangroves). If there were less than 1000 mangrove pixels within the 0.2 × 0.2 degree area then all mangrove pixels were defined as points and the number of mangrove reference points was reduced. The 2000 points were then split into 200 point sets (i.e., 100 mangrove and 100 non-mangrove) where the sets were assessed in turn until the 95% confidence interval for the macro F1-score was <5%. A minimum of 3 sets (i.e., 600 points) were assessed for each site, where typically 5 sets were required (1000 points) although 10 sets were used for one site. Points were manually annotated with a reference class through a combination of high-resolution Google Earth, Mapbox Satellite and Bing Maps imagery, the Sentinel-2 and 2010 ALOS PALSAR, and Landsat imagery data. In total, 50,750 points were assessed and used for the accuracy assessment. For sites where the mapping was updated, the points were also used to assess the improvement in map accuracy achieved through this study.

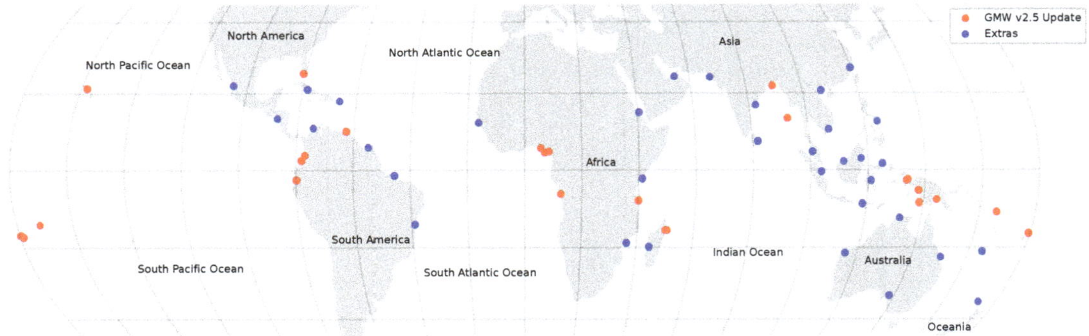

Figure 4. A map of the 60 sites used for the accuracy assessment. The 26 red points are over areas which have been mapped with Sentinel-2 as part of the GMW v2.5 analysis while the 34 blue points are further set of sites used to capture the global accuracy of the GMW v2.5 baseline rather than just the areas updated.

3. Results

3.1. Remapped Regions Comparison

To compare the accuracy of the updated v2.5 and v2.0 GMW 2010 baselines, the reference points for the 26 sites where the baseline has been updated were intersected with both layers. Summary statistics calculated were an overall accuracy, cohen kappa, and F1 score (per-class and overall), with summaries are provided in Tables 2 and 3. In addition, upper and lower confidence intervals for all metrics were calculated using bootstrapping. It was not possible to calculate metrics, such as the allocation and quantity disagreement as those metrics require a closed map where all pixels are allocated to a class such that the area of the whole region can be used to normalise. However, for this study we only have a single class of interest (i.e., mangroves).

As shown in Tables 2 and 3, the estimated accuracy of mapping in regions where the quality was identified previously as poor or missed increased from 82.6% (80.1–84.9) to 95.0% (93.7–96.4). The range for the individual site accuracies also decreased from 44.7% to 12.4%, demonstrating that the quality of mapping for these areas remapped resulted in a similar quality of mapping for all regions.

Table 2. For the region mapped to create v2.5, this table contains an overview of the accuracy statistics from the GMW v2.0 baseline, which can be compared to the statistics in Table 3.

Stats (GMW V2.0)	Overall	95th Confidence	Site Median	Site Min	Site Max
Overall (%)	82.6	80.1–84.9	84.8	51.6	96.3
Kappa	0.655	0.607–0.698	0.734	0.393	0.927
Macro F1-Score	0.824	0.799–0.847	0.866	0.334	0.963
Mangrove F1-Score	0.807	0.777–0.834	0.848	0.603	0.964
Mangrove Recall	0.706	0.665–0.746	0.774	0.437	0.976
Mangrove Precision	0.942	0.916–0.964	0.952	0.747	0.994
Other F1-Score	0.841	0.817–0.860	0.885	0.668	0.964
Other Recall	0.953	0.934–0.971	0.976	0.628	1.0
Other Precision	0.752	0.717–0.787	0.818	0.502	0.974

This improvement in mapping accuracy can be seen visually and is illustrated in Figures 5–7. Figure 5a provides a typical example of a region that was affected by the Landsat ETM+ striping but was remapped to improve the output Figure 5b. Figure 6a illustrates an area in Colombia where some areas of mangroves were missed but have now been mapped in GMW v2.5 (Figure 6b). Figure 7 illustrates an example where the habitat mask was too restricted in GMW v2.0 but has been improved within GMW v2.5 by expanding the habitat mask. In terms of the accuracy statistics (Table 3), the example shown in Figures 6 and 7 represents a region where the accuracy will have significantly improved, while Figure 5 resulted in only a modest statistical improvement but is visually much improved.

Table 3. For the region mapped to create v2.5, this table contains an overview of the accuracy statistics from the GMW v2.5 baseline, which can be compared to the statistics in Table 2.

Stats (GMW V2.5)	Overall	95th Confidence	Site Median	Site Min	Site Max
Overall (%)	95.0	93.7–96.4	96.3	87.4	99.8
Kappa	0.901	0.874–0.928	0.926	0.748	0.995
Macro F1-Score	0.951	0.937–0.964	0.963	0.872	0.997
Mangrove F1-Score	0.951	0.938–0.965	0.963	0.887	0.998
Mangrove Recall	0.936	0.901–0.915	0.982	0.803	1.0
Mangrove Precision	0.951	0.951–0.982	0.973	0.83	1.0
Other F1-Score	0.950	0.935–0.963	0.962	0.857	0.997
Other Recall	0.966	0.949–0.982	0.974	0.854	1.0
Other Precision	0.934	0.911–0.956	0.948	0.756	1.0

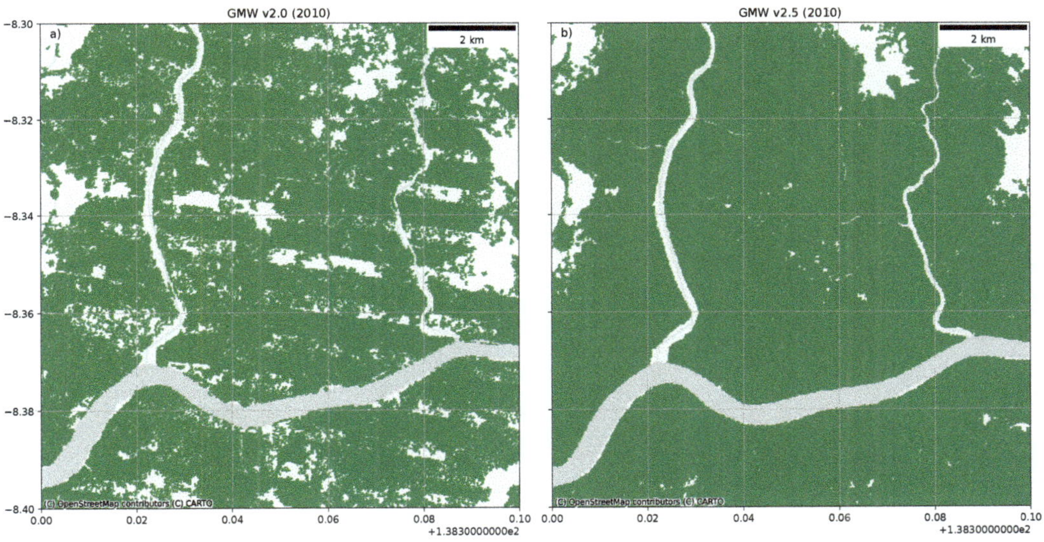

Figure 5. Comparison of GMW v2.0 (**a**) and v2.5 (**b**) products illustrated with an example from West Papua, Indonesia, where remapping with Sentinel-2 removed artefacts from Landsat ETM+ in GMW v2.0.

3.2. Overall Accuracy Assessment

Using all 60 sites, the overall accuracy statistics for the v2.5 map was calculated and presented in Table 4. The global assessment estimated an overall accuracy of 95.1% with a 95th confidence interval (i.e., 95% likelihood that the true value is within the range) of 93.8 and 96.5%. This was similar to those published by Bunting et al. [15] for v2.0, which estimated an overall accuracy of 94.0% with a 99th confidence interval of 93.6 and 94.5%. This is to be expected with only approximately 33% of the map having been remapped (i.e., replaced) and with only minor changes masking low backscatter pixels applied to all regions alongside the overall high estimated accuracy of the v2.0 map. However, as demonstrated in Table 2, the local accuracy of the v2.0 map could be as low as 51% where areas were missed.

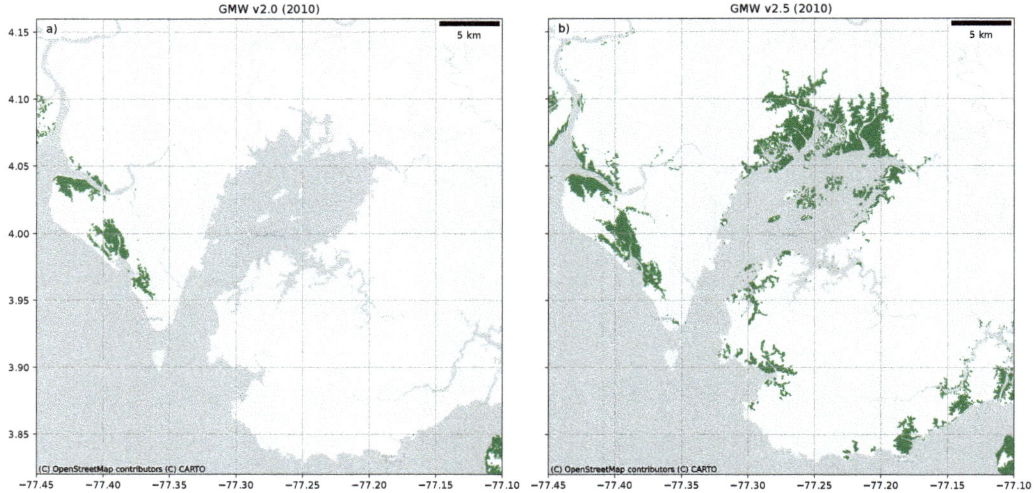

Figure 6. Comparison of GMW v2.0 (**a**) and v2.5 (**b**) products, illustrated with an example from Colombia, where regions had been omitted in GMW v2.0 but included in GMW v2.5.

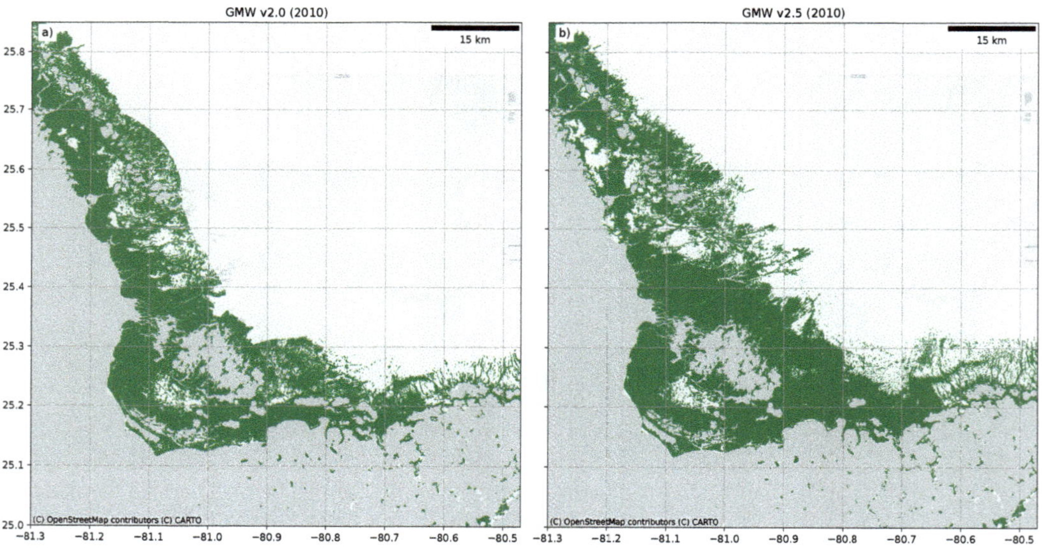

Figure 7. Comparison of GMW v2.0 (**a**) and v2.5 (**b**) products, illustrated with an example from Florida, USA, where the habitat mask was too restricted when used in the production of GMW v2.0 which has been improved for the GMW v2.5 product.

3.3. Area Statistics

The global mangrove extent mapped in v2.5 was 140,260 km^2, an increase of 2660 km^2 (2.5%) over the v2.0 GMW map, which had a global total of 137,600 km^2. Table 5 provides a range of example countries, some with significant changes in mangrove extent between v2.5 and 2.0. A full country table of mangrove extents for v2.5 and v2.0 has been provided in Appendix A Table A1.

Table 4. The overall GMW v2.5 accuracy assessment summary.

Stats (GMW V2.5)	Overall	95th Confidence	Site Median	Site Min	Site Max
Overall (%)	95.1	93.8–96.5	96.5	77.8	99.8
Kappa	0.902	0.876–0.930	0.930	0.556	0.995
Macro F1-Score	0.951	0.938–0.965	0.964	0.768	0.997
Mangrove F1-Score	0.951	0.937–0.964	0.964	0.720	0.998
Mangrove Recall	0.956	0.937–0.973	0.984	0.803	1.0
Mangrove Precision	0.947	0.926–0.964	0.969	0.570	1.0
Other F1-Score	0.952	0.938–0.965	0.966	0.817	0.997
Other Recall	0.947	0.923–0.966	0.968	0.696	1.0
Other Precision	0.956	0.938–0.973	0.986	0.756	1.0

Table 5. Example country statistics illustrating the changes between GMW v2.5 and v2.0. A full table has been provided in Appendix A (Table A1).

Country	GMW v2.5 (ha) 2010	GMW v2.0 (ha) 2010	Diff (ha)	Diff (%)
Angola	28,969	37,346	−8377	−22.4
Australia	988,842	1,006,021	−17,179	−1.7
Bahrain	59	82	−23	−28.4
Bangladesh	444,159	416,283	27,876	6.7
Benin	3390	84	3307	3957.0
Bermuda	8	0	8	–
Colombia	262,212	231,187	31,025	13.4
Fiji	49,984	51,166	−1182	−2.3
French Guiana	59,466	50,187	9279	18.5
Indonesia	2,801,795	2,688,955	112,840	4.2
Mauritius	345	703	−357	−50.9
Mozambique	298,841	301,125	−2284	−0.8
Nigeria	847,894	695,775	152,119	21.9
Papua New Guinea	445,785	476,167	−30,382	−6.4
United States	209,544	193,609	15,935	8.2
Global	**14,025,986**	**13,760,074**	**265,912**	**1.9**

Within the GMW v2.5 data, there are 121 countries with mangroves. Twelve countries, including Bermuda, were missing from the GMW v2.0, however they (and their areas) have now been added to the GMW v2.5 dataset. These were mostly small island nations where persistent cloud cover limits the acquisition of useable remote sensing data. The observed changes at a national level are variable, with 15 countries (e.g., Mozambique) having mangrove area differences of less than 1% between GMW v2.5 and v2.0, while 50 countries (e.g., Australia) had between 1–5% of change. The small changes between the two maps were attributed to the low backscatter pixel mask that was applied to all tiles. However, for countries with a small area of mangroves, these changes can be significant in percentage terms. For example, the area of mangroves mapped in Bahrain was 28% greater in the GMW v2.0.

Of the remaining 44 countries, 18 had a net change between the GMW v2.5 and v2.0 between 5–10% of their mangrove area, with 11 between 10–20% and 10 between 20–50% and 5 with a net change greater than 50% of the GMW v2.0 area. Many of the countries with the largest change area were those with small mangrove extents (e.g., Bahrain or Mauritius). However, in some regions remapped with Sentinel-2, substantial areas were either removed from the GMW v2.5 map (e.g., Angola had 8377 ha of fewer mangroves; Figure 8) or were added (e.g., Benin had 3307 ha more mangroves; Figure 9), with this improving the mangrove mask accuracy for these regions. In the GMW v2.0 map, a number of areas (e.g., Florida; Figure 7) were omitted because of the restricted GMW habitat mask, which was used to limit areas where mangroves could be classified. Improvements in this mask along with the remapping effort has allowed new areas to be included within the GMW v2.5 map.

However, some regions were found to have a mixture of substantial omissions and commissions within the v2.0 dataset. For example, Fiji, which was remapped with Sentinel-2, had an overall net change of −2.3% between v2.0 and v2.5. However, there were also significant regions of additional mangrove within Fiji in v2.5 as a processing error in the v2.0 product caused the mangroves in the west of the island nation (i.e., −180–−178) to be missed.

The improvement in mapping through the use of the Sentinel-2 data was significant in areas of high cloud cover and particularly in regions such as French Guiana (18.5%), Papua New Guinea (−6.4%), Nigeria (21.9%), and Colombia (13.4%; Figure 6). These areas had often significant striping artefacts present from the use of Landsat ETM+ data in the GMW v2.0 map (e.g., Figure 5).

These changes in the mapped mangrove area are not due to changes on the ground but rather to better input data (i.e., Sentinel-2) or new knowledge (e.g., improvements to the habitat mask) that have allowed us to generate a more accurate mangrove map for 2010.

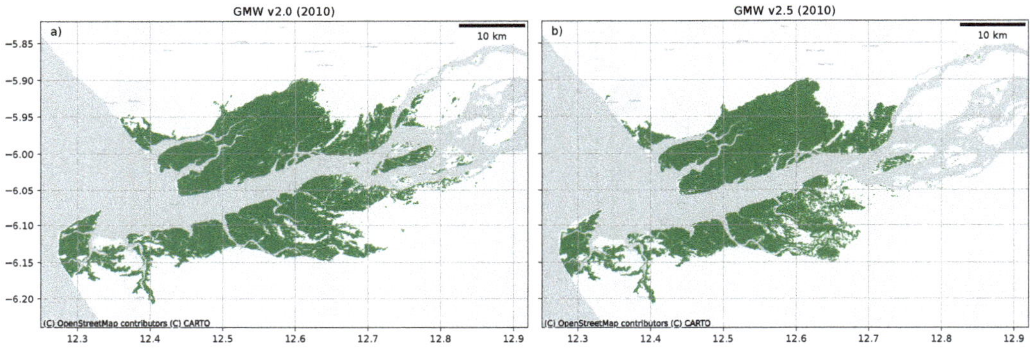

Figure 8. Comparison of GMW v2.0 (**a**) and v2.5 (**b**) products, illustrated with an example from Angola, where there were errors of commission within the GMW v2.0 product which were improved for the GMW v2.5 product.

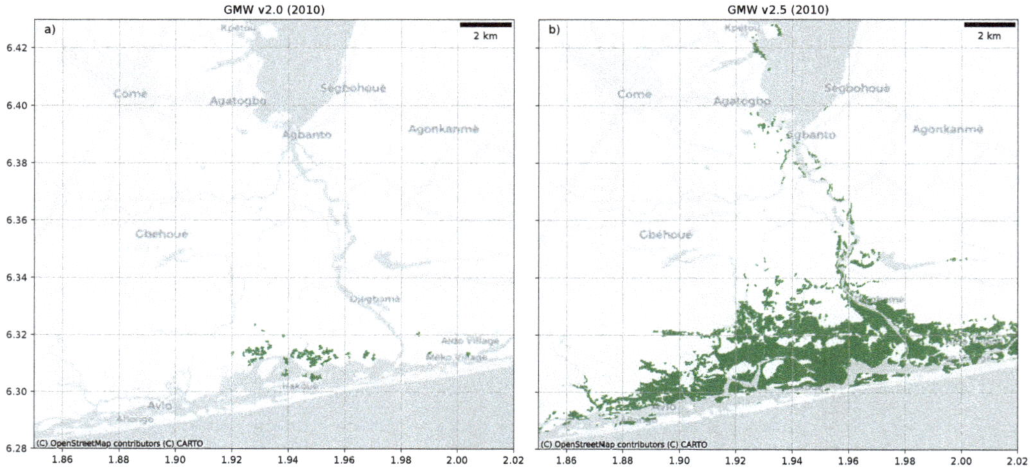

Figure 9. Comparison of GMW v2.0 (**a**) and v2.5 (**b**) products, illustrated with an example from Benin, where regions had been omitted in GMW v2.0 but included in GMW v2.5.

4. Discussion

To meet the requirements of global initiatives to achieve ambitious targets on the preservation and restoration of forested ecosystems, accurate and timely maps of the extent are critical. To date, the GMW has produced the most contemporary and comprehensive maps of global mangrove ecosystems. However, poorly mapped and omitted regions were present in the v2.0 dataset. We successfully identified 204 regions in need of updating or inclusion and proposed a new method to refine the maps in the selected locations. Our approach was able to increase the low accuracy of the map in these regions from 82.6% to 95.0%, bringing them in line with the level of overall accuracy of the global map. This updated map is better suited to meet the needs of the COP26 goal of ending deforestation by 2030 and the GMA goal of restoring 20% of mangroves by 2030. Accurate baselines are critical to measuring the success of such ambitious targets and ensuring accountability in reporting. With the accuracy of the updated regions mapped here increased by approximately 10%, GMW v2.5 is situated as the primary global scale mangrove extent product.

4.1. Data and Methods

As outlined by Thomas et al. [20], radar data used for the GMW v2.0 are limited in its ability to discern mangrove extent. Here, Sentinel-2 was relied upon for high-resolution high-cadence imagery, with spectral bands suited to wetland vegetation mapping. Sentinel-2 data are acquired as often as once every 5 days and do not suffer from the instrument degradations that impacted the Landsat 7 ETM+ imagery used for generating the GMW v2.0 2010 baseline. This provides a dense stack of imagery from which to derive a baseline map for afflicted regions. As an improvement over GMW v2.0, image composites were not used as these can result in image artefacts and inconsistent imaging conditions between images which can lead to classification confusion. Instead, each image was classified separately and a probabilistic approach was used to determine the mangrove extent. This is a more robust approach as it can provide additional minimum and maximum bounds and thus allows flexibility on definitions of extent and transparency on uncertainties. Despite this, errors do persist in specific locations. While the overall accuracy of the map was 95.1%, the ranges on a site-by-site basis was from 87.4% to 99.8%. In some locations, this was caused by the limitations of using moderate spatial resolution (i.e., 20–30-m resolution Sentinel-2, Landsat, and ALOS PALSAR) imagery to map very fine fringes and fragmented stands (e.g., Figure 10), that often are associated with human disturbance. These areas are challenging to identify and interpret and reliably differentiate from other vegetation types, even using very high spatial resolution imagery (i.e., <3 m). Access to local knowledge and field data is increasingly important for achieving high quality results in these locations.

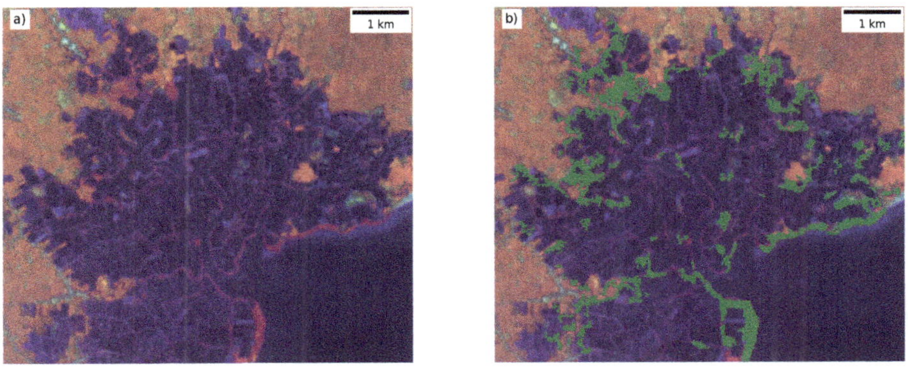

Figure 10. Example of fragmented mangroves from Sulawesi, Indonesia. (**a**) 30-m Landsat 8 imagery from 2016 (false colour: NIR, SWIR-1, and Red bands) and (**b**) 30-m Landsat 8 imagery from 2016 overlain with the GMW v2.5 baseline (green), illustrating the resulting v2.5 map for these fragmented areas of mangroves, which only maps the larger regions for mangroves and not the finer detail.

The use of the XGBoost classifier [32] provided the use of a gradient boosted decision tree, which can take advantage of larger training datasets than alternative methods such as random forests. This has advantages over the random forest algorithm used in GMW v2.0, by improving upon the single model through use of an iterative approach and training ensemble models in succession, with each new model correcting errors in the previous one. This is considered a more robust approach to ensemble learning. Our updated method also used a unique approach to classifier training, by using combinations of both local and global training data in order to derive results that are locally tuned but are also representative of the global mangrove extent.

4.2. Future Baseline Mapping

Going forward, we advocate that a new global baseline at a 10-m spatial resolution using Sentinel-2 be produced. Such a baseline would align well with other global datasets produced using Sentinel-2 and Landsat imagery. JAXA are also reprocessing the ALOS PALSAR and ALOS-2 PALSAR-2 products such that they align with this spatial baseline. Additionally, using Sentinel-2 data, this study and others (e.g., [21–25]) have demonstrated that optical remotely sensed data with a shortwave infrared channel (e.g., Landsat and Sentinel-2) can provide a reliable classification of mangroves and differentiation from other vegetation types in most regions. Regions where other wet tropic forests share a boundary with mangroves (e.g., Papua New Guinea and French Guiana) are however still challenging for classification. The additional spatial resolution (i.e., 10 m) should further help discriminate small mangrove patches, such as river edges and where disturbance and loss/gain patterns are complex. The spatial registration of the radar data used in GMW v2.0 has been found to have 1 or 2 pixels (i.e., 25–50 m) of mis-alignment with the latest Landsat (Collection 1 and 2) and Sentinel-2 datasets. The creation of a low backscatter mask is thought to align the remapped areas with the GMW v2.0 datasets. However, by using the ALOS PALSAR data alone, discrimination of mangroves from other vegetation types is poor, so the uncertainty in mapping the landward extent of mangroves is greater, with this also being the case for the GMW v2.0 dataset. JAXA are set to reprocess their L-band SAR data in order to alleviate this registration issue. Thus, future mapping efforts should maximise the use of these datasets when available.

5. Conclusions

The paper presented an updated version of the GMW mangrove extent baseline for 2010, producing version 2.5 of this dataset. The update focused on particular areas that were identified as being poor in quality (e.g., due to Landsat ETM+ stripping) or were missed in the GMW v2.0 product. The analysis demonstrated an increase in overall accuracy for updated regions from 82.61% (95th confidence interval 80.1–84.9%) for the GMW v2.0 product to 95.0% (95th confidence interval 93.7–96.4%) for the GMW v2.5 product. To our knowledge, this renders the GMW v2.5 baseline for 2010 as being the most complete global map of mangrove extent. This baseline will be used as the basis for an update to the GMW v3.0 change product, also extending the period from 1996–2016 to 1996–2020, which will be the subject of a forth-coming publication.

Author Contributions: Conceptualization, P.B., A.R., L.H., R.M.L. and N.T.; methodology, P.B., A.R. and N.T.; software, P.B.; validation, P.B., A.R., L.H., R.M.L. and N.T.; writing—original draft preparation, P.B.; writing—review and editing, P.B., A.R., L.H., R.M.L. and N.T. All authors have read and agreed to the published version of the manuscript.

Funding: The Global Mangrove Watch is funded by the Oak Foundation, the COmON Foundation, the National Philanthropic Trust, DOB Ecology, and the Dutch Postcode Lottery. This research was also funded by the Natural Environment Research Council (NERC) through the UKRI Newton Fund (NE/P014127/1) and the Japan Aerospace Exploration Agency (JAXA).

Institutional Review Board Statement: Not applicable.

Informed Consent Statement: Not applicable.

Data Availability Statement: The scripts for the data analysis are available on GitHub: https://github.com/globalmangrovewatch/gmw_gap_fill_2020 (accessed: 8 January 2022). The output GIS datasets are available here: https://doi.org/10.5281/zenodo.5828339 (accessed: 8 January 2022).

Acknowledgments: We acknowledge the support of the Supercomputing Wales project, which is partly funded by the European Regional Development Fund (ERDF) via the Welsh government, for providing the computing infrastructure for undertaking this study. We would also like to thank IUCN-France for the provision of the mangrove maps over the French overseas territories. We also thank all those who have provided feedback on the GMW v2.0 dataset. Tristram Irvine-Fynn is also thanked for reviewing the manuscript before submission.

Conflicts of Interest: The authors declare no conflict of interest.

Appendix A. National Mangrove Extent

Table A1. Country level 2010 mangrove extents for both GMW v2.5 and v2.0.

Country	GMW v2.5 (ha) 2010	GMW 2.0 (ha) 2010	Difference (ha)	Difference (%)
American Samoa	33	19	14	74.8
Angola	28,969	37,346	−8377	−22.4
Anguilla	1	1	0	−0.9
Antigua and Barbuda	863	889	−25	−2.9
Aruba	26	34	−7	−22.1
Australia	988,842	1,006,021	−17,179	−1.7
Bahamas	93,139	99,318	−6179	−6.2
Bahrain	59	82	−23	−28.4
Bangladesh	444,159	416,283	27,876	6.7
Barbados	14	14	0	−1.8
Belize	44,507	44,784	−277	−0.6
Benin	3390	84	3307	3957.0
Bermuda	8	0	8	–
Bonaire, Sint Eustatius and Saba	165	166	−2	−1.0
Brazil	1,081,106	1,107,207	−26,101	−2.4
British Virgin Islands	83	90	−7	−7.7
Brunei	11,491	11,163	328	2.9
Cambodia	58,517	59,230	−714	−1.2
Cameroon	199,109	191,326	7783	4.1
Cayman Islands	4148	4199	−50	−1.2
China	14,221	15,084	−863	−5.7
Colombia	262,212	231,187	31,025	13.4
Comoros	99	103	−3	−3.2
Cook Islands	3	0	3	–
Costa Rica	36,475	36,872	−397	−1.1
Côte d'Ivoire	5890	6131	−241	−3.9
Cuba	332,816	337,113	−4297	−1.3
Curaçao	7	7	0	−1.2
Democratic Republic of the Congo	24,017	25,810	−1793	−6.9
Djibouti	545	498	47	9.4
Dominica	2	2	0	−9.9
Dominican Republic	18,741	18,942	−201	−1.1
Ecuador	146,544	139,416	7128	5.1
Egypt	147	180	−33	−18.2
El Salvador	37,589	37,665	−76	−0.2
Equatorial Guinea	25,904	25,984	−80	−0.3
Eritrea	6918	7006	−88	−1.3
Fiji	49,984	51,166	−1182	−2.3
French Guiana	59,466	50,187	9279	18.5
French Polynesia	122	0	122	–
French Southern Territories	672	0	672	–
Gabon	176,632	177,091	−458	−0.3

Table A1. *Cont.*

Country	GMW v2.5 (ha) 2010	GMW 2.0 (ha) 2010	Difference (ha)	Difference (%)
Gambia	60,673	61,294	−621	−1.0
Ghana	20,021	20,557	−536	−2.6
Grenada	190	195	−5	−2.7
Guadeloupe	3713	3735	−22	−0.6
Guam	57	0	57	–
Guatemala	23,523	23,709	−187	−0.8
Guinea	222,286	223,849	−1562	−0.7
Guinea−Bissau	262,631	262,616	15	0.0
Guyana	28,640	27,472	1168	4.3
Haiti	14,432	14,635	−203	−1.4
Honduras	59,732	60,372	−640	−1.1
India	370,984	352,052	18,933	5.4
Indonesia	2,801,795	2,688,955	112,840	4.2
Iran	7587	7912	−325	−4.1
Jamaica	9411	9520	−109	−1.1
Japan	918	987	−69	−7.0
Kenya	52,888	53,788	−900	−1.7
Kiribati	135	0	135	–
Liberia	18,938	19,153	−215	−1.1
Madagascar	260,271	260,986	−715	−0.3
Malaysia	515,743	520,071	−4329	−0.8
Maldives	97	0	97	–
Marshall Islands	6	0	6	–
Martinique	2052	1700	352	20.7
Mauritania	177	104	73	70.4
Mauritius	345	703	−357	−50.9
Mayotte	702	589	113	19.2
Mexico	939,502	953,651	−14,150	−1.5
Micronesia	9084	8399	685	8.2
Mozambique	298,841	301,125	−2284	−0.8
Myanmar	496,686	501,116	−4430	−0.9
New Caledonia	33,593	29,625	3967	13.4
New Zealand	30,216	30,627	−411	−1.3
Nicaragua	73,988	74,988	−1000	−1.3
Nigeria	847,894	695,775	152,119	21.9
Oman	111	116	−5	−4.3
Pakistan	63,600	65,240	−1641	−2.5
Palau	6014	6183	−169	−2.7
Panama	153,337	154,477	−1140	−0.7
Papua New Guinea	445,785	476,167	−30,382	−6.4
Peru	4569	4083	486	11.9
Philippines	260,993	265,108	−4115	−1.6
Puerto Rico	8685	8869	−184	−2.1
Qatar	428	444	−16	−3.6
Republic of Congo	2063	0	2063	–
Saint Kitts and Nevis	28	29	−1	−3.0
Saint Lucia	164	166	−2	−1.1
Saint Vincent and the Grenadines	31	33	−2	−5.5
Saint−Martin	14	15	0	−3.0
Samoa	264	305	−40	−13.3
São Tomé and Príncipe	0	0	0	12.0
Saudi Arabia	5367	6824	−1457	−21.4
Senegal	128,077	128,934	−856	−0.7
Seychelles	385	107	277	258.2
Sierra Leone	160,038	127,759	32,279	25.3
Singapore	534	540	−6	−1.2
Solomon Islands	55,519	51,424	4094	8.0

Table A1. *Cont.*

Country	GMW v2.5 (ha) 2010	GMW 2.0 (ha) 2010	Difference (ha)	Difference (%)
Somalia	2253	2098	154	7.4
South Africa	2573	2637	−64	−2.4
Sri Lanka	18,941	20,164	−1223	−6.1
Sudan	433	366	67	18.2
Suriname	77,108	78,154	−1046	−1.3
Taiwan	159	171	−12	−7.0
Tanzania	107,775	113,101	−5326	−4.7
Thailand	223,137	225,770	−2634	−1.2
Timor−Leste	957	983	−26	−2.7
Togo	21	0	21	–
Tonga	1193	872	321	36.8
Trinidad and Tobago	7696	5523	2173	39.3
Turks and Caicos Islands	10,420	12,260	−1840	−15.0
Tuvalu	9	0	9	–
United Arab Emirates	6759	7766	−1008	−13.0
United States	209,544	193,609	15,935	8.2
Vanuatu	1724	1782	−58	−3.2
Venezuela	275,325	287,130	−11,805	−4.1
Vietnam	157,028	159,952	−2924	−1.8
Virgin Islands, U.S.	197	206	−8	−4.1
Wallis and Futuna	29	0	29	–
Yemen	1314	1753	−439	−25.0
Global	**14,025,986**	**13,760,074**	**265,912**	**1.9**

References

1. Hamilton, S.E.; Casey, D. Creation of a high spatio-temporal resolution global database of continuous mangrove forest cover for the 21st century (CGMFC-21). *Glob. Ecol. Biogeogr.* **2016**, *25*, 729–738. [CrossRef]
2. Goldberg, L.; Lagomasino, D.; Thomas, N.; Fatoyinbo, T. Global declines in human-driven mangrove loss. *Glob. Chang. Biol.* **2020**, *26*, 5844–5855. [CrossRef] [PubMed]
3. Thomas, N.; Lucas, R.; Bunting, P.; Hardy, A.; Rosenqvist, A.; Simard, M. Distribution and drivers of global mangrove forest change, 1996–2010. *PLoS ONE* **2017**, *12*, e0179302. [CrossRef] [PubMed]
4. Lacerda, L.D.D.; Ward, R.D.; Godoy, M.D.P.; Meireles, A.J.D.A.; Borges, R.; Ferreira, A.C. 20-Years Cumulative Impact From Shrimp Farming on Mangroves of Northeast Brazil. *Front. For. Glob. Chang.* **2021**, *4*, 653096. 653096. [CrossRef]
5. Giri, C.; Zhu, Z.; Tieszen, L.L.; Singh, A.; Gillette, S.; Kelmelis, J.A. Mangrove forest distributions and dynamics (1975–2005) of the tsunami-affected region of Asia. *J. Biogeogr.* **2008**, *35*, 519–528. [CrossRef]
6. Ai, B.; Ma, C.; Zhao, J.; Zhang, R. The impact of rapid urban expansion on coastal mangroves: A case study in Guangdong Province, China. *Front. Earth Sci.* **2020**, *14*, 37–49. [CrossRef]
7. Richards, D.R.; Friess, D.A. Rates and drivers of mangrove deforestation in Southeast Asia, 2000–2012. *Proc. Natl. Acad. Sci. USA* **2016**, *113*, 344–349. [CrossRef]
8. Duke, N.C.; Kovacs, J.M.; Griffiths, A.D.; Preece, L.; Hill, D.J.; Van Oosterzee, P.; Mackenzie, J.; Morning, H.S.; Burrows, D. Large-scale dieback of mangroves in Australia's Gulf of Carpentaria: A severe ecosystem response, coincidental with an unusually extreme weather event. *Mar. Freshw. Res.* **2017**, *68*, 1816–1829. [CrossRef]
9. Ermgassen, P.S.Z.; Mukherjee, N.; Worthington, T.A.; Acosta, A.; Araujo, A.R.d.R.; Beitl, C.M.; Castellanos-Galindo, G.A.; Cunha-Lignon, M.; Dahdouh-Guebas, F.; Diele, K.; et al. Fishers who rely on mangroves: Modelling and mapping the global intensity of mangrove-associated fisheries. *Estuar. Coast. Shelf Sci.* **2020**, *247*, 106975. [CrossRef]
10. Donato, D.C.; Kauffman, J.B.; Murdiyarso, D.; Kurnianto, S.; Stidham, M.; Kanninen, M. Mangroves among the most carbon-rich forests in the tropics. *Nat. Geosci.* **2011**, *4*, 293–297. [CrossRef]
11. Menéndez, P.; Losada, I.J.; Torres-Ortega, S.; Narayan, S.; Beck, M.W. The Global Flood Protection Benefits of Mangroves. *Sci. Rep.* **2020**, *10*, 4404. [CrossRef] [PubMed]
12. Spalding, M.; Parrett, C.L. Global patterns in mangrove recreation and tourism. *Mar. Policy* **2019**, *110*, 103540.. [CrossRef]
13. Costanza, R.; Groot, R.d.; Sutton, P.; Ploeg, S.v.d.; Anderson, S.J.; Kubiszewski, I.; Farber, S.; Turner, R.K. Changes in the global value of ecosystem services. *Glob. Environ. Chang.* **2014**, *26*, 152–158. [CrossRef]
14. Simard, M.; Fatoyinbo, L.; Smetanka, C.; Rivera-Monroy, V.H.; Castañeda-Moya, E.; Thomas, N.; Stocken, T.V.d. Mangrove canopy height globally related to precipitation, temperature and cyclone frequency. *Nat. Geosci.* **2019**, *12*, 40–45. [CrossRef]
15. Bunting, P.; Rosenqvist, A.; Lucas, R.; Rebelo, L.M.; Hilarides, L.; Thomas, N.; Hardy, A.; Itoh, T.; Shimada, M.; Finlayson, C. The Global Mangrove Watch—A New 2010 Global Baseline of Mangrove Extent. *Remote Sens.* **2018**, *10*, 1669. [CrossRef]

16. Bunting, P.; Rosenqvist, A.; Lucas, R.; Rebelo, L.M.; Hilarides, L.; Thomas, N.; Hardy, A.; Itoh, T.; Shimada, M.; Finlayson, M. *Global Mangrove Watch (1996–2016) Version 2.0*; Zenodo: Genève, Switzerland, 2019. [CrossRef]
17. Giri, C.; Ochieng, E.; Tieszen, L.L.; Zhu, Z.; Singh, A.; Loveland, T.; Masek, J.; Duke, N. Status and distribution of mangrove forests of the world using earth observation satellite data. *Glob. Ecol. Biogeogr.* **2011**, *20*, 154–159. [CrossRef]
18. Spalding, M. *World Atlas of Mangroves*; Routledge: London, UK, 2010. [CrossRef]
19. Thomas, N.; Lucas, R.; Itoh, T.; Simard, M.; Fatoyinbo, L.; Bunting, P.; Rosenqvist, A. An approach to monitoring mangrove extents through time-series comparison of JERS-1 SAR and ALOS PALSAR data. *Wetl. Ecol. Manag.* **2014**, *23*, 3–17.. [CrossRef]
20. Thomas, N.; Bunting, P.; Lucas, R.; Hardy, A.; Rosenqvist, A.; Fatoyinbo, T. Mapping Mangrove Extent and Change: A Globally Applicable Approach. *Remote Sens.* **2018**, *10*, 1466. [CrossRef]
21. Baloloy, A.B.; Blanco, A.C.; Ana, R.R.C.S.; Nadaoka, K. Development and application of a new mangrove vegetation index (MVI) for rapid and accurate mangrove mapping. *ISPRS J. Photogramm. Remote Sens.* **2020**, *166*, 95–117.. [CrossRef]
22. Li, H.; Han, Y.; Chen, J. Combination of Google Earth imagery and Sentinel-2 data for mangrove species mapping. *J. Appl. Remote Sens.* **2020**, *14*, 010501. [CrossRef]
23. Ghorbanian, A.; Zaghian, S.; Asiyabi, R.M.; Amani, M.; Mohammadzadeh, A.; Jamali, S. Mangrove Ecosystem Mapping Using Sentinel-1 and Sentinel-2 Satellite Images and Random Forest Algorithm in Google Earth Engine. *Remote Sens.* **2021**, *13*, 2565. [CrossRef]
24. Cissell, J.R.; Canty, S.W.J.; Steinberg, M.K.; Simpson, L.T. Mapping National Mangrove Cover for Belize Using Google Earth Engine and Sentinel-2 Imagery. *Appl. Sci.* **2021**, *11*, 4258. [CrossRef]
25. Liu, X.; Fatoyinbo, T.E.; Thomas, N.M.; Guan, W.W.; Zhan, Y.; Mondal, P.; Lagomasino, D.; Simard, M.; Trettin, C.C.; Deo, R.; Barenblitt, A. Large-Scale High-Resolution Coastal Mangrove Forests Mapping Across West Africa With Machine Learning Ensemble and Satellite Big Data. *Front. Earth Sci.* **2021**, *8*, 560933. [CrossRef]
26. Awty-Carroll, K.; Bunting, P.; Hardy, A.; Bell, G. Using Continuous Change Detection and Classification of Landsat Data to Investigate Long-Term Mangrove Dynamics in the Sundarbans Region. *Remote Sens.* **2019**, *11*, 2833. [CrossRef]
27. Awty-Carroll, K.; Bunting, P.; Hardy, A.; Bell, G. Evaluation of the Continuous Monitoring of Land Disturbance Algorithm for Large-Scale Mangrove Classification. *Remote Sens.* **2021**, *13*, 3978. [CrossRef]
28. Zhu, Z.; Zhang, J.; Yang, Z.; Aljaddani, A.H.; Cohen, W.B.; Qiu, S.; Zhou, C. Continuous monitoring of land disturbance based on Landsat time series. *Remote Sens. Environ.* **2019**, *238*, 111116. [CrossRef]
29. Bunting, P.; Clewley, D.; Lucas, R.M.; Gillingham, S. The Remote Sensing and GIS Software Library (RSGISLib). *Comput. Geosci.* **2014**, *62*, 216–226. [CrossRef]
30. Bunting, P.; Gillingham, S. The KEA image file format. *Comput. Geosci.* **2013**, *57*, 54–58. [CrossRef]
31. Available online: https://www.remotesensing.info/pbprocesstools/ (accessed on 8 January 2022).
32. Chen, T.; Guestrin, C. XGBoost: A Scalable Tree Boosting System. In Proceedings of the 22nd ACM SIGKDD International Conference on Knowledge Discovery and Data Mining; San Francisco, CA, USA, 13–17 August 2016; pp. 785–794. [CrossRef]
33. Available online: https://cloud.google.com/storage/docs/public-datasets/sentinel-2 (accessed on 8 January 2022).
34. Available online: https://www.remotesensing.info/arcsi/ (accessed on 8 January 2022).
35. John, E.; Bunting, P.; Hardy, A.; Silayo, D.S.; Masunga, E. A Forest Monitoring System for Tanzania. *Remote Sens.* **2021**, *13*, 3081. [CrossRef]
36. Vermote, E.; Tanre, D.; Deuze, J.; Herman, M.; Morcrette, J. Second Simulation of the Satellite Signal in the Solar Spectrum, 6S: An overview. *IEEE Trans. Geosci. Remote Sens.* **1997**, *35*, 675–686. [CrossRef]
37. Wilson, R.T. Py6S: A Python interface to the 6S radiative transfer model.*Comput. Geosci.* **2013**, *51*, 166–171. 2012.08.002. [CrossRef]
38. Shepherd, J.D.; Dymond, J.R. Correcting satellite imagery for the variance of reflectance and illumination with topography. *Int. J. Remote Sens.* **2003**, *24*, 3503–3514. [CrossRef]
39. Zhu, Z.; Woodcock, C.E.; Olofsson, P. Continuous monitoring of forest disturbance using all available Landsat imagery. *Remote Sens. Environ.* **2012**, *122*, 75–91. [CrossRef]
40. Zhu, Z.; Wang, S.; Woodcock, C.E. Improvement and expansion of the Fmask algorithm: Cloud, cloud shadow, and snow detection for Landsats 4–7. 8, and Sentinel 2 images. *Remote Sens. Environ.* **2015**, *159*, 269–277. [CrossRef]
41. Available online: http://www.pythonfmask.org (accessed on 8 January 2022).
42. Available online: https://github.com/sentinel-hub/sentinel2-cloud-detector (accessed on 8 January 2022).
43. Ke, G.; Meng, Q.; Finley, T.; Wang, T.; Chen, W.; Ma, W.; Ye, Q.; Liu, T.Y. LightGBM: A Highly Efficient Gradient Boosting Decision Tree. In Proceedings of the 31st Conference on Neural Information Processing Systems (NIPS 2017), Long Beach, CA, USA, 4–9 December 2017.
44. Reseau National D'observation et D'aide a la Gestion des Mangroves. Les Surfaces de Mangroves an France en 2020. Available online: https://uicn.fr/wp-content/uploads/2017/07/plaquette-rom-060717.pdf (accessed on 8 January 2022).
45. Pekel, J.F.; Cottam, A.; Gorelick, N.; Belward, A.S. High-resolution mapping of global surface water and its long-term changes. *Nature* **2016**, *540*, 418–422. [CrossRef]

remote sensing

MDPI

Article

Multiscale Diagnosis of Mangrove Status in Data-Poor Context Using Very High Spatial Resolution Satellite Images: A Case Study in Pichavaram Mangrove Forest, Tamil Nadu, India

Shuvankar Ghosh [1,2], Christophe Proisy [1,3,4,*], Gowrappan Muthusankar [1], Christiane Hassenrück [5,6], Véronique Helfer [5], Raphaël Mathevet [1,7], Julien Andrieu [1,8], Natesan Balachandran [1] and Rajendran Narendran [9]

1 French Institute of Pondicherry, Pondicherry 605001, India; shuvankar.ghosh@ifpindia.org (S.G.); muthusankar@ifpindia.org (G.M.); raphael.mathevet@cefe.cnrs.fr (R.M.); julien.andrieu@ifpindia.org (J.A.); balachandran.n@ifpindia.org (N.B.)
2 Iora Ecological Solutions Pvt. Ltd., New Delhi 110030, India
3 AMAP, University Montpellier, IRD, CNRS, INRAE, 34000 Montpellier, France
4 AMAP, Centre IRD de Cayenne, 97300 Cayenne, France
5 Leibniz Centre for Tropical Marine Research (ZMT), 28359 Bremen, Germany; christiane.hassenrueck@io-warnemuende.de (C.H.); veronique.helfer@leibniz-zmt.de (V.H.)
6 Leibniz Institute for Baltic Sea Research Warnemünde (IOW), 18119 Rostock, Germany
7 CEFE, University Montpellier, CNRS, EPHE, IRD, 34090 Montpellier, France
8 ESPACE, University Cote d'Azur, 06103 Nice, France
9 Centre of Advanced Study in Marine Biology, Faculty of Marine Science, Annamalai University, Parangipettai 608002, India; narenbios666@gmail.com
* Correspondence: christophe.proisy@ird.fr

Citation: Ghosh, S.; Proisy, C.; Muthusankar, G.; Hassenrück, C.; Helfer, V.; Mathevet, R.; Andrieu, J.; Balachandran, N.; Narendran, R. Multiscale Diagnosis of Mangrove Status in Data-Poor Context Using Very High Spatial Resolution Satellite Images: A Case Study in Pichavaram Mangrove Forest, Tamil Nadu, India. *Remote Sens.* **2022**, *14*, 2317. https://doi.org/10.3390/rs14102317

Academic Editor: Chandra Giri

Received: 6 April 2022
Accepted: 27 April 2022
Published: 11 May 2022

Abstract: Highlighting spatiotemporal changes occurring within mangrove habitats at the finest possible scale could contribute fundamental knowledge and data for local sustainable management. This study presents the current situation of the Pichavaram mangrove area, a coastal region of Southeast India prone to both cyclones and reduced freshwater inflow. Based on the supervised classification and visual inspection of very high spatial resolution (VHSR) satellite images provided with a pixel size of <4 m, we generated time-series maps to analyze the changes that occurred in both the natural and planted mangroves between 2003 and 2019. We achieved a high mapping accuracy (>85%), which confirmed the potential of classification techniques applied to VHSR images in capturing changes in mangroves on a very fine scale. Our diagnosis reveals variable expansion rates in plantations made by the local authorities. We also report an ongoing mangrove dieback and confirm progressive shoreline erosion along the coastline. Despite a lack of field data, VHSR images allowed for the multiscale diagnosis of the ecosystem situation, thus constituting the first fine-scale assessment of the fragile Pichavaram mangrove area upon which the coastal community is dependent.

Keywords: remote sensing-based monitoring; plantation; restoration; change detection; dieback; Bay of Bengal

1. Introduction

Each mangrove region has its own history, which reflects how species diversity and structural patterns adapt to local and evolving geomorphic and climatic settings [1,2]. The status of mangroves, however, remains insufficiently documented in many places of the world despite a consensus on their rates of global decline over decadal timescales [3–9]. Even if annual rates of mangrove loss tended to decrease between 2000 and 2012 (0.16–0.39%) compared to the 1980–2000 period (0.99%) [10–14], the present situation in South Asia and Southeast Asia remains illustrative of the major drivers of global mangrove loss, which include land-use conversion to aquaculture, timber harvesting, pollution, freshwater deficits, low silt deposition, shoreline erosion, and other environmental events, such as dieback, floods, and tropical storm landfalls [8,12,13]. Beyond the awareness of the importance to protect and restore mangroves raised among the public at large including policy makers [13] after the Indian Ocean tsunami in 2004, local and regional administrations must be continuously and at the earliest stage cautioned on the mangrove status for better ecosystem management [15–17].

Mangrove environments are complex forest landscapes where spatiotemporal changes continuously occur at fine spatial scales as a consequence of bio-geomorphological processes operating at different timescales, and the ecosystem transformation remains highly situational. More than anything, progressive small-scale changes occurring within forest landscape area, such as stand structure degradation, dieback, or loss due to sediment erosion, must be monitored to detect stress factors as early as possible [17–19].

As evident from Figure 1, mangrove species distribution, forest structural attributes, or landscape fragmentation characterized by forest patches interspersed with water bodies and channels may hinder moderate spatial resolution (MSR) images, provided with a pixel size of >4 m (such as Landsat or Sentinel-2 images) when mapping mangrove areas with accuracy and robustness. Small-scale changes might not get captured using MSR images until the fine-scale changes accumulate and result in major changes; at this point, it might be too late to mitigate.

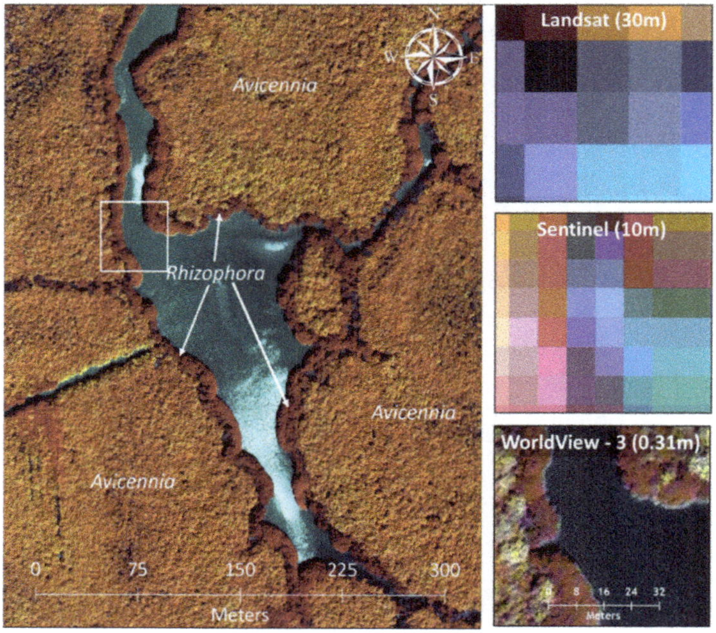

Figure 1. Importance of spatial resolution for mapping changes in mangrove cover. Mangrove subsets from MSR images from Landsat-8 (15 March 2019) and Sentinel-2 (31 May 2019), along with very high spatial resolution image Worldview-3 (28 January 2019).

For the better analysis of fine-scale changes and the implementation of rehabilitation or mitigation measures, the periodic monitoring of mangrove ecosystems using very high spatial resolution images (VHSR) satellite images delivered with a pixel size of finer than 4 m is necessary. Efforts to map mangrove habitats and analyze spatiotemporal changes using VHSR (Quickbird, GeoEye, Worldview) images are rare, but, when achieved, valuable information about mangrove status has been obtained [18,20–25]. First, VHSR remote sensing supports the setting up of field experiments. Second, when combined with field surveys, VHSR remote sensing images are able to provide the fine-scale mapping of species distribution or biomass dynamics through the development of quantitative and robust remote sensing methods based on spectral and textural image characteristics [26,27]. Third, time series of high resolution images facilitate the accurate diagnoses of changes in mangrove habitats and allow for the questioning of the reasons for local demise or the establishment of mangroves [18,28,29].

In India, efforts to monitor spatiotemporal changes in mangrove habitats using VHSR images have been even less frequent [30,31]; mangrove rehabilitation measures have been generally assessed as successful solely from the monitoring of the surface, which have been estimated using moderate resolution images such as Landsat or IRS LISS-III [32,33]. The relevance of the plantation techniques, hastily referred to restoration practices, remains insufficiently questioned though multiple studies drawing attention towards the limitations of this approach [18,34], especially when the post-planting in situ monitoring of planted mangrove attributes is not carried out. This is the case of the fishbone plantation, a typical approach that has been in practice since the 1980s in India to (re)introduce mangrove on degraded lands [32]. This plantation method diverts water from existing creeks and channels through the construction of new channels, thus converting the barren land into an area that tides are expected to reach more easily and frequently, with expected impacts on the growth and survival of planted mangrove seedlings.

Here, we demonstrate the potential of VHSR remote sensing to study the Pichavaram mangrove region located in Southeast India based on the use of a time series of five VHSR satellite images acquired over a period of 16 years (2003–2019). This mangrove region is part of a major restoration program of wetlands of the east coast of India initiated in the 1990s by national and local authorities [32]. So far, changes in Pichavaram mangroves have been analyzed with moderate resolution images such as Landsat and IRS LISS-III [32,33]. As few ancillary data are available for scientific analysis, the current status of this mangrove area must be questioned. We therefore attempted to map mangrove habitats through an analysis of changes in the mangrove extent with the aim to identify and understand the environmental and anthropogenic driving forces, a number of them being exogenous to the mangrove area itself. A second objective was to assess the performance of the mangrove rehabilitation efforts that have been undertaken by the state forest department. In addition, as we identified a mangrove dieback event in Pichavaram that has resulted in massive tree mortality, we discuss the current status of the Pichavaram mangroves and provide recommendations to improve knowledge and monitoring of this fragile area.

2. Materials and Methods

2.1. Study Area

The Pichavaram mangrove is located almost 15 km northeast of Chidambaram in the Cuddalore District of Tamil Nadu. Based on the Köppen–Geiger system [35], climate in Pichavaram is tropical wet and dry (tropical savannah sub-type), characterized by hot and humid summers, with maximum temperatures reaching almost 42–45 degrees Celsius followed by mild and pleasant winters (average high temperatures of 25 degrees Celsius).

The mangrove region is bordered by the Vellar estuary in the north and the Kollidam estuary in the south (Figure 2). The total estuarine area is about 23 km^2, of which mangroves cover about 1100 ha [33,36]. The area is characterized by fertile alluvium deposits in the west that gradually change to fluviomarine deposits and beach sands towards the eastern area close to the Bay of Bengal. The mangroves receive most of their freshwater from the Kollidam River (rather than the Vellar River) due to the south–north elevation gradient. The region receives a majority of rainfall from the northeast monsoon (October–December), and to a much lesser extent from the southwest monsoon. From October to December, the freshwater input can generally increase considerably while the region remains dry [37], with salinity levels being very high (35–45 ppt) during the rest of the year [36].

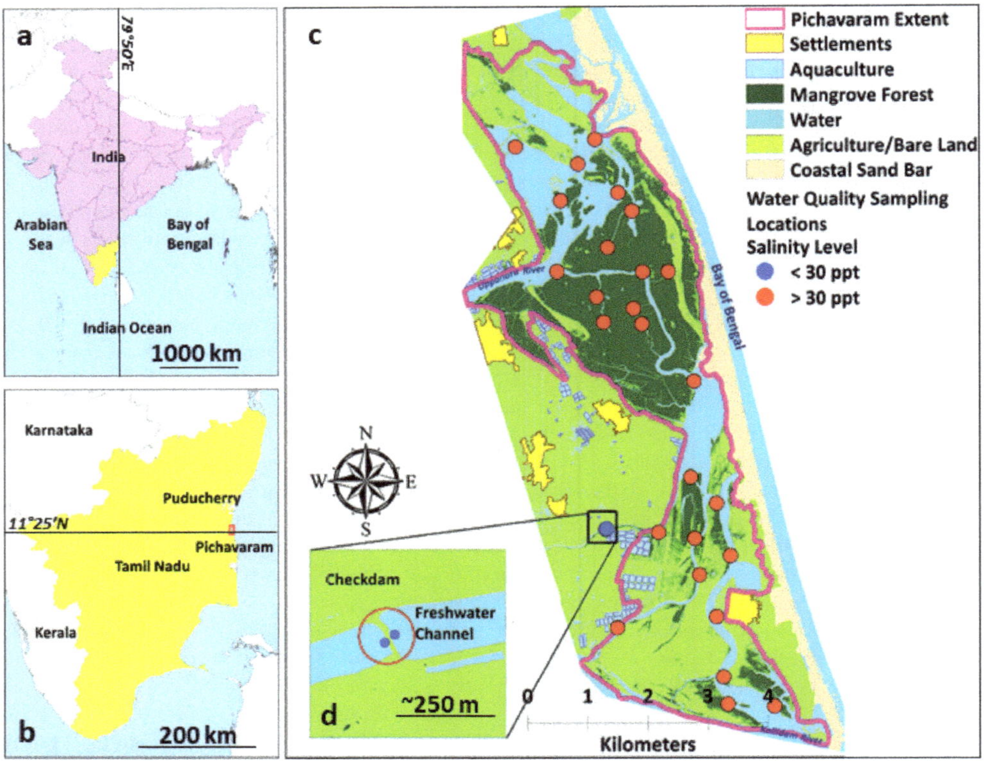

Figure 2. Location of the Pichavaram mangroves in Tamil Nadu, Southeast India (**a**,**b**). The mangrove area, along with the adjacent land-use and land-cover such as agriculture, aquaculture, check-dams on freshwater streams, and settlements, is illustrated (**c**). Locations of water quality samples are also shown along with the observed salinity levels (**d**).

The Pichavaram mangrove ecosystem is surrounded inland by fishermen hamlets, agriculture, and aquaculture ponds that have shown significant increases in numbers, especially in the last decade. The mangroves serve as lifeline for the surrounding coastal fishermen community. The fishery production harvested from the mangrove wetland has been reported to be almost 245 tons (prawns constituting almost 85% of the reported harvest) [36,38,39]. Almost 1000 fishermen seasonally fish in the mangrove; in addition, 800–900 cattle graze the mangrove wetland area, although the rates of cattle grazing have been reported to decline over the years [36]. The Pichavaram mangroves played a critical role in mitigating the devastating effects of the inundation resulting from the Indian Ocean Tsunami of 2004 on the hamlets that were guarded by the mangroves on the seaward side [31].

Today, the habitat is mainly dominated by *Avicennia marina* in the interior parts, while *Rhizophora apiculata* and *Rhizhophora mucronata* form the periphery of mangrove patches. Apart from *Avicennia* and *Rhizophora*, a few other mangrove plants have been intermittently observed in the ecosystem, such as *Bruguiera cylindrica*, *Excoecaria agallocha*, *Ceriops decandra*, *Avicennia officinalis*, *Aegiceras corniculatum*, *Rhizophora annamalayana*, *Acanthus ilicifolius*, and *Lumnitzera racemosa*, with some rare observations of *Xylocarpus granatum* and *Sonneratia apetala* [39]. No detailed data on mangrove forest structure are available. Based on personal visual inspection, *Rhizophora* formations along the river banks can reach heights of up to 6–7 m, while old and probably non-planted *Avicennia marina* (>20 years) may reach heights of 10–12 m. Tree height in plantations, where mangroves are mostly shrubby in nature, seems to be lower (<3 m). The state forest department has been implementing fishbone plots in Pichavaram since the 1990s in an attempt to (re)introduce mangroves in the area, even on barren land (Figure 3). It is worth noting that, to our knowledge, the biogeochemistry, topography, and elevation of barren land are not documented prior or even after planting.

Table 1. Specifications and dates of acquisition of the VHSR images used in this study. The angles θs and θv denote the sun zenith and viewing zenith angles, respectively, while $\phi s\text{–}v$ is the relative sun-viewing azimuth angle. The tide levels during the time of acquisition, obtained from NOAA Tides and Currents (http://tidesandcurrents.noaa.gov, accessed on 12 November 2021) are also indicated. P = panchromatic; MS = multi-spectral.

Satellite and Format	Acquisition Date	Pixel Size (m)	θs (°)	θv (°)	$\phi s\text{–}v$ (°)	Tide Level (cm)
QBD [P, 4 MS]	4 May 2003	[0.6, 2.4]	22.8	6.5	292.6	7
	5 January 2005		38.7	32.9	230.3	46
GEO [P, 4 MS]	22 March 2011	[0.4, 1.6]	26.1	19.6	262.3	0
	30 January 2016		38.8	26.3	86.1	20
WV3 [P, 8 MS]	28 January 2019	[0.4, 1.6]	34.5	19.2	169.2	44

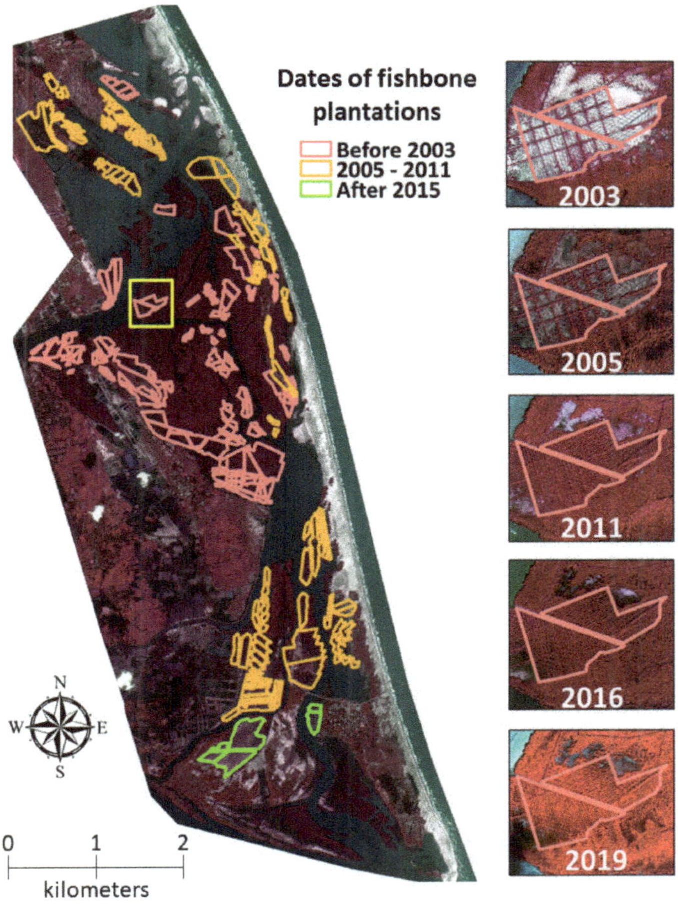

Figure 3. Pichavaram mangroves in Tamil Nadu along with the locations of the fishbone plots (**left**). Changes in the mangrove cover inside a 400 × 400 m fishbone plot (yellow square) implemented before 2003 are illustrated on the **right** using the VHSR image dataset described in Table 1.

2.2. Data

2.2.1. VHSR Satellite Imagery

We searched for cloud-free VHSR satellite images covering the Pichavaram mangroves on the Maxar® web-catalog. We shortlisted five images and purchased them; these comprised Quickbird (QBD) images from the years 2003 and 2005 (acquired a few days after the Indian Ocean Tsunami that had heavily impacted the region on 26 December 2004), GeoEye-1 (GEO) images from 2011 and 2016, and a Worldview-3 (WV3) image from 2019 (Table 1). All the images were captured at low tides. Each image was delivered in the GeoTIFF format as a set of two files, one comprising the panchromatic band (P) and the second one containing a multichannel stack generated by combining either four or eight multispectral (MS) bands, depending on the sensor. Panchromatic and multispectral bands were provided with pixel sizes ranging from 0.4 to 0.6 m and from 1.2 to 2.4 m, respectively. We selected the GeoEye image of 30 January 2016 as a reference image for the geo-registration of all the other images to Universal Transverse Mercator (Zone 44N). We estimated the spatial registration accuracy as 1–2 m for the satellite images.

The pixel intensity (or digital number) of satellite images was transformed into a TOA reflectance value, without correction for atmospheric effects, as follows:

$$\rho = \frac{\pi \cdot L_\lambda \cdot D^2}{E_{sun} \cdot \cos\theta_s} \tag{1}$$

where ρ is the pixel reflectance, L_λ is the radiance obtained using sensor-specific equations provided in the image metadata files, D is the Sun–Earth distance (expressed in astronomical units), E_{sun} is the corresponding mean solar exo-atmospheric spectral irradiance ($mW.cm^{-2}.\mu m^{-1}$), and θ_s is the solar zenith angle [40]. In addition to the five raw VHSR multispectral images, we used RGB composites provided by the Google Earth platform at different dates between 2003 and 2019.

2.2.2. In Situ Water Quality

We conducted two field visits in the study area during the pre-monsoon season of 2019–2020 to measure in situ surface water salinity levels across a number of locations in the mangrove area (Figure 2) using the HI-98192 professional waterproof salinity meter. The instrument was calibrated prior to the field visits using an EC calibration solution. No field visits were possible during the post-monsoon season and the monsoon season of 2020–2021 due to travel restrictions imposed in response to the COVID-19 pandemic.

2.2.3. Meteorological Observations

We obtained monthly rainfall (0.25-degree grid) and air temperature (1-degree grid) data for the 1901–2019 and 1951–2019 periods, respectively, using meteorological observations from the Indian Meteorological Department (IMD) (https://mausam.imd.gov.in/ accessed on 12 November 2021) for the Cuddalore district in Tamil Nadu. From these long-term records, we investigated possible anomalies in precipitation and temperature regimes during our study period of 2001–2019 through a comparison of short-term and long-term linear trends.

2.3. Methods

Our main objective was to distinguish mangrove areas from non-mangrove areas to capture changes in the mangrove habitat. For this, we developed our analysis through a supervised classification process based on all panchromatic and multispectral bands available for each image, as described below. We did not use any vegetation indices.

2.3.1. Supervised Classification

Supervised classification was performed on each of the VHSR image using the nearest neighbor algorithm (see [41] for a review of image classification methods) in ERDAS Imagine 2016 (Hexagon Geospatial; https://www.hexagongeospatial.com/products/power-portfolio/erdas-imagine accessed on 31 March 2022).

Due to lack of geo-registered in situ data on vegetation for the whole study period, species-level classification was avoided, while a generic land-cover level classification was conducted. From a visual inspection of VHSR images consolidated by our ground experience, training and validation polygons of an area ranging from 0.1 to 4 ha were delineated for three predominant land-covers, i.e., mangrove, water, and other non-mangrove areas (e.g., barren sandy and dry areas, sandy vegetation, aquaculture ponds, areas, roads, and buildings) in each image, mostly independently of each other. User's and producer's accuracy values were generated along with the kappa statistic, as well as overall accuracy. New training and validation polygons for each of the three land-covers were created and added to the classification process for each image until the maximum variability of spectral responses for each class was captured and the accuracy levels reached stable values. This was achieved through a few iterations for each image.

2.3.2. Land-Cover Change Detection

Comparisons were performed to detect changes in the whole Pichavaram region with respect to the initial land-cover of 2003 (such as comparisons between 2003 and 2005, 2003 and 2011, 2003 and 2016, and 2003 and 2019), as well progressive changes in land-cover since 2003 (comparisons between 2003 and 2005, 2005 and 2011, 2011 and 2016, and 2016 and 2019). The change classes generated through these comparisons comprised unchanged mangrove, water, and non-mangrove classes and the following changes (initial land-cover to subsequent land-cover): 'mangrove to non-mangrove', 'mangrove to water', 'water to mangrove', 'water to non-mangrove', 'non-mangrove to mangrove', and 'non-mangrove to water'. Progressive and cumulative mangrove gains and losses were derived since the beginning of the study period (2003). The analysis of the 'mangrove to water' or 'water to mangrove' classified areas along the 15 km of Pichavaram shoreline was expected to provide a fine-scale diagnosis of the eventual impact of coastal erosion or seaward mangrove progradation.

It was also crucial to document the changes within the non-mangrove areas in proximity to Pichavaram in order to highlight any natural or anthropogenic features or phenomena that might threaten the mangrove ecosystem health and resilience. For that, we visually inspected the changes within the Pichavaram adjoining areas using the raw VHSR scenes, along with images available through Google Earth at different dates.

2.3.3. Mangrove Cover Change in Fishbone Plantations

We wanted to analyze the success and/or failure rates in the fishbone plots implemented just prior to the Quickbird image acquisition of 2003 (Figure 3). For this, we visually delineated fishbone units on all VHSR images; plots were segregated based on the time of their implementation (i.e., plots laid prior to 2003, between 2005 and 2011, between 2011 and 2016, and between 2016 and 2019). Some fishbone plantation units were implemented close/adjacent to one another: if it was during the same timeframe, we gathered them into a singular unit assuming that the geographical and environmental conditions were likely to be identical.

For further analysis, we only retained the fishbone plots that were laid prior to 2003 and between 2005 and 2011, as only two areas with fishbone plots were observed between 2011 and 2016 (see "after 2016" in Figure 3) and the plots implemented after 2016 were captured only in the VHSR acquired in 2019. Based on the observed rates of growth, we classified the performance of fishbone units into three categories, i.e., 'high growth', 'moderate growth', and 'low growth' plots.

2.3.4. Delineation of Dieback Areas

In 2019, during a field experiment, we noticed tree mortality in fringing *Rhizophora* trees across a certain mangrove area of Pichavaram. We visually inspected all VHSR images available from 2016 to 2019, including satellite images provided in Google Earth to estimate when dieback came up and how many hectares of mangroves were affected. This objective could not be achieved using MSR satellite images.

3. Results

3.1. Supervised Classification

The supervised classification of the VHSR images yielded a very high level of accuracy (Table 2). For most of the images, an overall mapping accuracy of >90% was achieved with kappa values of >0.85. For 2005, though the mapping accuracy levels were observed to be relatively lower (85% and kappa = 0.77). Within the overall accuracy, the mangrove mapping accuracy was observed to be mostly >90% (both producer's and user's accuracy), apart from 2005 and 2019 when the user's accuracy reached 84.8% and 86.8%, respectively, which could still be considered reasonably high.

Table 2. Classification accuracies of individual land-cover classes for Pichavaram, as obtained from the supervised classification of the VHSR images (QBD = Quickbird; GEO = GeoEye-1; WV3 = Worldview-3).

Year	Class Names	Producer's Accuracy	User's Accuracy	Kappa
2003 (QBD)	Mangrove	96.2%	93.8%	0.91
	Water	90.8%	98.8%	0.98
	Non-Mangrove	97.3%	91.2%	0.87
		Total Accuracy = 95%		*Kappa = 0.92*
2005 (QBD)	Mangrove	84.8%	97.5%	0.96
	Water	88.9%	70.0%	0.59
	Non-Mangrove	82.4%	87.5%	0.81
		Total Accuracy = 85%		*Kappa = 0.77*
2011 (GEO)	Mangrove	94.1%	100.0%	1
	Water	100.00%	97.5%	0.96
	Non-Mangrove	97.4%	93.8%	0.91
		Total Accuracy = 97%		*Kappa = 0.96*
2016 (GEO)	Mangrove	93.0%	100.0%	1
	Water	100.0%	98.8%	0.98
	Non-Mangrove	98.6%	92.5%	0.89
		Total Accuracy = 97%		*Kappa = 0.96*
2019 (WV3)	Mangrove	86.8%	98.8%	0.98
	Water	97.1%	82.5%	0.76
	Non-Mangrove	90.1%	91.3%	0.87
		Total Accuracy = 91%		*Kappa = 0.86*

The confidence attached to the temporal analysis of VHSR images was judged sufficient to map (Figure 4) and document spatiotemporal changes at fine scales of the main land-covers within the mangrove area itself and its surroundings.

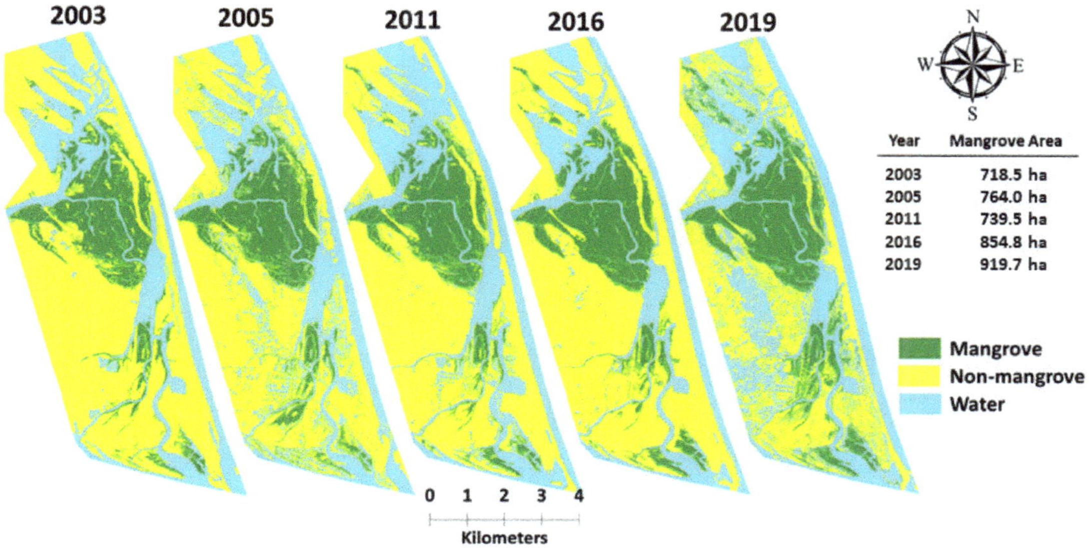

Figure 4. Land-cover maps produced from VHSR images (Quickbird for 2003 and 2005, GeoEye-1 for 2011 and 2016, and Worldview-3 for 2019) after supervised classification and visual adjustment; mangrove cover (in hectares) for individual years are mentioned.

Between 2003 and 2019, the overall mangrove cover of Pichavaram expanded by about 2 km^2, from 718.5 ha to 919.7 ha, (Figure 5a). The area of natural mangroves increased by 62 ha from 600 to 662 ha while the area of mangroves planted in fishbone plots expanded by about 139 ha from 118.5 to 257.7 ha.

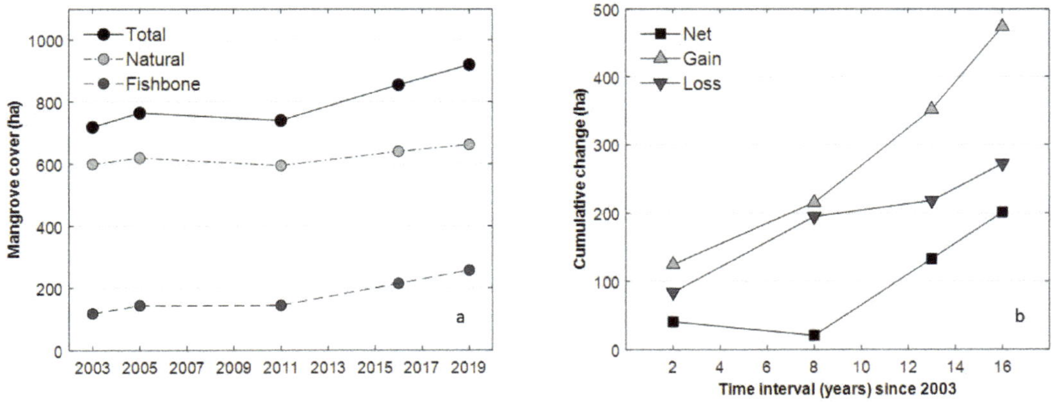

Figure 5. Change in cover among total, natural, and artificial mangrove area (**a**). Cumulative gross gain, gross loss, and net gain of mangrove cover at different time intervals since 2003 (**b**).

3.2. Change Detection and Analysis

The gain and loss in mangrove habitat have been gradual, with mangrove gross gains of 17.4% (124.8 ha), 30.0% (215.6 ha), 49.3% (354.1 ha), and 65.9% (473.6 ha) and mangrove gross losses of 11.7% (84.0 ha), 27.1% (195.0 ha), 30.4% (218.4 ha), and 37.9% (272.4 ha) at 2, 8, 13, and 16 year intervals, respectively (Figure 5b). The mangrove gross loss was always lower than the mangrove gross gain, and a net gain of mangrove areas of 28.0% (201.2 ha) relative to the mangrove cover of 2003 was assessed.

Looking more closely at the different change classes, it appears that the gain in mangrove areas mostly resulted from the conversion of non-mangrove areas to mangrove areas (Figure 6).

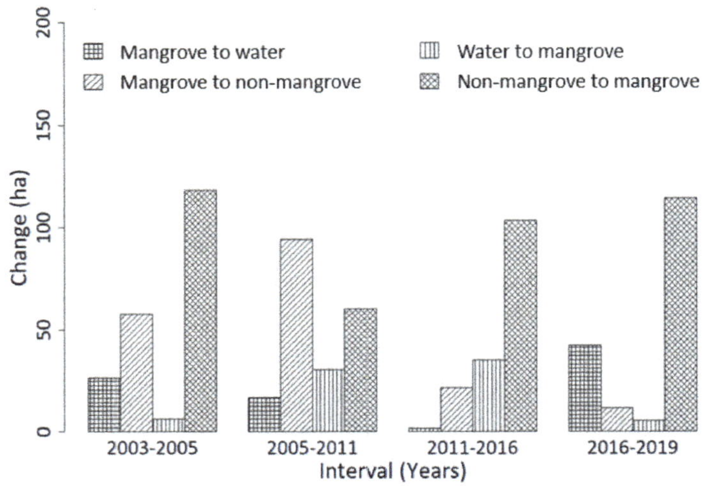

Figure 6. Progressive change in land-cover (in hectares) at image acquisition intervals.

Non-mangrove areas overall contributed to 396.4 ha of the mangrove growth between 2003 and 2019 with progressive increments of 118.5 ha (2003–2005), 60.3 ha (2005–2007), 103.4 ha (2011–2016), and 114.2 ha (2016–2019). Conversion from water to mangrove areas was also observed, albeit negligible, with an overall conversion between 2003 and 2019 of 2%. Conversely, we observed an overall combined loss of mangrove areas to non-mangrove or water areas with progressive losses of 84.1 ha (2003–2005), 81.9 ha (2003–2011), 61.8 ha (2003–2016), and 59.1 ha (2003–2019), this latter value corresponding to 8.2% of the mangrove area in 2003.

Looking at the change classes 'mangrove to water' and 'non-mangrove to water', we could detect changes in the shoreline (Figure 7).

Notably, the Quickbird image of early January 2005, 10 days after the Indian Ocean tsunami, revealed the disappearance of sandbars close to the mouth of the Uppanaru River (Figure 7a,b) and along the coastline near central Pichavaram (Figure 7f,g). However, in the subsequent images, no lingering effects of the apparent flood event were visible (Figure 7c–e,h–j) New sandbars visible in 2011–2019 images appeared near the mouth of the Kollidam River in the south (Figure 7k–o). Furthermore, a substantial extent of non-mangrove inland areas were apparently flooded, thus creating temporary water bodies in the 2005 image (waterlogging as an aftereffect of irrigation; discussed later) (Figure 7p–t) that were converted back to non-mangrove areas between 2005 and 2011 (waterlogging was not observed in 2011).

In addition, we observed a constant and progressive conversion of non-mangrove areas to water areas along the coast, indicating shoreline erosion, while new non-mangrove areas were being formed in the southern part of Pichavaram (particularly near the mouth of the Kollidam River) as a probable response of sediment deposition. These latter areas were not colonized by mangroves in 2019.

The construction of artificial channels within the mangrove area itself, as well as the construction of artificial pond-like structures (Figure 8a–e) or artificial barriers (Figure 8f–j) across upstream channels, could be observed along the periphery of the Pichavaram mangrove forests and further inland.

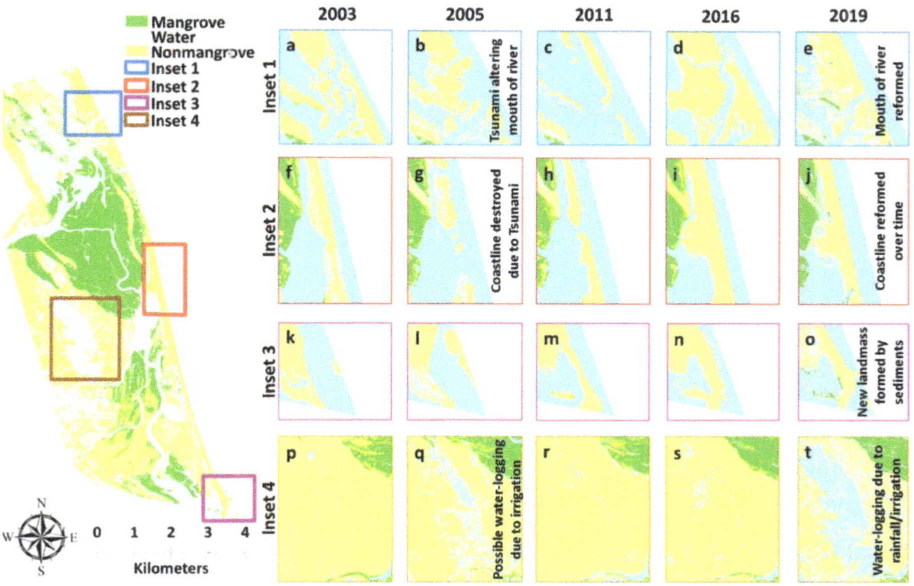

Figure 7. Changes, with a focus on the land (mangrove or non-mangrove) to water changes, in the land-cover of Pichavaram area and surrounding areas such as (**a–e**) the mouth of the Uppanaru River, (**f–j**) coastline near the central Pichavaram forest, (**k–o**) the mouth of the Kollidam River, and (**p–t**) inland area.

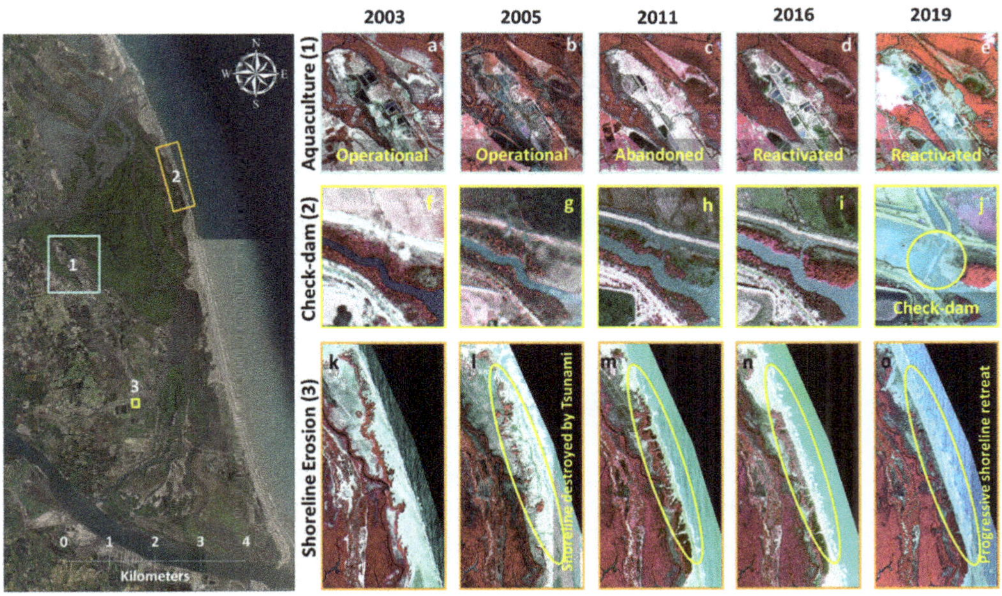

Figure 8. Changes observed in the land-cover of Pichavaram mangroves and the surrounding areas from the visual inspection of the VHRS images (image excerpts on the **right**) in combination with a 2016 color-composite image provided by Google Earth (**left**). Aquaculture expansion in the vicinity of the mangrove forest (**a–e**), check-dam construction in freshwater channel to Pichavaram (**f–j**), and progressive shoreline erosion (**k–o**) can be easily identified and monitored.

3.2.1. Change in Mangrove Cover within Fishbone Plantations

Overall, we observed a gradual increase in the mangrove cover within the fishbone areas (Figure 9a). The annual expansion rate of the mangroves within the fishbone plots implemented prior to or in 2003 was about 1.7% but could reach 4% for a number of plots where mangrove cover was between 20 and 40% in 2003. However, a number of plots showed minor loss in mangrove cover, and a single fishbone plot, located close to the mouth of the Uppanaru River, did not show an expansion of its mangrove cover during the entire study period (Figure 9a, bottom evolution profile).

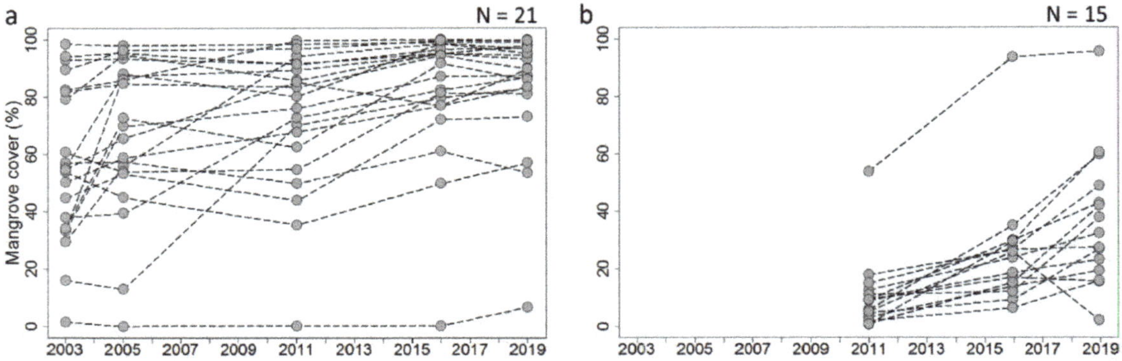

Figure 9. Changes in the mangrove cover (% relative to fishbone area) within individual fishbone plots implemented prior to or in (**a**) 2003 and (**b**) between 2005 and 2011.

A similar trend (Figure 9b) was observed for the fishbone plots implemented between 2005 and 2011, with most of the plots reaching between 20 and 60% of mangrove cover by 2019. The annual mangrove expansion rates observed within plots ranged from 0.4% to 3.5%, with an average growth rate of 1.7%—values very similar to those observed for the plots implemented prior to or in 2003.

3.2.2. Dieback in Natural Mangroves

Mangrove dieback in the central west Pichavaram area (Figure 10a), first reported by the forest department in late 2008, was confirmed by our in situ observations in 2019 (Figure 10d).

We observed that the phenomenon was affecting only a number of river-fringing *Rhizophora apiculata* trees along about 1.5 km of river banks. From a closer visual inspection of the GeoEye image of January 2016 with support from the Google Earth VHSR satellite images of April and November 2017, we estimated that tree defoliation started in 2017 (between April and November) along the river banks of a 2.5 ha water basin located 700 m eastward of active aquaculture ponds, impacting a mangrove area of 0.8 ha. The majority of the dieback was noticed along three interior channels, two southward and one northward from a branch of the Uppanaru river, and mortality was observed along both sides of the channels. In total, the phenomenon was observed along 2 km of fringe mangrove area along several channels (Figure 11a).

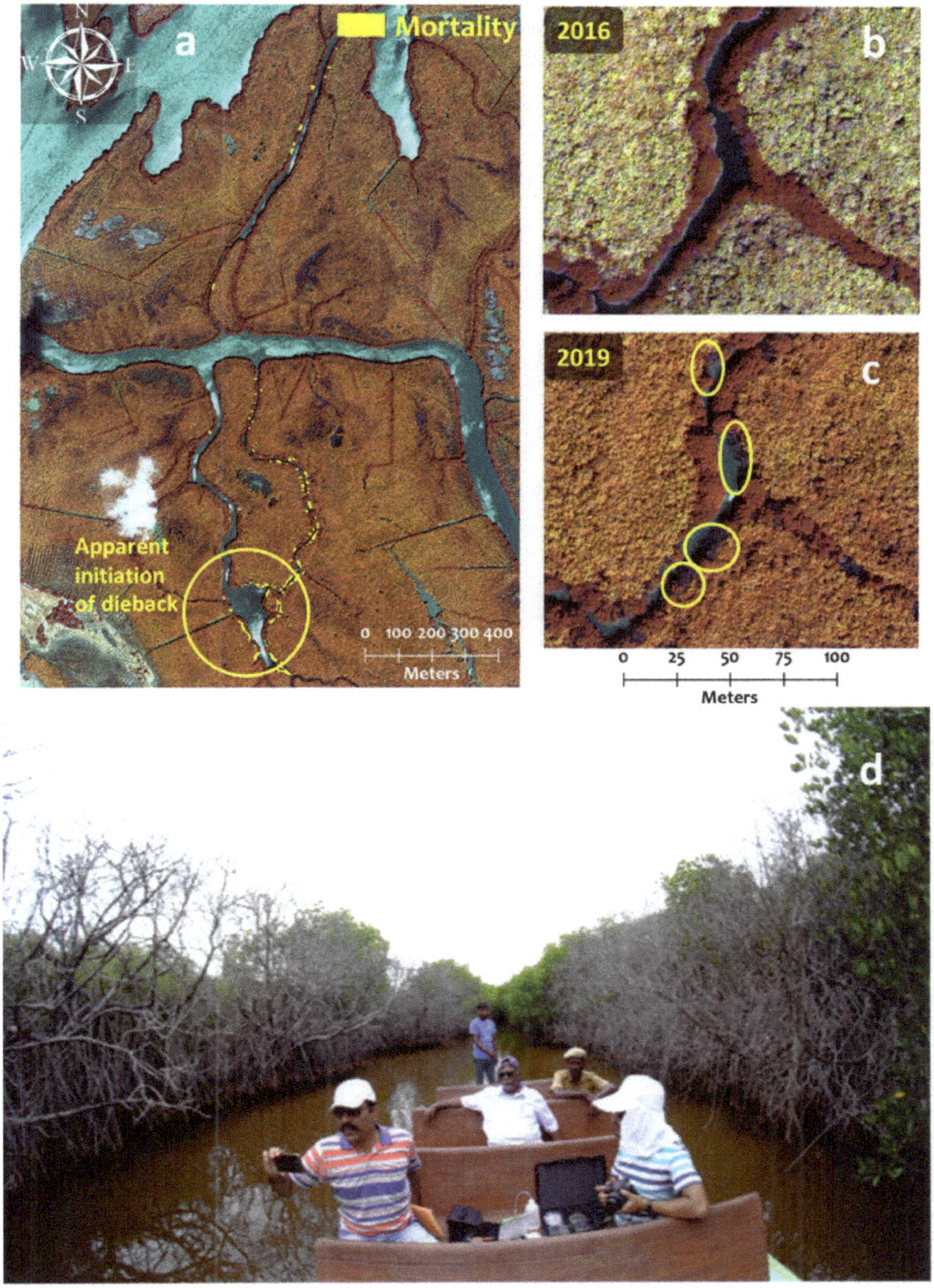

Figure 10. *Rhizophora* dieback spots in Pichavaram highlighted from the 2019 Worldview-3 image (**a**). Inset images show areas with presence of *Rhizophora* in 2016 (**b**) and subsequently dieback in 2019 (**c**), as confirmed by in situ observations (**d**).

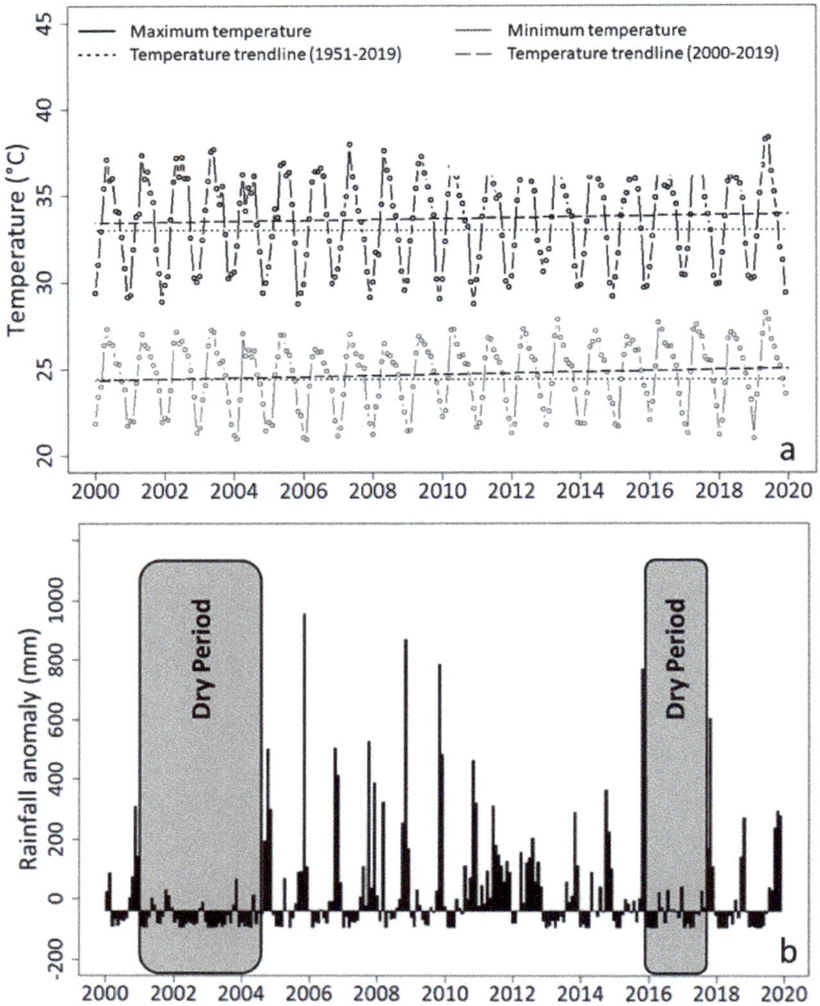

Figure 11. Air temperature for Cuddalore, near Pichavaram. Monthly maximum and minimum temperatures are provided for the time period of 2000–2019. Linear trends for both the 1951–2019 and 2000–2019 time periods are also indicated (**a**). Median rainfall anomalies for Pichavaram from 2000 to 2019 (encompassing the study period) are presented. Abnormal dry periods from 2001 to 2005 and from 2016 to 2017 are highlighted (**b**).

3.2.3. Additional Diagnostics Based on Meteorological Observations and Water Salinity

Overall, maximum air temperature ranged between 27.3 and 38.3 degrees Celsius for the time period of 1951–2019 and between 29.7 and 38.2 degrees Celsius for the time period of 2000–2019 (Figure 11a). Monthly minimum and maximum temperatures increased by an average of 1.37 and 1.77 degrees Celsius, respectively, over the 1951–2019 period in the Pichavaram region. Both the minimum and maximum air temperature trends showed gradual increases over the two time periods (with a steeper increase in minimum temperatures over the last 20 years than over the long-term period), as measured by the upward slopes of the trend lines that were 0.0356 and 0.02573, respectively.

Regarding median monthly rainfall data (Figure 11b), the analysis showed two exceptionally dry periods between 2001 and 2005 and between 2016 and 2017. A rainfall surplus was observed throughout 2011.

Salinity levels (data not shown) recorded from the water samples collected during our field visits (see location of samples in Figure 2) ranged from 30 parts per thousand (ppt) to 38 ppt, which is comparable to sea-level salinity. Low salinity levels (<1 ppt; comparable to freshwater salinity level) were observed only in two locations, close to each other, in the vicinity of a check-dam that was recently constructed upstream of the mangrove area (Figure 2d).

4. Discussion

4.1. Mangrove Cover Estimates Using Very High Spatial versus Moderate Spatial Resolution Images

The VHSR-based mangrove cover estimates (from 718 ha in 2003 to 919 ha in 2019) were consistently lower than the Landsat-derived mangrove cover estimates (from 800 ha to 1078 ha) for the same study area (Figure 12).

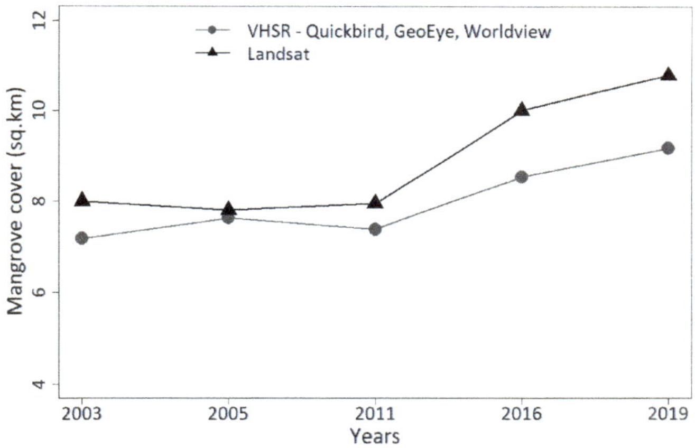

Figure 12. Comparison of mangrove cover as estimated by VHSR images (Quickbird, GeoEye, and Worldview) and Landsat.

The Landsat-image-based mangrove cover was in much greater agreement with the estimates of mangrove cover provided in earlier studies [30,36] that used moderate resolution sensors for mapping mangrove vegetation in Pichavaram. This difference may be attributed to the more accurate level of detail captured by VHSR images compared to that of moderate resolution images. The geomorphology of mangroves is often characterized by fragmentation, with patches of vegetation interlaced by water channels and small open water bodies. Unlike VHSR images, moderate resolution sensors often classify otherwise mixed pixels as vegetation, resulting in overestimates of mangrove vegetation (Figure 1). This study emphasizes the need to map complex and fragmented habitats such as mangroves using VHSR images, especially while studying very localized mangrove habitats such as Pichavaram. The methodology used in this study may not be novel and the results demonstrating the robustness of VHSR may not be unexpected. However, in the case of the Pichavaram mangrove forest, an analysis of time series of VHSR images acquired over 16 years proved to be particularly interesting to highlight the range of fine-scale positive or negative evolution trends of the mangrove habitats through an up-to-date diagnosis of the mangrove situation, where a lack of in situ data has been an impediment.

4.2. Monitoring Spatio-Temporal Changes in Mangroves Using Time Series of VHSR Images

The majority of the VHSR images were classified with a >90% accuracy. From the confusion matrix, it is clear that the mapping accuracies of the individual land-covers were affected by the presence of water (particularly in the Quickbird image from 2005), which may have been either residual moisture or actual sea-water logged in both open-mangrove and non-mangrove areas. This is not entirely unexpected due to the contrasting spectral responses of aquatic and non-aquatic surfaces, which hinders the performance of both supervised and unsupervised methods of classification. Satellite images of coastal regions and ecosystems in particular are often influenced by tidal water fluctuations. Therefore, image acquisitions at low tide may guarantee desired levels of accuracy. As suggested by the high-level accuracy reached for the classification of the GeoEye image of 2011, a year with rainfall surplus, steeper viewing zenith angles along with adequate sun-viewing azimuth angles may help mangrove areas to be discriminated from water areas. Sun-frontward angle configurations must be particularly avoided to limit sun backscattering from water areas, as observed and explained in [18].

Further, this analysis highlights the robustness of the VHSR images in capturing details of the changes that have occurred in the mangrove ecosystem including progressive gain and loss of mangrove habitat since 2003. The ecosystem itself is very dynamic, with constant changes taking place at a very fine spatial scale. As evident from Figure 13, changes that were captured by the VHSR images in this study would otherwise be difficult to observe and analyze using moderate or coarse resolution images.

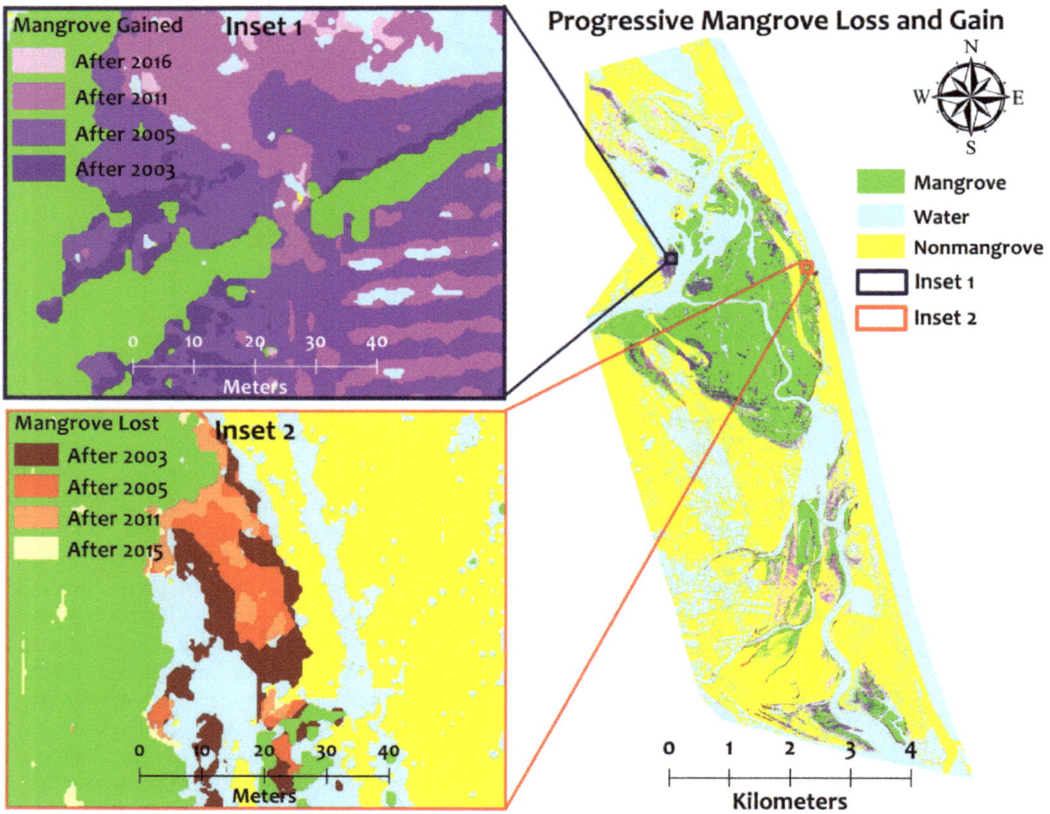

Figure 13. Mangrove gains and losses as captured by VHSR images at different time intervals in Pichavaram.

The visual inspection of VHSR satellite images provided by Google Earth is often helpful and complementary to the analysis of any multispectral images. The performance of the supervised classification of mangroves areas based on Google Earth VHSR false-color composites deserves further analysis.

Despite the robustness of the VHSR images in capturing spatial details, the lack of in situ observations has been a great impediment towards mapping the Pichavaram mangroves at the species level. Although the existing literature provides some basic information about the existing mangrove vegetation community [42,43], it is very difficult to match this information with the VHSR image data without any accurate spatial and ground reference of species identification and distribution. From our in situ observations and literature records, we can safely assert the domination of *Rhizophora* sp. along the river edges and *Avicennia marina* in the interior mangroves. However, mapping and identifying the co-existing species from multispectral VHSR satellite images remains difficult without accurate in situ observations. Even from the structural point of view, mapping species distribution in Pichavaram remains complicated. The forest canopy in the interior mangroves is formed by tree crowns of few square meters, either densely distributed with uniform tree height or sparsely located in the case of fishbone plantations. In such a scenario, even the use of sophisticated textural or combined spectral–textural approaches [26,29] applied on VHSR images for mapping and monitoring species has its limitations [44]. This aspect could be addressed in the future using a combination of active sensors, such as LiDAR, along with extensive in situ field observations of species location and canopy structure and size. Utilizing LiDAR may also facilitate understanding coastal geomorphology, more specifically erosion and the accumulation of sediments, shoreline retreat, and sea-level rise, factors which will determine the current and future sustainability management of the Pichavaram mangroves.

4.3. *Preliminary Diagnosis for the Pichavaram Mangroves*

4.3.1. Land-Cover Conversion to Mangrove Area

We observed a considerable extent of non-mangrove areas being converted to mangrove from our change analysis over a period of seventeen years. This conversion was mostly the result of the efforts undertaken by the state forest department to (re)introduce mangroves in Pichavaram through the implementation of fishbone plantations.

4.3.2. Mangrove Cover Fluctuations

However, in addition to the increment in mangrove cover, losses in mangrove areas were also observed, albeit to a lesser extent. Most of the losses in mangrove areas were observed from 2003 to 2005 and then from 2016 to 2019. The loss in mangroves from 2003 to 2005 may be attributed to the effects of the Indian Ocean Tsunami in December 2004. The mangroves also witnessed the landfall of Cyclone Thane in late 2011. However, they may have recovered in the subsequent five-year gap, as no mangrove decline was observed between the 2011 and 2016 images. Since no VHSR images were captured immediately after the landfall of the cyclone, the estimation of any mangrove damages whatsoever due to the landfall has not been possible. The loss in mangrove between 2016 and 2019 may be partly attributed to the mangrove dieback that was reported in the later part of 2017, which resulted almost 1 ha of *Rhizophora* being lost by 2019.

We did observe mangrove losses in years between 2005 and 2016 in our analysis; however, we did not have any in situ observations or studies on mangrove losses in Pichavaram at such a fine spatial scale that allowed us ascertain the causal factors behind the observed losses. Therefore, we can probably attribute the losses to possible localized environmental disturbances.

4.3.3. Effects of Tsunami and Shoreline Erosion

The destruction of the sandbars at the mouth of the Uppanaru River and the coastline in central Pichavaram can be clearly attributed to the devastating effect of the Indian Ocean

Tsunami of 2004 that was captured by the Quickbird image acquired just 10 days after the incident. The visual inspection of the 2005 image showed the effect of the tsunami, mostly along the coastal areas and to a lesser extent on the inland areas (Figures 4 and 7a–e). These sandbars were reformed in time, most likely as a result of the deposition of both riverine and marine sediments. In addition, we also observed progressive shoreline erosion along the eastern boundary of the Pichavaram mangroves and sediment accumulation leading to shoreline creation along the mouth of the Kollidam River, which borders the mangrove ecosystem in the south (Figure 7f–o) The progressive shoreline retreat might be disastrous to the mangroves of Pichavaram, as it might lead to sandy bars encroaching on mangrove habitat or the creation of a perennial hypersaline environment due to seawater inflow that may lead to salinity-induced stress, particularly among the emergent mangroves, both natural and artificially planted in fishbone plots. The Pichavaram mangroves are already facing significant long-term threats from climate-change-induced sea-level rises in the northern Indian Ocean coastline [45].

4.3.4. Agriculture, Aquaculture and Other Developmental Activities

We also observe inland non-mangrove areas being converted to water from 2003 to 2005; this temporary conversion could be attributed to effects of irrigation practices. In addition, a lot of inland non-mangrove areas were converted to water bodies from 2016 to 2019. On a closer visual inspection of the raw image, these areas were found to be mostly agriculture and dry aquaculture ponds that were water-logged, probably due to rainfall accumulation or irrigation practices, just before the acquisition of the Worldview image of 2019. In addition, we observed the progressive expansion of aquaculture ponds close to the mangrove boundary throughout the 16-year study period.

While some of the aquaculture ponds may have been abandoned over time (hence, we observed these ponds to be colonized with natural vegetation), the majority of them remained apparently active. These anthropogenic practices should be regularly monitored, and any possible evidences of effluent discharge from these systems into the mangrove ecosystem should be investigated. The legal prohibition of such activities should also be considered if such activities are found to affect the mangrove ecosystem. In terms of any possible effect of the aquaculture practices on the Pichavaram mangrove ecosystem, opinions have been ambiguous [46,47] (discussed in Section 4.3.6).

In addition, we observed the construction of recent check-dams across freshwater streams flowing into the Pichavaram mangrove forest. These dams were constructed with the intention of restricting saltwater intrusion into upstream agriculture, though, in the process, these structures have also impeded the flow of freshwater into the mangrove ecosystem, which otherwise is critical for mangrove ecosystem functioning. Therefore, a mechanism for regulating the flow of freshwater into the mangrove ecosystem should be implemented.

4.3.5. On the Mangrove Rehabilitation Efficiency

Most of the fishbone plots implemented in 2003 showed gradual improvements in mangrove cover, with a gain of about 1.3 km^2 between 2003 and 2019. The mangrove cover in the fishbone plots implemented between 2005 and 2011 had also improved and will hopefully achieve complete mangrove cover in due course if environmental conditions remain favorable.

The particular fishbone plot that showed negligible mangrove growth during the sixteen-year time period was very close to the mouth of the Uppanaru River, an area most likely directly impacted from the Indian Ocean Tsunami in 2004 that may have destroyed any mangrove saplings planted there. Any chance of a quicker recovery may have been hampered by the landfall of Cyclone Thane in late 2011. The fishbone plots that had not shown any significant improvement in mangrove cover need to be revisited and periodically monitored, and any possible reason for the reduced mangrove growth rate should be investigated.

This improvement in overall mangrove cover highlights the apparent success of the rehabilitation and conservation measures that have been implemented so far by the state Forest Department of Tamil Nadu. However, there has been no attempt to assess the efficiency of the restoration measures implemented in Pichavaram so far. This is the first study to provide a quantitative assessment of the efficiency of the restoration efforts at the finest possible scale, and it further emphasizes the importance of the periodic monitoring of the Pichavaram mangroves for the future assessment of the success/failure of the current/ongoing restoration efforts and their possible effects on the mangrove ecosystem. Whether the monoculture plantations by the Forest Department contribute to the long-term sustainability management of the Pichavaram mangroves or to reducing biodiversity leading to ecological complications in the future remains to be ascertained.

Therefore, we argue that the increase in mangrove cover cannot be the sole objective of future coastal management plans by the state forest department. The Pichavaram mangrove environment is already hypersaline, possibly owing to erratic rainfall-induced freshwater fluctuation and deficits along with rising minimum temperatures, resulting in more frequent hot and dry periods [45]. Further, the construction of check-dams upstream for preventing salt-water intrusion has also reduced the inflow of freshwater into the mangroves. Unless a policy for regulating freshwater inflow into the mangroves is implemented, such persistent hypersaline conditions may prove to be detrimental to newly planted mangrove saplings, resulting in reduced mangrove growth or even sapling mortality. Historically, hypersaline conditions have severely affected the existing plant diversity in Pichavaram, with extreme halophytic plants becoming dominant in the mangrove landscape, thus reducing the distribution of the less halophytic plant species [42,43,48]. Low plant diversity may be a precursor for ecosystem collapse, especially in an event of a widespread dieback or disease, from which mangroves may not recover from.

Furthermore, the sites for implementing restoration measures need to be carefully chosen, considering all the environmental factors that may affect mangrove growth and recovery and, ultimately, the success of the restoration effort. Care should also be taken in terms of choosing the right species for plantation [49] and restoring plant diversity along with vegetation cover to ensure proper ecosystem structure and functioning. Alternatively, just restoring the environmental conditions suitable for mangrove regrowth and survival may foster natural mangrove recovery [50].

4.3.6. Mangrove Dieback

Possible causes of mangrove dieback are plenty; studies have linked mangrove dieback to a multitude of factors, such as local herbivory [51], hurricane landfall [52], gale storm and wind gusts [53], iron toxicity [54], sea-level rise [55], drought [56], and even localized herbicide usage [57]. Therefore, it is difficult to ascertain the exact cause of the dieback observed in Pichavaram. A persistent hypersaline environment may be a reason; we have observed water salinity levels in Pichavaram comparable to sea level salinity in the dry season. However, salinity levels are usually lower in the wet season in this region, when the mangroves receive substantial freshwater from precipitation and river discharge. Earlier water quality studies in Pichavaram [37], along with careful observations of the site-specific meteorological data (i.e., temperature and rainfall) indicated severe fluctuations and progressive reductions in the level of freshwater coming into the mangrove ecosystem owing to erratic rainfall patterns and increasing temperatures, something that can be traced back to the first decade of the 21st century.

Again, we do not have any in situ observations of salinity from the wet season, as fieldwork was not possible due to the restriction of activities as a result of COVID-19-induced restriction measures. Additional in situ observations and estimations of water and soil bio-geochemistry are required to analyze whether there are any possible connections between salinity levels and/or freshwater discharge and mangrove mortality/dieback.

Further, as observed during the change analysis, human activities, especially aquaculture, have gradually increased in the vicinity of the Pichavaram mangrove ecosystem.

Although early observations had ruled out any negative effects of aquaculture practices on the mangrove ecosystem [46], current observations indicate the presence of shrimp farms and subsequent farming practices (sometimes illegal) responsible for the discharge of effluents and chemicals leading to fish and bird population declines, along with the introduction of invasive species in the ecosystem, all of which have caused damage to the ecosystem itself and the economy of the fishermen [47]. Therefore, a possibility of these chemical discharges into the adjacent mangrove ecosystem causing tree mortality cannot be ruled out. However, we need more evidence in the form of in situ bio-geochemical observations to substantiate this possibility.

4.4. Preliminary Recommendations for the Pichavaram Mangroves

We insist on the necessity of the continuous monitoring of the Pichavaram mangrove ecosystem to highlight the hotspots of mangrove degradation, as well as the subsequent prioritization of appropriate fine-scale restoration and conservation measures. For this purpose, the present analysis confirms the robustness of VHSR images in capturing details of the changes that have occurred in the mangrove ecosystem compared to previous studies conducted using coarser spatial resolution imagery [36]. Implementing a supervised classification of VHSR images remains quite intuitive because image interpretation matches the scale of human perception of mangrove attributes and environment features in the field. For the large-scale analysis of change in mangroves, the use of MSR images remains complementary.

However, as the tremendous potential of VHSR images in studying mangroves cannot compensate for the lack of in situ observations, initiatives to carry out regular in situ scientific forest inventories for describing forest structures and species distribution, along with abiotic parameters such as soil and water quality parameters (salinity and other biochemical properties), should be supported and encouraged by the local authorities to provide early warning in case of stress events [17], as experienced in Pichavaram [32,37].

The implementation of socio-ecological regulations for the integrated management and exploitation of mangrove resources is of prime urgency for the sustainability of both the mangrove forest and the livelihood of the fishermen community [38]. Indeed, progressive shoreline erosion, freshwater deficits, rising temperatures, and massive dieback are factors that are already threatening the mangroves of Pichavaram. Furthermore, any major environmental disturbance in the future such as a cyclone landfall or a tsunami could make the recovery of the mangroves very difficult, resulting in the unfortunate collapse of the mangrove ecosystem. This would, in turn, severely affect the livelihood of the surrounding coastal community (mostly fishermen), which is dependent on the sustenance and well-being of the mangrove ecosystem for their survival.

Finally, conservation decisions need to consider not only the environmental dynamics within the mangrove ecosystem but also the developments taking place outside of it [58]. The threat of developmental projects such as hydrocarbon exploration has already been perceived as a significant threat to the Pichavaram mangrove ecosystem [59]. Therefore, a holistic approach towards the conservation of coastal wetlands such as the Pichavaram mangroves is critical to the resilience of both the mangrove ecosystem and the livelihood of the local coastal community [60].

5. Conclusions

This study is the first known attempt of utilizing VHSR images for mapping and analyzing spatiotemporal changes in the Pichavaram mangroves. Our work also provides the first fine-scale in-depth assessment of the success of the fishbone plantations implemented by the forest department for restoring mangroves in Pichavaram. We also report widespread dieback and tree mortality among the fringe mangroves of Pichavaram, more specifically in *Rhizophora apiculata*. Our approach can be replicated to study and analyze similar mangrove ecosystems within coastal India and beyond, where the lack of ancillary data might be an impediment. Through the study of the Pichavaram mangrove region, we illustrated how VHSR images can provide robust and crucial information on fragile mangrove regions in a simple manner, especially where there are an apparent lack of forest data and infrequent monitoring. This remains a prerequisite to diagnose ecosystem status, identify stressful conditions, and warn against rapid degradation.

Author Contributions: Conceptualization, S.G., C.P. and G.M.; Investigation, S.G., C.P. and G.M.; Writing—original draft preparation, S.G. and C.P.; Writing—review and editing, All authors; Image interpretation, All authors; Supervision, C.P.; Project administration, G.M., C.P. and R.M. All authors have read and agreed to the published version of the manuscript.

Funding: This research was funded by the Agence Nationale de la Recherche (ANR) via the Economic and Social Research Council, ESRC (UK) and EU–India Platform for Social Sciences and Humanities (EqUIP) program (ES/R010404/1) in the frame of the FISHERCOAST project (Coastal transformations and fishing community wellbeing—synthesized perspectives from India and Europe). The travel and research stay of C.H. and V.H. were supported by the Leibniz Centre for Tropical Marine Research (ZMT) from the Institute's core funding from the Federal Ministry of Education and Research and the State of Bremen.

Data Availability Statement: The data presented in this study are available upon request to the corresponding author.

Acknowledgments: The authors acknowledge K. Kathiresan, R. A. James, C. Lakshumanan, and T. Usha for kindly providing the independent validation data set of this project. The authors thank the Tamil Nadu Forest Department for allowing us to undertake this study.

Conflicts of Interest: The authors declare no conflict of interest.

References

1. Ellison, J.C. (Ed.) *Biogeomorphology of Mangroves*; Elsevier: Amsterdam, The Netherlands, 2019; pp. 687–715. [CrossRef]
2. Rovai, A.S.; Riul, P.; Twilley, R.R.; Castañeda-Moya, E.; Rivera-Monroy, V.H.; Williams, A.A.; Simard, M.; Cifuentes-Jara, M.; Lewis, R.R.; Crooks, S.; et al. Scaling mangrove aboveground biomass from site-level to continental-scale. *Glob. Ecol. Biogeogr.* **2016**, *25*, 286–298. [CrossRef]
3. Bunting, P.; Rosenqvist, A.; Hilarides, L.; Lucas, R.M.; Thomas, N. Global Mangrove Watch: Updated 2010 Mangrove Forest Extent (v2.5). *Remote Sens.* **2022**, *14*, 1034. [CrossRef]
4. Duke, N.C.; Meynecke, J.O.; Dittmann, S.; Ellison, A.M.; Anger, K.; Berger, U.; Cannicci, S.; Diele, K.; Ewel, K.C.; Field, C.D.; et al. A world without mangroves? *Science* **2007**, *317*, 41–42. [CrossRef]
5. Giri, C.; Ochieng, E.; Tieszen, L.L.; Zhu, Z.; Singh, A.; Loveland, T.; Masek, J.; Duke, N. Status and distribution of mangrove forests of the world using earth observation satellite data. *Glob. Ecol. Biogeogr.* **2011**, *20*, 154–159. [CrossRef]
6. Spalding, M. *World Atlas of Mangroves*, 1st ed.; Routledge: London, UK, 2010. [CrossRef]
7. Spalding, M.D.; Blasco, F.; Field, C.D. *World Mangrove Atlas*; The International Society for Mangrove Ecosystems: Okinawa, Japan, 1997; p. 178. Available online: http://www.archive.org/details/worldmangroveatl97spal (accessed on 1 April 2022).
8. Thomas, N.; Lucas, R.; Bunting, P.; Hardy, A.; Rosenqvist, A.; Simard, M. Distribution and drivers of global mangrove forest change, 1996–2010. *PLoS ONE* **2017**, *12*, e0179302. [CrossRef]
9. Valiela, I.; Bowen, J.L.; York, J.K. Mangrove forests: One of the world's threatened major tropical environments. *BioScience* **2001**, *51*, 807–815. [CrossRef]
10. Hamilton, S.E.; Casey, D. Creation of a high spatio-temporal resolution global database of continuous mangrove forest cover for the 21st century (CGMFC-21). *Glob. Ecol. Biogeogr.* **2016**, *25*, 729–738. [CrossRef]
11. Goldberg, L.; Lagomasino, D.; Thomas, N.; Fatoyinbo, T. Global declines in human-driven mangrove loss. *Glob. Chang. Biol.* **2020**, *26*, 5844–5855. [CrossRef]
12. Giri, C.; Long, J.; Abbas, S.; Murali, R.M.; Qamer, F.M.; Pengra, B.; Thau, D. Distribution and dynamics of mangrove forests of South Asia. *J. Environ. Manag.* **2015**, *148*, 101–111. [CrossRef]
13. Richards, D.R.; Friess, D.A. Rates and drivers of mangrove deforestation in Southeast Asia, 2000–2012. *Proc. Natl. Acad. Sci. USA* **2016**, *113*, 344–349. [CrossRef]
14. Giri, C.; Zhu, Z.; Tieszen, L.L.; Singh, A.; Gillette, S.; Kelmelis, J.A. Mangrove forest distributions and dynamics (1975–2005) of the tsunami-affected region of Asia. *J. Biogeogr.* **2008**, *35*, 519–528. [CrossRef]
15. Nordhaus, I.; Toben, M.; Fauziyah, A. Impact of deforestation on mangrove tree diversity, biomass and community dynamics in the Segara Anakan lagoon, Java, Indonesia: A ten-year perspective. *Estuar. Coast. Shelf Sci.* **2019**, *227*, 106300. [CrossRef]
16. Queiroz, L.D.; Rossi, S.; Calvet-Mir, L.; Ruiz-Mallen, I.; Garcia-Betorz, S.; Salva-Prat, J.; Meireles, A.J.D. Neglected ecosystem services: Highlighting the socio-cultural perception of mangroves in decision-making processes. *Ecosyst. Serv.* **2017**, *26*, 137–145. [CrossRef]

17. Lewis, R.R.; Milbrandt, E.C.; Brown, B.; Krauss, K.W.; Rovai, A.S.; Beever Iii, J.W.; Flynn, L.L. Stress in mangrove forests: Early detection and preemptive rehabilitation are essential for future successful worldwide mangrove forest management. *Mar. Pollut. Bull.* **2016**, *109*, 764–771. [CrossRef]
18. Proisy, C.; Viennois, G.; Sidik, F.; Andayani, A.; Enright, J.A.; Guitet, S.; Gusmawati, N.; Lemonnier, H.; Muthusankar, G.; Olagoke, A.; et al. Monitoring mangrove forests after aquaculture abandonment using time series of very high spatial resolution satellite images: A case study from the Perancak estuary, Bali, Indonesia. *Mar. Pollut. Bull.* **2018**, *131*, 61–71. [CrossRef]
19. Wang, L.; Jia, M.; Yin, D.; Tian, J. A review of remote sensing for mangrove forests: 1956–2018. *Remote Sens. Environ.* **2019**, *231*, 111223. [CrossRef]
20. Rahman, M.M.; Lagomasino, D.; Lee, S.; Fatoyinbo, T.; Ahmed, I.; Kanzaki, M. Improved assessment of mangrove forests in Sundarbans East Wildlife Sanctuary using WorldView 2 and TanDEM-X high resolution imagery. *Remote Sens. Ecol. Conserv.* **2019**, *5*, 136–149. [CrossRef]
21. Everitt, J.H.; Yang, C.; Sriharan, S.; Judd, F.W. Using High Resolution Satellite Imagery to Map Black Mangrove on the Texas Gulf Coast. *J. Coast. Res.* **2008**, *2008*, 1582–1586. [CrossRef]
22. Neukermans, G.; Dahdouh-Guebas, F.; Kairo, J.G.; Koedam, N. Mangrove species and stand mapping in Gazi bay (Kenya) using Quickbird satellite imagery. *J. Spat. Sci.* **2008**, *53*, 75–86. [CrossRef]
23. Lee, T.-M.; Yeh, H.-C. Applying remote sensing techniques to monitor shifting wetland vegetation: A case study of Danshui River estuary mangrove communities, Taiwan. *Ecol. Eng.* **2009**, *35*, 487–496. [CrossRef]
24. Wang, L.; Sousa, W.P.; Gong, P.; Biging, G.S. Comparison of IKONOS and QuickBird images for mapping mangrove species on the Caribbean coast of Panama. *Remote Sens. Environ.* **2004**, *91*, 432–440. [CrossRef]
25. Proisy, C.; Féret, J.-B.; Lauret, N.; Gastellu-Etchegorry, J.-P. (Eds.) *Mangrove Forest Dynamics Using Very High Spatial Resolution Optical Remote Sensing*; Elsevier: Paris, France, 2016; Volume 7, pp. 269–295. [CrossRef]
26. Proisy, C.; Couteron, P.; Fromard, F. Predicting and mapping mangrove biomass from canopy grain analysis using Fourier-based textural ordination of IKONOS images. *Remote Sens. Environ.* **2007**, *109*, 379–392. [CrossRef]
27. Taureau, F.; Robin, M.; Proisy, C.; Fromard, F.; Imbert, D.; Debaine, F. Mapping the Mangrove Forest Canopy Using Spectral Unmixing of Very High Spatial Resolution Satellite Images. *Remote Sens.* **2019**, *11*, 367. [CrossRef]
28. Wang, T.; Zhang, H.; Lin, H.; Fang, C. Textural–spectral feature-based species classification of mangroves in Mai Po Nature Reserve from Worldview-3 imagery. *Remote Sens.* **2016**, *8*, 24. [CrossRef]
29. Proisy, C.; Walcker, R.; Blanchard, E.; Gardel, A.; Anthony, E.J. (Eds.) *Mangroves: A Natural Early Warning System of Erosion on Open Muddy Coasts in French Guiana*; Elsevier: Amsterdam, The Netherlands, 2021; pp. 47–63. [CrossRef]
30. Gnanappazham, L.; Selvam, V. The dynamics in the distribution of mangrove forests in Pichavaram, South India—Perception by user community and remote sensing. *Geocarto Int.* **2011**, *26*, 475–490. [CrossRef]
31. Olwig, M.F.; Sørensen, M.K.; Rasmussen, M.S.; Danielsen, F.; Selvam, V.; Hansen, L.B.; Nyborg, L.; Vestergaard, K.B.; Parish, F.; Karunagaran, V.M. Using remote sensing to assess the protective role of coastal woody vegetation against tsunami waves. *Int. J. Remote Sens.* **2007**, *28*, 3153–3169. [CrossRef]
32. Selvam, V.; Ravichandran, K.K.; Gnanappazham, L.; Navamuniyammal, M. Assessment of community-based restoration of Pichavaram mangrove wetland using remote sensing data. *Curr. Sci.* **2003**, *85*, 794–798.
33. Vani, M.; Rama Chandra Prasad, P. (Eds.) *Geospatial Assessment of Spatio-Temporal Changes in Mangrove Vegetation of Pichavaram Region, Tamil Nadu, India*; Springer International Publishing: Cham, Switzerland, 2018; pp. 89–102. [CrossRef]
34. Löf, M.; Madsen, P.; Metslaid, M.; Witzell, J.; Jacobs, D.F. Restoring forests: Regeneration and ecosystem function for the future. *New For.* **2019**, *50*, 139–151. [CrossRef]
35. Beck, H.E.; Zimmermann, N.E.; McVicar, T.R.; Vergopolan, N.; Berg, A.; Wood, E.F. Present and future Köppen-Geiger climate classification maps at 1-km resolution. *Sci. Data* **2018**, *5*, 180214. [CrossRef]
36. Selvam, V.; Ravichandaran, K.K.; Karunakaran, V.M.; Mani, K.G.; Beula, E.J.; Gnanappazham, L. *Pichavaram Mangrove Wetland: Situation Analysis*; MS Swaminathan Research Foundation: Chennai, India, 2010; p. 39.
37. Sathyanathan, R.; Thattai, D.; Selvam, V. The Coleroon river flow and its effect on the Pichavaram mangrove ecosystem. *J. Coast. Conserv.* **2014**, *18*, 309–322. [CrossRef]
38. Kathiresan, K. A review of studies on Pichavaram mangrove, southeast India. *Hydrobiologia* **2000**, *430*, 185–205. [CrossRef]
39. Sathya, T.; Sekar, C. Stakeholder Preference, Dependence and Attitude towards Conservation of Mangrove Eco-System in South-East Coast of India. *Innovare J. Soc. Sci.* **2014**, *2*, 15–25.
40. Kuester, M.A. Absolute Radiometric Calibration: 2016 v0, Digital Globe. 2017. Available online: https://dgv4-cms-production.s3.amazonaws.com/uploads/document/file/136/ABSRADCAL_FLEET_2016v0_Rel20170606.pdf (accessed on 1 April 2022).
41. Wilkinson, G.G. Results and implications of a study of fifteen years of satellite image classification experiments. *IEEE Trans. Geosci. Remote Sens.* **2005**, *43*, 433–440. [CrossRef]
42. Kathiresan, K. Globally threatened mangrove species in India. *Curr. Sci.* **2010**, *98*, 1551.
43. Kathiresan, K.; Rajendran, N. Fishery resources and economic gain in three mangrove areas on the south-east coast of India. *Fish. Manag. Ecol.* **2002**, *9*, 277–283. [CrossRef]
44. Viennois, G.; Proisy, C.; Féret, J.-B.; Prosperi, J.; Sidik, F.; Suhardjono; Rahmania, R.; Longépé, N.; Germain, O.; Gaspar, P. Multitemporal analysis of high spatial resolution satellite imagery for mangrove species mapping, Bali, Indonesia. *IEEE J. Sel. Top. Appl. Earth Observ. Remote Sens.* **2016**, *9*, 3680–3686. [CrossRef]

45. Unnikrishnan, A.S.; Nidheesh, A.G.; Lengaigne, M. Sea-level-rise trends off the Indian coasts during the last two decades. *Curr. Sci.* **2015**, *108*, 966–971.
46. Jayanthi, M.; Thirumurthy, S.; Nagaraj, G.; Muralidhar, M.; Ravichandran, P. Spatial and temporal changes in mangrove cover across the protected and unprotected forests of India. *Estuar. Coast. Shelf Sci.* **2018**, *213*, 81–91. [CrossRef]
47. Senthalir, S. Shrimp Farms Threatening Pichavaram Forest. *The Hindu*. 9 August 2017. Available online: https://www.thehindu.com/news/cities/puducherry/shrimp-farms-threatening-pichavaram-forest/article19453172.ece (accessed on 1 April 2022).
48. Srivastava, J.; Farooqui, A.; Hussain, S.M. Vegetation history and salinity gradient during the last 3700 years in Pichavaram estuary, India. *J. Earth Syst. Sci.* **2012**, *121*, 1229–1237. [CrossRef]
49. Bosire, J.O.; Dahdouh-Guebas, F.; Walton, M.; Crona, B.I.; Lewis III, R.R.; Field, C.; Kairo, J.G.; Koedam, N. Functionality of restored mangroves: A review. *Aquat. Bot.* **2008**, *89*, 251–259. [CrossRef]
50. Kamali, B.; Hashim, R. Mangrove restoration without planting. *Ecol. Eng.* **2011**, *37*, 387–391. [CrossRef]
51. Rossi, R.E.; Archer, S.K.; Giri, C.; Layman, C.A. The role of multiple stressors in a dwarf red mangrove (*Rhizophora mangle*) dieback. *Estuar. Coast. Shelf Sci.* **2020**, *237*, 106660. [CrossRef]
52. Lagomasino, D.; Fatoyinbo, T.; Castañeda-Moya, E.; Cook, B.D.; Montesano, P.M.; Neigh, C.S.R.; Corp, L.A.; Ott, L.E.; Chavez, S.; Morton, D.C. Storm surge and ponding explain mangrove dieback in southwest Florida following Hurricane Irma. *Nat. Commun.* **2021**, *12*, 4003. [CrossRef] [PubMed]
53. Servino, R.N.; Gomes, L.E.d.O.; Bernardino, A.F. Extreme weather impacts on tropical mangrove forests in the Eastern Brazil Marine Ecoregion. *Sci. Total Environ.* **2018**, *628–629*, 233–240. [CrossRef]
54. Sippo, J.Z.; Santos, I.R.; Sanders, C.J.; Gadd, P.; Hua, Q.; Lovelock, C.E.; Santini, N.S.; Johnston, S.G.; Harada, Y.; Reithmeir, G.; et al. Reconstructing extreme climatic and geochemical conditions during the largest natural mangrove dieback on record. *Biogeosciences* **2020**, *17*, 4707–4726. [CrossRef]
55. Lovelock, C.E.; Feller, I.C.; Reef, R.; Hickey, S.; Ball, M.C. Mangrove dieback during fluctuating sea levels. *Sci. Rep.* **2017**, *7*, 1680. [CrossRef]
56. Duke, N.C.; Kovacs, J.M.; Griffiths, A.D.; Preece, L.; Hill, D.J.E.; van Oosterzee, P.; Mackenzie, J.; Morning, H.S.; Burrows, D. Large-scale dieback of mangroves in Australia's Gulf of Carpentaria: A severe ecosystem response, coincidental with an unusually extreme weather event. *Mar. Freshw. Res.* **2017**, *68*, 1816–1829. [CrossRef]
57. Duke, N.C.; Bell, A.M.; Pederson, D.K.; Roelfsema, C.M.; Bengtson Nash, S. Herbicides implicated as the cause of severe mangrove dieback in the Mackay region, NE Australia: Consequences for marine plant habitats of the GBR World Heritage Area. *Mar. Pollut Bull.* **2005**, *51*, 308–324. [CrossRef]
58. Mathevet, R.; Targowla, S.; Munisamy, A.; Govindan, V.; Narayanan, A.; Bautès, N. Wetlands for a sustainable urban future: Insights from Pondicherry, South India. *Grassroots J. Nat. Resour.* **2020**, *3*, 74–93. [CrossRef]
59. Prasad, S. Hunt for Hydrocarbons Could Sink Pichavaram, Fear Activists. *The Hindu*. 13 June 2019. Available online: https://www.thehindu.com/news/cities/puducherry/hunt-for-hydrocarbons-could-sink-pichavaram-fear-activists/article27891952.ece (accessed on 1 April 2022).
60. Mathevet, R.; Poulin, B. From conservation biology to conservation geography. *Bull. L'assoc. Géogr. Fr.* **2006**, *83*, 341–354. [CrossRef]

remote sensing

MDPI

Article

Mapping Multi-Decadal Mangrove Extent in the Northern Coast of Vietnam Using Landsat Time-Series Data on Google Earth Engine Platform

Thuy Thi Phuong Vu [1], Tien Dat Pham [2,*], Neil Saintilan [2], Andrew Skidmore [2,3], Hung Viet Luu [4], Quang Hien Vu [1], Nga Nhu Le [5], Huu Quang Nguyen [6] and Bunkei Matsushita [7]

1 Forest Inventory and Planning Institute (FIPI), Ministry of Agricultural and Rural Development (MARD), Vinh Quynh, Thanh Tri, Hanoi 100000, Vietnam
2 School of Natural Sciences, Faculty of Science and Engineering, Macquarie University, Sydney, NSW 2109, Australia
3 Faculty of Geo-Information Science and Earth Observation (ITC), University of Twente, 7522 NB Enschede, The Netherlands
4 Centre of Multidisciplinary Integrated Technologies for Field Monitoring (FIMO), The University of Engineering and Technology, Vietnam National University (VNU), 144 Xuan Thuy, Cau Giay, Hanoi 100000, Vietnam
5 Department of Marine Mechanics and Environment, Institute of Mechanics, Vietnam Academy of Science and Technology (VAST), 264 Doi Can Street, Ba Dinh District, Hanoi 100000, Vietnam
6 Department of Policy and Planning Sciences, University of Tsukuba, 1-1-1 Tennoudai, Tsukuba 305-8573, Ibaraki Prefecture, Japan
7 Faculty of Life and Environmental Sciences, University of Tsukuba, 1-1-1 Tennoudai, Tsukuba 305-8572, Ibaraki Prefecture, Japan
* Correspondence: tiendat.pham@mq.edu.au

Citation: Vu, T.T.P.; Pham, T.D.; Saintilan, N.; Skidmore, A.; Luu, H.V.; Vu, Q.H.; Le, N.N.; Nguyen, H.Q.; Matsushita, B. Mapping Multi-Decadal Mangrove Extent in the Northern Coast of Vietnam Using Landsat Time-Series Data on Google Earth Engine Platform. *Remote Sens.* **2022**, *14*, 4664. https://doi.org/10.3390/rs14184664

Academic Editor: Guangsheng Chen

Received: 22 July 2022
Accepted: 16 September 2022
Published: 19 September 2022

Abstract: A pixel-based algorithm for multi-temporal Landsat (TM/ETM+/OLI/OLI-2) imagery between 1990 and 2022 monitored mangrove dynamics and detected their changes in the three provinces (i.e., Thai Binh, Nam Dinh and Hai Phong), which are located on the Northern coast of Vietnam, through the Google Earth Engine (GEE) cloud computing platform. Results showed that the mangrove area in the study area decreased from 2960 ha in 1990 to 2408 ha in 1995 and then significantly increased to 4435 ha in 2000 but later declined to 3502 ha in 2005. The mangrove areas experienced an increase from 4706 ha in 2010 to 10,125 ha in 2020 and reached a highest peak of 10,630 ha in 2022. In 2022, Hai Phong province had the largest area of mangrove (3934 ha), followed by Nam Dinh (3501 ha) and Thai Binh (3195 ha) provinces. The overall accuracies for 2020 and 2022 were 94.94% and 91.98%, while the Kappa coefficients were 0.90 and 0.84, respectively. The mangrove restoration programs and policies by the Vietnamese government and local governments are the key drivers of this increase in mangroves in the three provinces from 1990 to 2022. The results also demonstrated that the combination of Landsat time series images, a pixel-based algorithm, and the GEE platform has a high potential for monitoring long-term change of mangrove forests during 32 years in the tropics. Moreover, the obtained mangrove forest maps at a 30-m spatial resolution can serve as a useful and up-to-date dataset for sustainable management and conservation of these mangrove forests in the Red River Delta, Vietnam.

Keywords: mangrove; remote sensing; Landsat; Google Earth Engine; Red River Delta; Vietnam

1. Introduction

Mangrove forests are trees and shrubs found in tidal wetlands and located in the tropical and sub-tropical region between 30°N and 30°S latitude [1]. They cover only 0.1% of Earth's continental surface, yet they provide a wide range of ecosystem services, including water purification, natural hazards reduction, soil and water conservation, shoreline protection and enhanced local livelihood and are considered as natural-based solutions in dealing with climate change impact [2]. The areas of mangrove forests have been changed significantly on a global scale due to anthropogenic disturbance (i.e., urbanization and increased agricultural production) and climate change [2]. However, recent studies pointed out that the rate of deforestation has been decreasing [3], and mangroves have expanded in some Southeast Asian nations and in Australia [4,5].

Among approximately 3260 km of the total coastal length of Vietnam, 2365 km are covered by mangroves representing 29 coastal provinces. Mangrove ecosystems, therefore, play an important role in protecting the Vietnamese coastline against flooding and erosion, providing biodiversity and livelihood for coastal communities as well as sequestering carbon, known as 'blue carbon' [6,7]. The Vietnamese mangroves are mainly distributed in the two deltas, the Red River Delta (RRD) in the north and the Mekong River Delta in the south [8,9]. However, the mangrove forest area in Vietnam has decreased dramatically over the past 70 years, falling from 408,500 ha in 1943 to 178,000 ha in 2000, and then continuously shrinking to 138,318 ha in 2016 [10–14]. Therefore, it is essential to obtain accurate information about mangrove forests in the past and current state that is useful to manage and effectively protect mangrove ecosystems across the Vietnamese coastline. However, there is no up-to-date map of mangroves in Vietnam; thus, mapping mangrove forests and detecting their dynamics are vital for sustainable conservation and management of mangrove resources.

Mapping mangroves at a large scale remains challenging due to the costs and labour intensiveness in field measurements for large areas. In recent years, remote-sensing-based techniques have become widely used for monitoring the Earth's surface including mangrove forests and has proven to be a key tool to effectively map mangrove dynamics in large areas in Southeast Asia [12,15–18]. Pixel-based and object-based approaches are the most common techniques for mapping mangroves and detecting their changes. These approaches can provide the most frequently updated data at a low cost [19,20]. For instance, the distribution of mangroves at a global scale using a multi-temporal Landsat dataset and a supervised Maximum Livelihood Classification (MLC) was reported by Giri et al. [17] with an overall accuracy ranging from 79% to 86%. In Vietnam, Nguyen-Thanh et al. [12] used the Landsat time series data and an object-based image analysis to monitor mangrove extent in the Ca Mau Peninsula, Vietnam, whereas Pham and Brabyn [13] used the SPOT imagery and a support vector machine (SVM) classifier to map mangrove dynamics in the Can Gio biosphere reserve region with overall accuracies of 77 and 83%. However, to date there is no spatial distribution map of mangroves along the RRD, and there is a complete lack of reliable updated statistical data of mangroves in the three coastal provinces of the RRD.

More recently, with the development of open-source software and cloud computing platforms such as the Google Earth Engine (GEE), the applications of remote sensing techniques in monitoring mangrove changes have become more popular. Previous studies have widely applied the GEE platform to map mangrove changes using multispectral sensors such as the Thematic Mapper (TM), the Enhanced Thematic Mapper Plus (ETM+), the Operational Land Imager (OLI), and the Operational Land Imager-2 (OLI-2) in Landsat and the Multi-Spectral Instrument (MSI) in Sentinel-2 [21–25]. However, to the best of our knowledge, the up-to-date mangrove forests maps and their change detection using time series Landsat (TM/ETM+/OLI/OLI-2) imagery between 1990 and 2022 have not been reported in Vietnam. Thus, this study aims to fill the gap in the current literature by investigating a pixel-based algorithm: (1) to map multi-decadal mangrove dynamics using Landsat time series data between 1990 and 2022 through the GEE platform, (2) to provide up-to-date statistical analysis of areas of mangrove forests in the Northern coast of Vietnam for the first time in 2022 using Landat-9 OLI-2 as an important national mangrove database, and (3) to provide a useful tool for decision makers in supporting the mangrove conservation and management in Vietnam.

2. Materials and Methods

2.1. Study Area

In the current study, three coastal provinces in the RRD were selected to test the performance of the pixel-based algorithm. They are Hai Phong province (20°51′54.5004″N, 106°41′1.7880″E), Thai Binh province (20°27′0″N, 106°20′24.07″E) and Nam Dinh province (20°16′45.048″N, 106°12′18.533″E), which are shown in Figure 1.

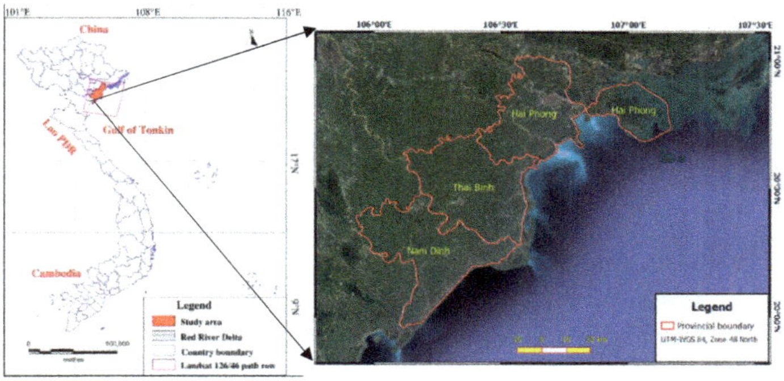

(a) (b)

Figure 1. Study area in the northern coast of Vietnam: (**a**) map of Vietnam; (**b**) three coastal provinces in the RRD.

The RRD, consisting of nine provinces (i.e., Hai Duong, Bac Ninh, Vinh Phuc, Hung Yen, Thai Binh, Nam Dinh, Ha Nam, Ninh Binh and Quang Ninh) is the second-largest delta and is located in the northern region of Vietnam with a total area of 15,000 km^2. With a population of 22 million, the RRD is the most densely populated region in Vietnam [26]. In 2019, the population density of the RRD had reached 1064 inhabitants per km^2 [27]. The total area of the three provinces is approximately 473,700 ha, of which Nam Dinh province is the largest with 166,800 ha, followed by Thai Binh province and Hai Phong province with an area of 153,400 ha and 152,300 ha, respectively [8].

The mangrove ecosystems in the RRD play a key role in protecting coastal habitats, supporting biodiversity, and providing coastal resources for local people. In the RRD, the Xuan Thuy National Park was listed as the first Ramsar site in Southeast Asia in 1989 to promote the sustainable conservation of wetlands [28]. The Ramsar site was defined as "the sustainable utilization of wetlands for the benefit of mankind in a way compatible with the maintenance of the natural properties of the ecosystem" [29]. There are five dominant mangrove species observed in this park being *Sonneratia caseolaris*, *Kandelia obovata*, *Aegiceras corniculatum*, *Rhizophora stylosa* and *Avicennia marina* [30]. Furthermore, this park is the habitat to 116 flora species and 106 fish and has significantly contributed to wetland biodiversity protection on the northern coastline of Vietnam [30]. A nature reserve, which is located in Thai Binh Province, is well-known as the Bird Conservation area, and there are several rare species listed in the Vietnamese Red Book [27]. There are four seasons in the RRD with a mean annual temperature of approximately 28 °C. The annual precipitation recorded in the last ten years is around 1800–2000 mm. In recent years, the RRD has been seriously affected by climate variability including higher temperatures, storms and flooding [28]. In particular, 2020 was recorded as the hottest year over the last 45 years and likely resulted in a dieback of mangroves [31,32].

2.2. Materials

Satellite Data

Multi-decadal Landsat surface reflectance (SR) data obtained through the GEE platform was used to map mangrove dynamics in the study area (Table 1). We used Collection 2, which were atmospherically corrected SR data with a single-channel algorithm developed by the National Aeronautics and Space Administration (NASA) Jet Propulsion Laboratory (JPL). All Landsat time series Collection 2 SR data used in the current study (Table 1) were acquired using the Java scripts on the GEE.

Table 1. Time-series Landsat imagery used for 32 years in the study area.

Sensor	Spatial Resolution (m)	Path/Row	Year	Band Used
Landsat-5 TM	30	126/46	1990, 1995, 2005, 2010	Blue, Green, Red, NIR, SWIR
Landsat-7 ETM+	30	126/46	2000	
Landsat-8 OLI	30	126/46	2015, 2020	
Landsat-9 OLI-2	30	126/26	2022	

Considering the seasonality changes of mangrove forests and their species, a total of 82 cloud-free Landsat scenes between 1990 and 2022 were used to map mangrove dynamics in the RRD. We applied an image normalisation technique to make all images consistent during the pre-processing process. To minimize the effects of tidal and water levels, we selected the datasets acquired in early morning when the tidal level was the lowest.

2.3. Methods

2.3.1. Generation of Training and Validation Datasets for the Study Area

In this study, the training and the validation data were obtained from very high spatial resolution images in Google Earth Pro imagery (2020). A total of 2370 points were randomly selected, of which 1896 points (80%) were used for the training set and 474 points (20%) were used for the validation set (Figure 2).

As shown in Figure 2, a polygon including mangrove forest and non-mangrove was created from the spatial data and consisted of a sea dyke and river layers with a total area of 35,566 ha. A 3 km buffer generated from the high-resolution images of Google Earth Pro imagery in 2020 was used to capture the entire mangrove forests area as suggested by Thomas et al. [33] and Bunting et al. [34].

2.3.2. Computation of Spectral Indices

Four indices were calculated from the SR data of Landsat (5/7/8/9) images to identify vegetation and open surface water bodies as suggested by Wang et al. [22] and Pham et al. [35]. They are the Normalized Difference Vegetation Index (NDVI) [36], the Enhanced Vegetation Index (EVI) [37], the Land Surface Water Index (LSWI) [38] and the modified Normalized Difference Water Index (mNDWI) [39]. The equations are shown below:

$$\text{NDVI} = \frac{\rho_{nir} - \rho_{red}}{\rho_{nir} + \rho_{red}} \quad (1)$$

$$\text{EVI} = 2.5 \times \frac{\rho_{nir} - \rho_{red}}{\rho_{nir} + 6 \times \rho_{red} - 7.5 \times \rho_{blue} + 1} \quad (2)$$

$$\text{LSWI} = \frac{\rho_{nir} - \rho_{swir}}{\rho_{nir} + \rho_{swir}} \quad (3)$$

$$\text{mNDWI} = \frac{\rho_{green} - \rho_{swir}}{\rho_{green} + \rho_{swir}} \quad (4)$$

where ρ_{red}, ρ_{green}, ρ_{blue} and ρ_{swir} are the surface reflectance at red (band 3 for TM/ETM+ or band 4 for OLI and OLI-2), green (TM/ETM+ band 2 or OLI/OLI-2 band 3), blue (TM/ETM+ band 1 or OLI/OLI-2 band 2) and short-wave infrared (SWIR: TM/ETM+ band 5 or OLI/OLI-2 band 6) bands, respectively.

We proposed a framework using a pixel-based mapping algorithm to map mangrove forests and automatically detect their changes using time series Landsat images from 1990 to 2022 through the GEE platform as shown in Figure 3. We developed the Python scripts using the geemap package (https://github.com/giswqs/geemap, accessed on 16 July 2021) to map mangrove extent in the RRD.

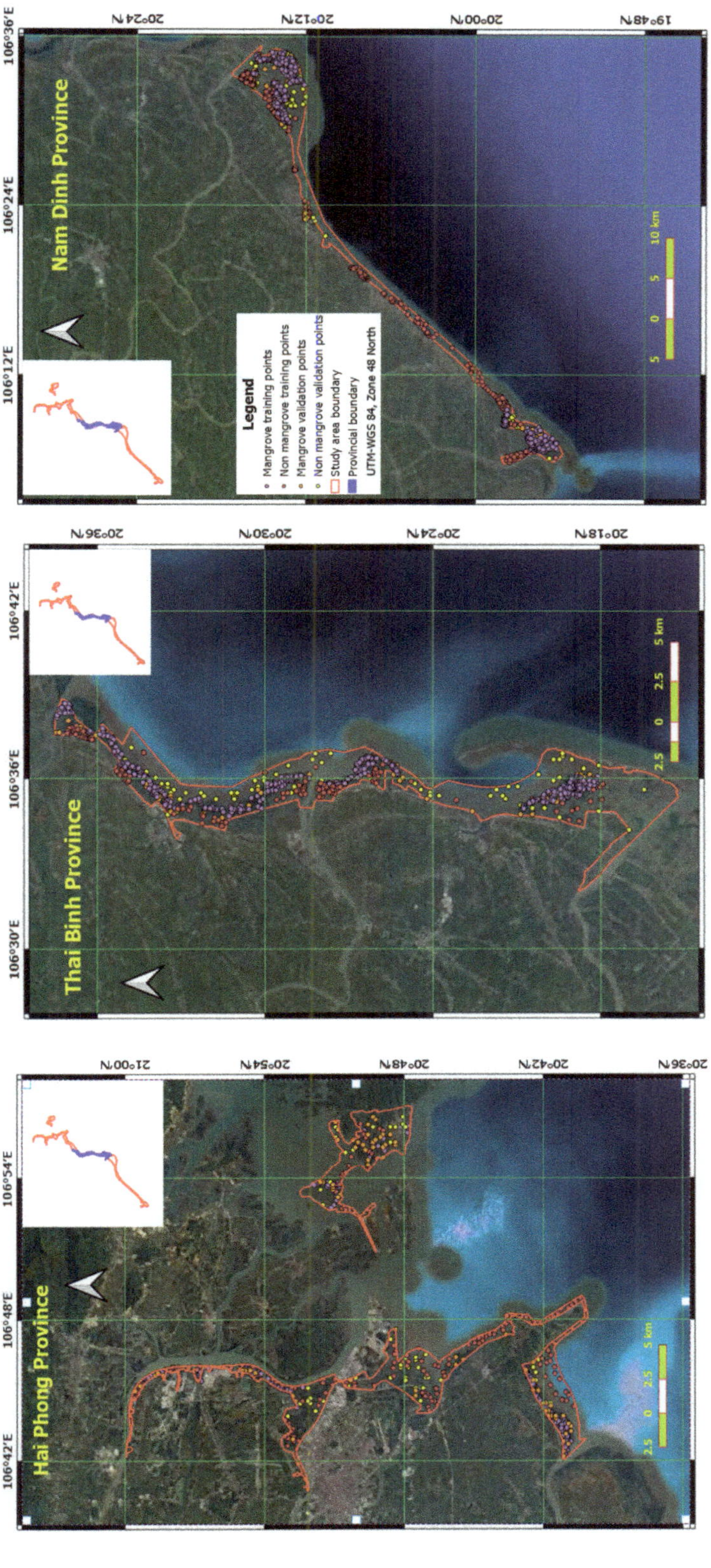

Figure 2. Training and validation points for mangrove mapping during the year 2020 in the RRD, Vietnam: (**a**) Hai Phong; (**b**) Thai Binh; (**c**) Nam Dinh.

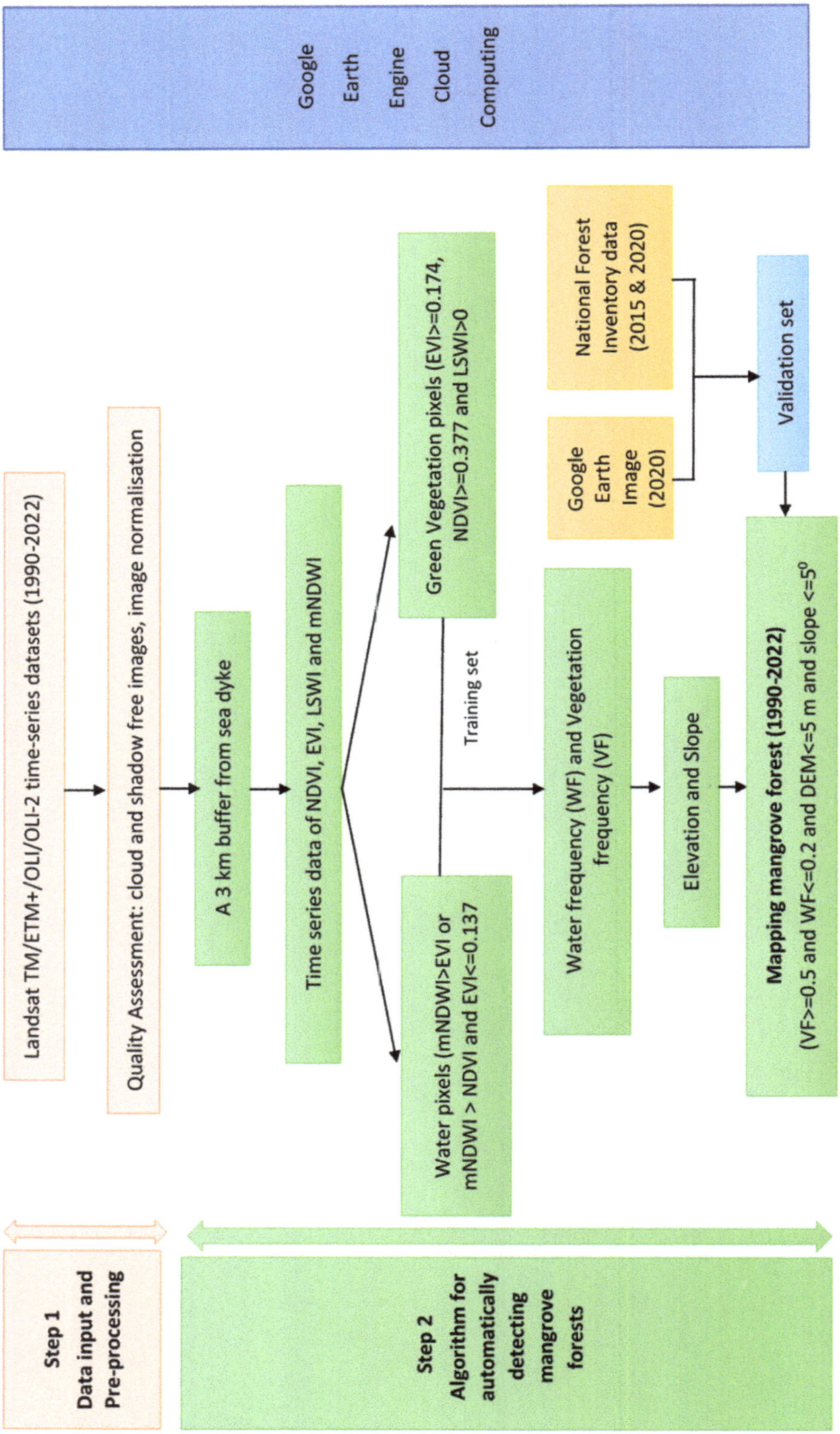

Figure 3. A framework developed in this study for mangrove mapping and change detection using a pixel-based algorithm.

2.3.3. Mangrove Mapping Algorithm

In this study, we used a pixel-based algorithm, which was developed by Wang et al. [22] for mapping coastal wetlands using time series Landsat datasets in 2018 to generate annual maps of mangrove forests between 1990 and 2022. The algorithm includes three steps for processing each pixel: (1) identifying open surface water body and green vegetation, (2) calculating annual frequency for surface water body and vegetation, and (3) generating annual maps of mangrove forest. The present study used data in 2020 and in 2022 to check and modify the thresholds provided by the original study and then used these modified thresholds to estimate the mangrove area for other years.

We used a frequency-based approach from Landsat time series to mitigate the effect of periodical tidal dynamics and bad-quality observations as suggested by Wang et al. [22]. The open surface water body and vegetation frequencies in a year were calculated using the following equations:

$$\mathrm{WF} = \frac{\mathrm{Nwater}}{\mathrm{Ngood}} \tag{5}$$

$$\mathrm{VF} = \frac{\mathrm{Nvegetation}}{\mathrm{Ngood}} \tag{6}$$

where

WF and VF are the frequencies of surface water body and vegetation, respectively (-1 to 1).

N_{water} and $N_{vegetation}$ are the numbers observations identified as water and vegetation in a year, respectively, while

N_{good} is the number of observations with good quality, which was characterised as cloud and shadow-free during the observed year.

We defined the thresholds based on the training data collected from the high spatial resolution Google Earth images in 2020 to identify evergreen wetlands as follows:

$$\text{Evergreen} = \mathrm{VF} \geq 0.9 \text{ and } \mathrm{WF} \leq 0.2 \text{ and } \mathrm{DEM} \leq 5 \text{ m and Slope} \leq 5^{\circ} \tag{7}$$

where WF and VF are the frequencies of surface water body and vegetation, respectively. These indices values are ranked between -1 and 1, while the Shuttle Radar Topography Mission (SRTM) Digital Elevation Model (DEM) data were used to generate a slope layer in the current study [22].

To identify open surface water bodies and green vegetation, a combination of mNDWI and two vegetation indices (EVI and NDVI) was employed to reduce the errors in mixed pixels of the water body and vegetation [22]. In this study, almost all pixels identified as water body in 2020 have an EVI $\leq$ 0.137 and mNDWI > EVI or mNDWI > NDVI.

The NDVI and the EVI are two popular indices to detect vegetation as suggested in prior studies [19,27]. Their values are defined between -1 to 1 in which the negative values indicate no vegetation, while greater positive values indicate available green vegetation cover. However, a given pixel is often mixed on vegetation, water and soil. The LSWI is an alternative useful index to identify water content in vegetation and soil, and its values are also between -1 to 1. The current study found that most vegetation pixels have EVI $\geq$ 0.174, NDVI $\geq$ 0.377 and LSWI $\geq$ 0. The final formulations to identify surface water body and green vegetation in 2020 are shown as below:

- Pixels of surface water body: EVI $\leq$ 0.137 and (mNDWI > EVI or mNDWI > NDVI);
- Pixels of green vegetation: EVI $\geq$ 0.174, NDVI $\geq$ 0.377 and LSWI > 0.

2.3.4. Annual Maps of Mangrove Forest

As shown in Figure 3, almost all vegetation pixels have values of VF $\geq$ 0.5, while water pixels have WF values ranging from 0.05 to 0.85. In addition, both mangrove and non-mangrove pixels have DEM $\leq$ 5 m and Slope $\leq 5^{\circ}$. Therefore, the criteria for mangrove mapping in 2020 and in 2022 was described as follows:

$$\text{Mangrove forest} = \text{VF} \geq 0.5 \text{ and WF} \leq 0.2 \text{ and DEM} \leq 5 \text{ m and Slope} \leq 5^{\circ} \quad (8)$$

$$\text{Non-mangrove forest} = \text{VF} \leq 0.15 \text{ and } 0.05 \leq \text{WF} \leq 0.2 \text{ and DEM} \leq 5 \text{ m and Slope} \leq 5^{\circ} \quad (9)$$

2.3.5. Accuracy Assessment

The mangrove maps of the three provinces in 2020 and in 2022 were generated from Landsat-8 OLI and Landsat-9 OLI-2 data using the modified algorithm through the GEE cloud computing platform. In this study, the stratified random sampling approach was employed to generate the verification samples, and very high-resolution images were used to evaluate the accuracy of the classification maps in 2020 and in 2022. The size of verification points for each class (mangrove or non-mangrove) was identified by Cochran's formula (the confidence level was set to 0.95 in this study):

$$n_0 = \frac{Z^2 pq}{e^2} \quad (10)$$

where

n_0 is the sample size,
Z is derived from the standard normal distribution,
e is the desired level of precision,
p is the required accuracy, and
$q = 1 - p$.

In this study, a total of 474 validation points (20% of the total points) were selected for evaluating the accuracy of mangrove forest mapping in 2020 and in 2022. The random sampling points include 243 mangrove samples and 231 non-mangrove samples. Then, each sample was checked with its location, which was identified from very high spatial resolution Google Earth images by visual interpretation. With the validation samples, the user's accuracy, the producer's accuracy, the overall accuracy, and the Kappa coefficient were calculated in this study [35,40].

2.3.6. Analysis and Statistical Method

The mangrove distribution maps and the mangrove statistical areas in the study sites were automatically computed. In addition, the present study used QGIS 3.10.2 software to produce the spatial distribution of mangroves in the RRD of Vietnam.

3. Results

3.1. Mangrove Classification and Accuracy Assessment

As shown in Figure 4, the total area of mangrove forest was estimated as 10,125 ha and 10,630 ha in 2020 and in 2022, respectively. The largest mangrove forest area was observed in Hai Phong province (3790 ha), followed by Nam Dinh province (3325 ha) and Thai Binh province (3010 ha) in 2020.

The results in Tables 2 and 3 show that the overall accuracies obtained from the stratified random sampling points were 94.94% in 2020 and 91.98% in 2022, while the Kappa coefficients of classification maps for 2020 and 2022 were 0.90 and 0.84, respectively, indicating a good-of-fit agreement between the classification result and reference data. The Landsat-8 OLI sensor produced relatively higher accuracy for 2020 than that of the Landsat-9 OLI-2 sensors for 2022.

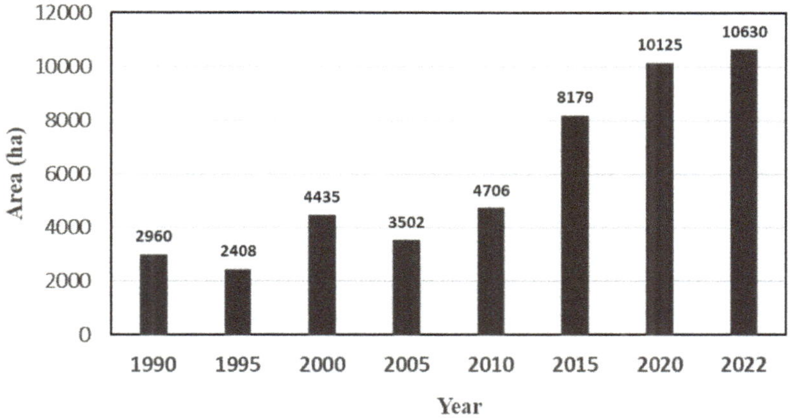

Figure 4. The estimated mangrove area in the RRD from 1990 to 2022.

Table 2. The confusion matrix for accuracy assessment of mangrove forest using Landsat 8-OLI in 2020.

		Reference Pixels			
	Class	Mangrove	Non-mangrove	Total	User's accuracy
Predicted pixels	Mangrove	221	22	243	90.95%
	Non-mangrove	2	229	231	99.13%
	Total	223	251	474	
	Producer's Accuracy	99.10%	88.18%		
	Overall accuracy	**94.94%**			
	Kappa coefficient	**0.90**			

Table 3. The confusion matrix for accuracy assessment of mangrove forest using Landsat-9 OLI-2 in 2022.

		Reference Pixels			
	Class	Mangrove	Non-mangrove	Total	User's accuracy
Predicted pixels	Mangrove	210	33	243	86.42%
	Non-mangrove	5	226	231	97.84%
	Total	215	259	474	
	Producer's Accuracy	97.70%	87.26%		
	Overall accuracy	**91.98%**			
	Kappa coefficient	**0.84**			

3.2. Mangrove Dynamics from 1990 to 2022

By using our defined thresholds in 2020 and the proposed framework in Figure 3, we generated mangrove maps in the three provinces (Hai Phong, Nam Dinh and Thai Binh) in the RRD between 1990 and 2015 together with mangrove maps in 2022 (See Figures A1–A3). We also estimated the areas of mangrove forests in the three provinces for other years (1990, 1995, 2000, 2005, 2010, 2015 and 2022). The mangrove distribution maps and the statistical areas in the study sites were automatically computed using the Java scrips on the GEE cloud computing platform. As shown in Figure 4, the mangrove forest area increased in the three provinces across the RRD over the 32 years (1990–2022). The change of mangrove area in each province and each period can be found in Table 4. Figure 5 shows the spatial distribution of mangrove in the RRD of Vietnam in 1990 (Figure 5a) and in 2022 (Figure 5b). Figures 6 and 7 represent the mangrove maps of each province in the RRD in 1990 (Figure 6) and in 2022 (Figure 7). Mangrove forests are mainly distributed in the river mouth of the three provinces in the RRD, and they are found in front of the sea dykes (Figures 5–7).

As shown in Figure 4, the area of mangrove forests in the RRD significantly increased from 1990 to 2022. The mangrove area decreased from 2960 ha in 1990 to 2408 ha in 1995 and then significantly increased to 4435 ha in 2000. Notably, the area of mangrove forests decreased to 3502 ha in 2005. In contrast, the mangrove area experienced an increase from 4706 ha in 2010 to 8179 ha in 2015 and continued its upward trend to 10,125 ha in 2020 and reached the highest peak value of 10,630 ha in 2022.

Table 4 shows the change detection of the mangrove area in Hai Phong, Thai Binh and Nam Dinh provinces over 32 years. Overall, the mangrove area across the three provinces increased considerably since 2010. Hai Phong province had the largest area of mangrove in 2022 with 3934 ha, followed by Nam Dinh province (3591 ha). In contrast, the mangrove area in Thai Binh province was the lowest with 3195 ha.

Table 4. The change detection of the mangrove area in the three provinces over 32 years.

Year/Province	Hai Phong (ha)	Nam Dinh (ha)	Thai Binh (ha)	Total (ha)
1990	1433	459	1068	2960
1995	1190	776	442	2408
1990–1995	−243	317	−626	−552
2000	1495	1335	1605	4435
1995–2000	305	559	1163	2027
2005	1061	1287	1154	3502
2000–2005	−434	−48	−451	−933
2010	1628	1564	1514	4706
2005–2010	567	277	360	1204
2015	3065	2781	2333	8179
2010–2015	1437	1217	819	3473
2020	3790	3325	3010	10,125
2015–2020	725	544	677	1946
2022	3934	3591	3195	10,630
2020–2022	144	176	185	505

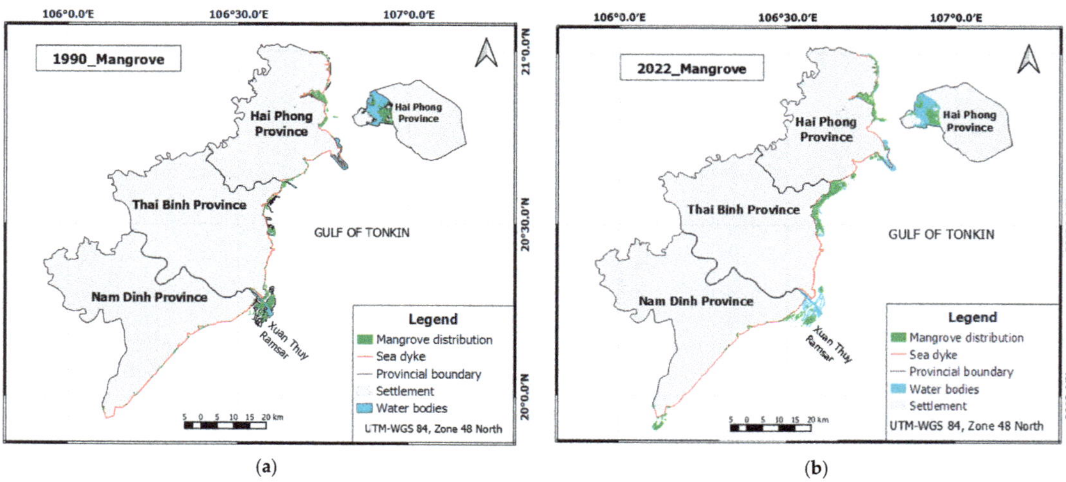

(**a**) (**b**)

Figure 5. The spatial distribution map of mangrove in the RRD of Vietnam: (**a**) mangrove map in 1990; (**b**) mangrove map in 2022.

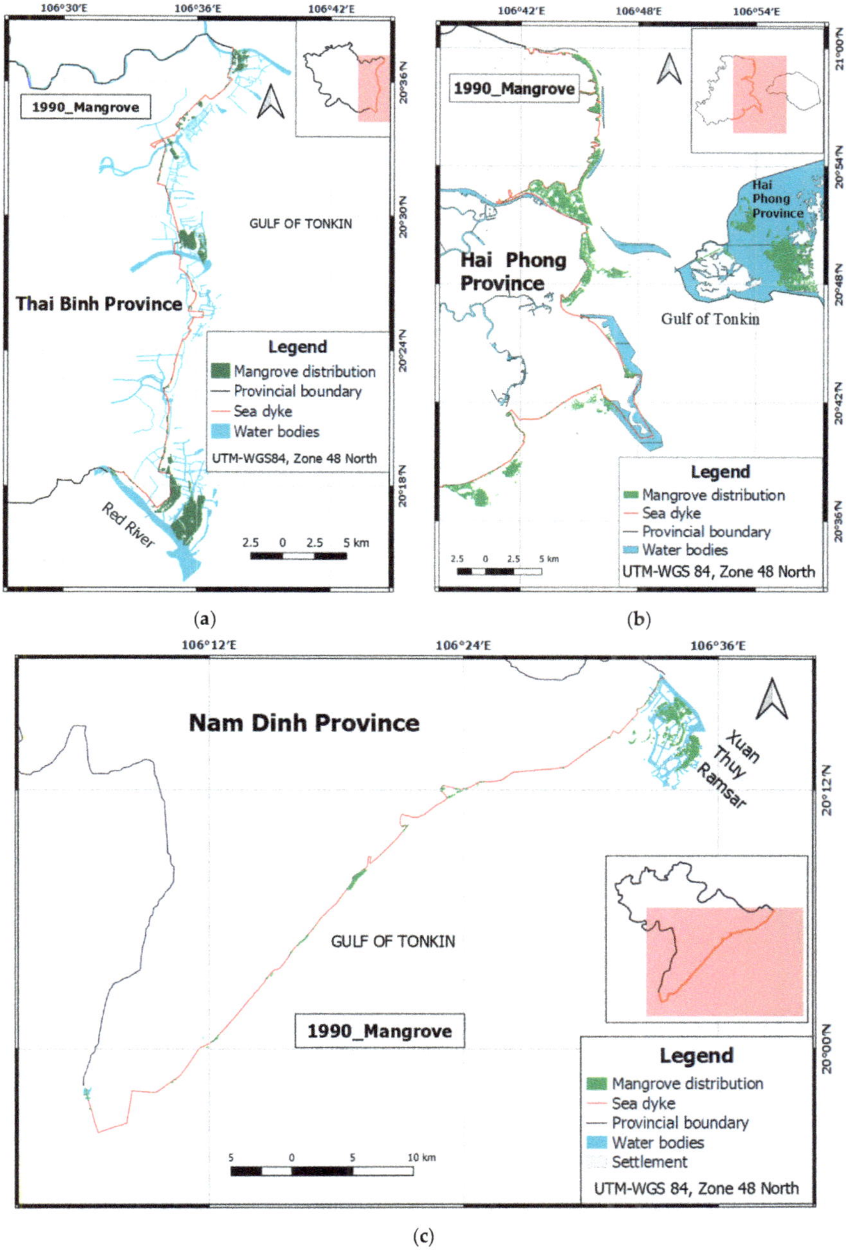

Figure 6. Mangrove maps in the three provinces in 1990 across the RRD, northern Vietnam: (**a**) Thai Binh; (**b**) Hai Phong; (**c**) Nam Dinh.

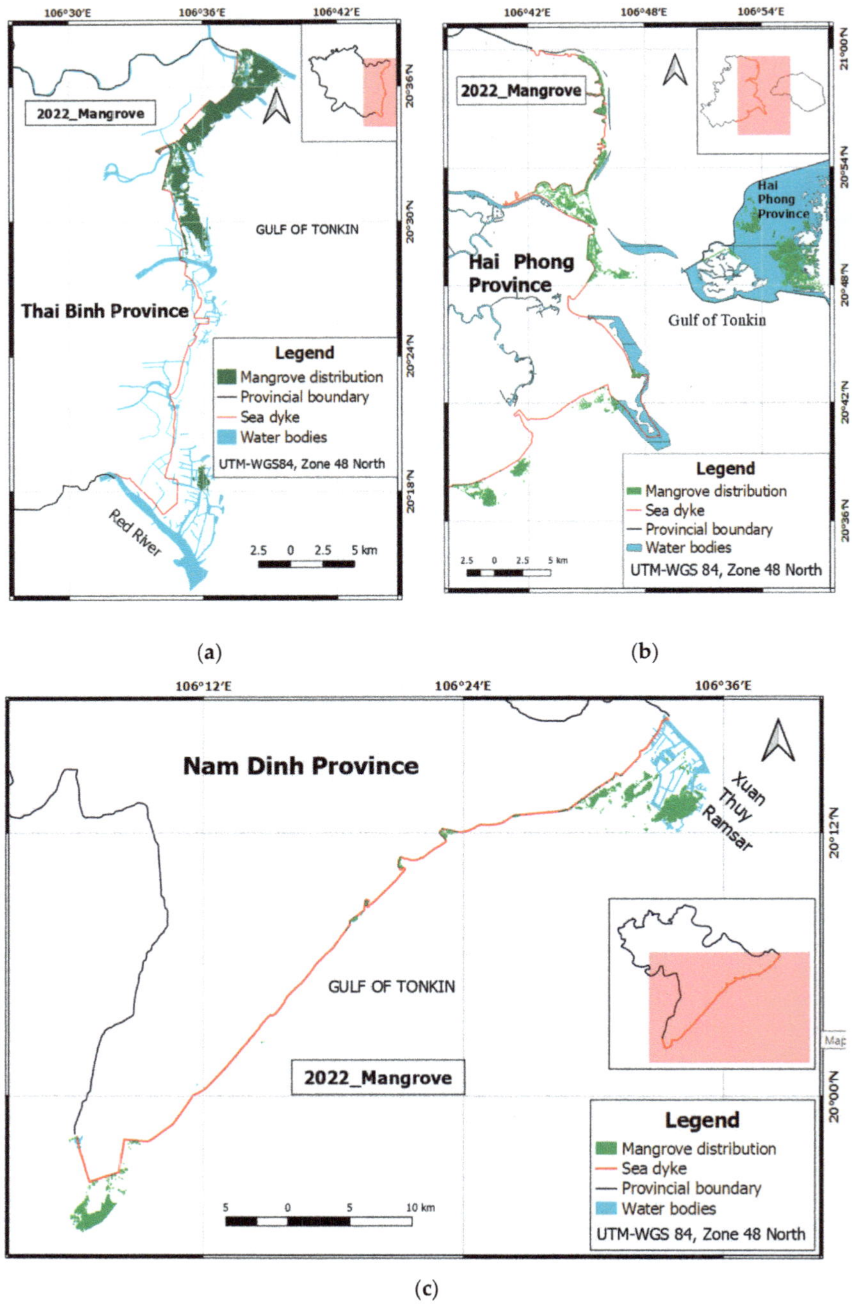

Figure 7. Mangrove maps in the three provinces in 2022 across the RRD, northern Vietnam: (**a**) Thai Binh; (**b**) Hai Phong; (**c**) Nam Dinh.

4. Discussion

4.1. Uncertainty of Mangrove Mapping and Change Detection

The overall accuracies (OA) of the mangrove maps in 2020 and in 2022 were 94.94% and 91.98% with Kappa coefficients of 0.90 and 0.84, respectively. These values indicate the successful use of the pixel-based algorithm for mapping mangrove forests and detecting change using multi-temporal Landsat datasets on the GEE cloud computing platform. The Landsat-8 OLI sensor produced better results than those obtained from Landsat-9 OLI-2 (Tables 2 and 3). It is likely due to more available multi-temporal Landat-8 datasets in 2020 with 14 time series scenes compared to only 3 cloud-free scenes available during 2022. The number of available cloud-free time series Landsat data would influence the overall accuracy and produce better results when mapping mangrove forests using the pixel-based algorithm. Future studies applying our framework and thresholds should be further tested in other mangrove regions with more available Landsat-9 OLI-2. Our results suggested satisfactory accuracies for mapping mangrove forests during 2020 and 2022. These results are relatively higher than those reported by the previous studies in Vietnam using SPOT-7 imagery with a higher spatial resolution of 6 m (OA = 92.9%) [41] and using time series Landsat data (OA ranged from 85% to 92%) [42]. Our results are similar to Hauser et al. [43] with an attempt to detect mangrove dynamics on the southern coast of Vietnam using GGE with an overall accuracy ranging from 94 to 96%. However, there is an uncertainty involved in the mangrove classification and change detection. There are several factors that could affect the accuracy of mangrove mapping in the study area during 2020. As shown in Figure A4, this study only obtained about 83% of the pixels in 2020 with seven good-quality observations. Therefore, it can be considered that the acceptable uncertainty [40,41] in mangrove area estimation probably resulted from the lower quality of available Landsat time series (TM/ETM+/OLI/OLI-2) data obtained in the current study area between 1990 and 2022. In addition, the mixed pixels of mangrove and other vegetation in the study area may also affect the accuracy of the generated mangrove maps. For example, *Casuarina* spp. sites were misclassified as mangrove forests because several *Casuarina* spp. species have a quite similar reflectance spectrum with other mangrove species observed in the RRD [30] such as *Sonneratia caseolaris*, *Kandelia obovata*, *Aegiceras corniculatum*, *Rhizophora stylosa* and *Avicennia marina*. Importantly, in the RRD, many mixed small and shrub species are often observed and reported in the previous studies [30,35].

In this study, the defined thresholds were created based on the calibration data collected from the high spatial resolution Google Earth images in 2020 to automatically map and detect mangrove canopy changes across the RRD. As shown in Figure 4, the estimated mangrove area in 2015 was about 8179 ha. This number is close to the estimate as reported in the National Forest Inventory (NFI) in Vietnam during 2015 (8225 ha), fitting well with the model developed in the current study using Landsat data on the GEE.

We observed an increase in the extent of mangroves across the three provinces in the RRD located on the northern coast of Vietnam from 1990 to 2022. The trend is consistent with the forest coverage change in Vietnam, which was reported in recent studies [44,45] and is similar to those observed in other Southeast Asian countries by Goldberg et al. [3] in the southeast and northern Australia by Saintilan et al. [4]. The increase in forest coverage benefited from the efforts of the Vietnamese government in mangrove planting, restoration, and protection. The total forest area in Vietnam was slightly increased between 1990 and 2020 and includes both inland forest and mangrove forest in Vietnam [44]. Overall, the mangrove forest area increase over 32 years (1990–2022) can be automatically detected and mapped by using Landsat 5/7/8/9 time series images through the GEE platform as a result of a number of mangrove restoration projects and programs by the Vietnamese government and policy recommendations based on policy measures from many research studies [11,46].

4.2. Driving Factors for Mangrove Dynamic in Three Provinces from 1990 to 2022

As exhibited in Table 4 and Figure 4, the mangrove area changed during the period of 1990–2022. Key drivers that caused changes of mangrove forests in each period are considered and discussed as follows:

Between 1990 and 1995: The total mangrove area of three provinces decreased from 2960 ha in 1990 to 2408 ha in 1995. This decline was caused by the mangrove deforestation in the Hai Phong province and the Thai Binh province during the period. Specifically, the mangrove area in Hai Phong province declined from 1433 ha to 1190 ha, and in Thai Binh province it reduced from 1068 ha to 442 ha. This period witnessed the smallest mangrove area during the 32-year period. The reasons behind the decrease were the consequence of a new policy, the Reform Policy, initiated in 1986 and officially launched in 1991 [47]. During the period, natural resources, including forest resources, were exhaustively exploited for economic development.

During this period, many regions were converted to aquaculture farms, significantly destroying mangrove forests in Thai Binh and Hai Phong provinces [10]. In contrast, Nam Dinh province had a mangrove area increase of 317 ha from 1990 to 1995 thanks to strict protection and constant expansion of Xuan Thuy National Park [48].

Period of 1995–2000: This period witnessed an increase in the mangrove area in three provinces (Table 4). The mangrove areas of Hai Phong, Nam Dinh and Thai Binh provinces in 2000 reached 1495 ha, 1335 ha and 1605 ha, respectively. This increase was due mainly to the efforts of planting and protecting mangrove forests through various programs and projects implemented in such provinces. During this period, the Five Million Hectare Reforestation Program (661 program) was carried out between 1998 and 2010 at the national level to increase forest coverage. The percentage of forest coverage was up to 43% of the total land cover in the final year of the program. In addition, other programs and projects were also implemented. Several projects such as Red Cross of Japan, PAM5325, ACTMANG, the 661 programs (Mangrove Plantation and Disaster Risk Reduction project) were undertaken in such provinces. These projects significantly contributed to the increase of mangrove cover in the RRD [49,50].

Period of 2000–2005: In this period, the mangrove areas decreased from 4435 ha in 2000 to 3502 ha in 2005 (Figure 4). The main cause for mangrove loss in 2005 may probably be explained by the negative impacts of natural disasters. In 2005, an extreme typhoon event, typhoon "Damrey", hit the northern region of Vietnam [46] and damaged the mangrove forest in these areas, especially young mangrove forests. This typhoon was also reported by Hong, Avtar and Fujii [9] as the amongst the strongest tropical cyclones impacting the coastal zones of Vietnam during the last 30 years.

Period 2005–2020: This period witnessed a continuous increase in the area of mangroves in such provinces sustained for 15 years. As shown in Table 4, the total area of mangrove reached 10,125 ha in 2020. This number was three-times higher than that in 2005. The mangrove restoration received priority attention and investment by the Vietnamese government in this period and enhancement of community-based mangrove management [51,52]. In addition to the 661 Program implemented from 1998 to 2010, many other projects and programs funded by the Vietnamese government and other organizations were implemented in the whole country [52], especially in the Red River Delta [53]. Further sustainable mangrove conservation and management across the Vietnamese coastline should be carefully considered in protecting existed mangrove forests and restoring degraded mangroves as well as planting new ones to enhance not only the mangrove area but also quality and biodiversity in the context of climate change issues.

Period 2020–2022: This short period was characterised by an increased upward trend in mangrove area in the RRD. The Vietnamese government continued to support mangrove conservation and management schemes in dealing with climate change impact.

5. Conclusions

Mangrove forests play an important role in mitigating climate change impacts and are able to sequester blue carbon for their protection and restoration. Mapping mangrove extent at a large scale remains challenging due to cloud coverage and spatial limitations of single satellite sensors. This study developed a framework using the pixel-based algorithm applied to Landsat TM/ETM+/OLI/OLI-2 time series data on the Google Earth Engine cloud computing platform to automatically map and quantify mangrove forest changes in the three provinces of Hai Phong, Nam Dinh and Thai Binh across the RRD over 32 years.

The results showed that the mangrove area has increased considerably in the RRD over 32 years in response to the mangrove restoration programs and policies by the Vietnamese government and local governments. The mangrove areas were 2960 ha, 2408 ha, 4435 ha, 3502 ha, 4706 ha, 8179 ha, 10,125 ha and 10,630 ha in 1990, 1995, 2000, 2005, 2010, 2015, 2020 and 2022, respectively.

The overall accuracies of the Landsat-8 OLI and the Landsat-9 OLI-2 image processing for 2020 and 2022 were 94.94% and 91.98%, respectively, while the Kappa coefficients were 0.90 and 0.84, indicating promising results for mapping mangrove forest cover in the tropics using the GEE platform associated with free open-source code. It could be said that the pixel-based algorithm and Landsat time series images on the GEE cloud computing are suitable for long-term monitoring of mangrove change in tropical regions. The Landsat family has shown the potential use in mapping mangrove dynamics in the tropics and should be further used worldwide.

Author Contributions: Conceptualization, T.T.P.V., T.D.P. and B.M.; methodology, T.T.P.V., T.D.P., H.V.L. and Q.H.V.; software, T.T.P.V. and H.Q.N.; validation, T.T.P.V., Q.H.V. and T.D.P.; formal analysis, T.T.P.V. and H.Q.N.; investigation, T.T.P.V.; resources, T.T.P.V. and N.N.L.; data curation, T.T.P.V.; writing—original draft preparation, T.T.P.V. and T.D.P.; writing—review and editing, T.D.P., N.S., N.N.L. and A.S.; visualization, T.D.P. and H.V.L.; supervision, B.M. All authors have read and agreed to the published version of the manuscript.

Funding: This research received no external funding.

Data Availability Statement: Not applicable.

Acknowledgments: The authors are thankful to the JICE (Japanese International Cooperation Center) for financial support to the first author studying a master course at the University of Tsukuba, Japan for 2 years. In addition, we also would like to thank the Forest Inventory and Planning Institute (FIPI), Vietnam for providing useful data for the current study. T.D.P. is supported by a Macquarie University Research Fellowship (Grant No. MQRF0001124-2021).

Conflicts of Interest: The authors declare no conflict of interest.

Appendix A

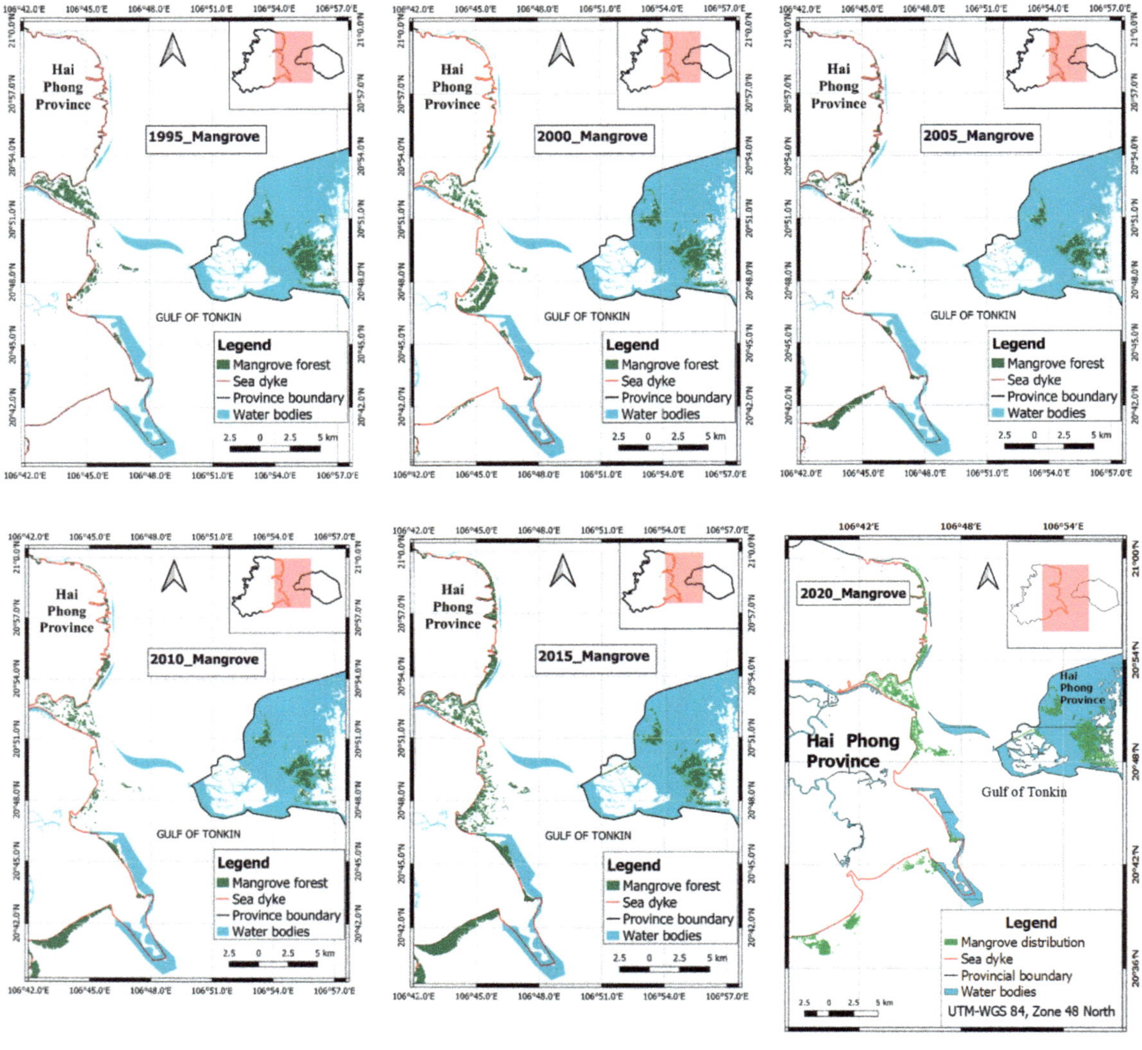

Figure A1. Mangrove distribution maps in Hai Phong between 1995 and 2020.

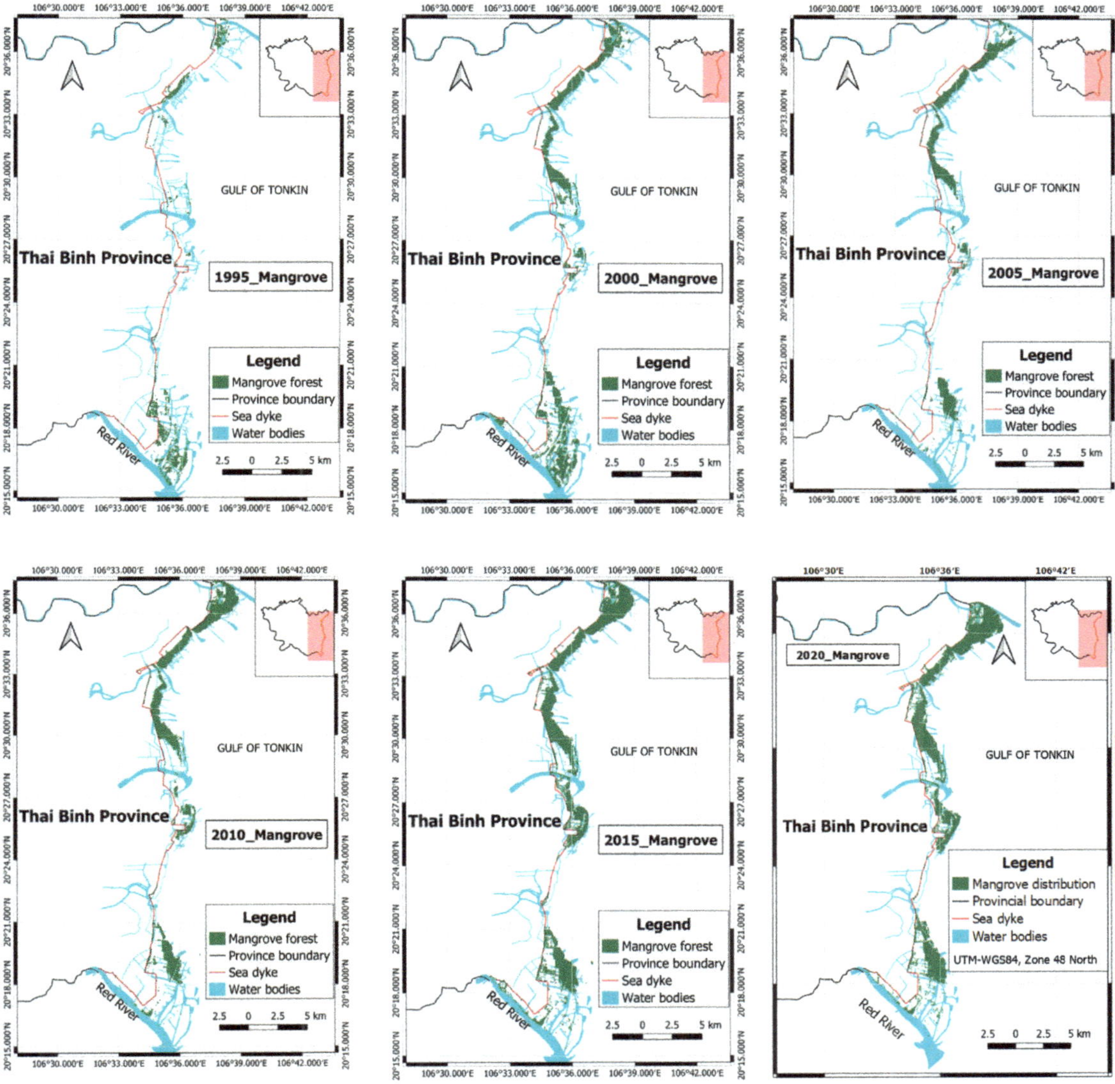

Figure A2. Mangrove distribution maps in Thai Binh between 1995 and 2020.

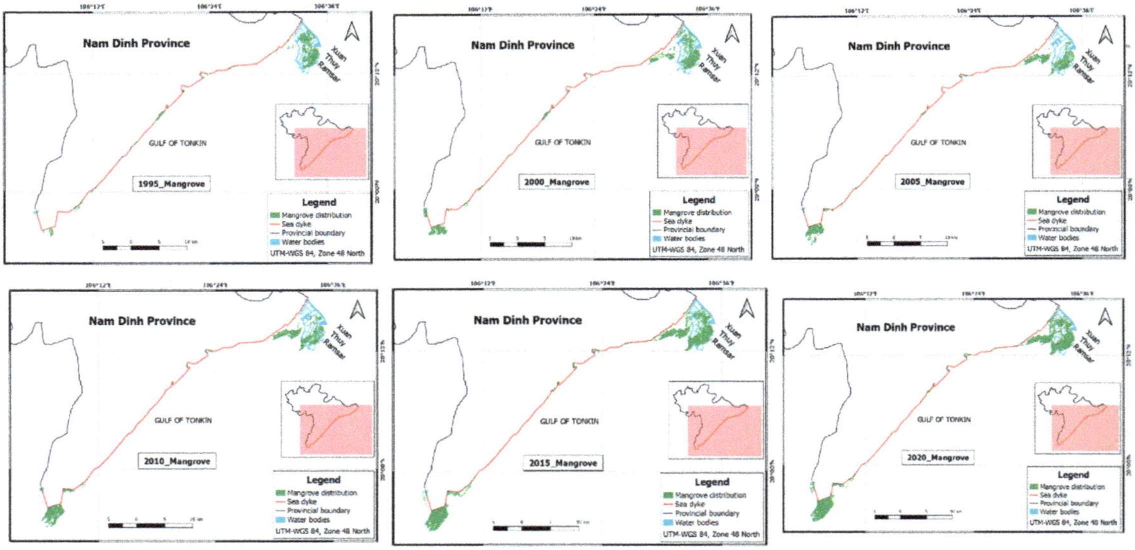

Figure A3. Mangrove distribution maps in Nam Dinh between 1995 and 2020.

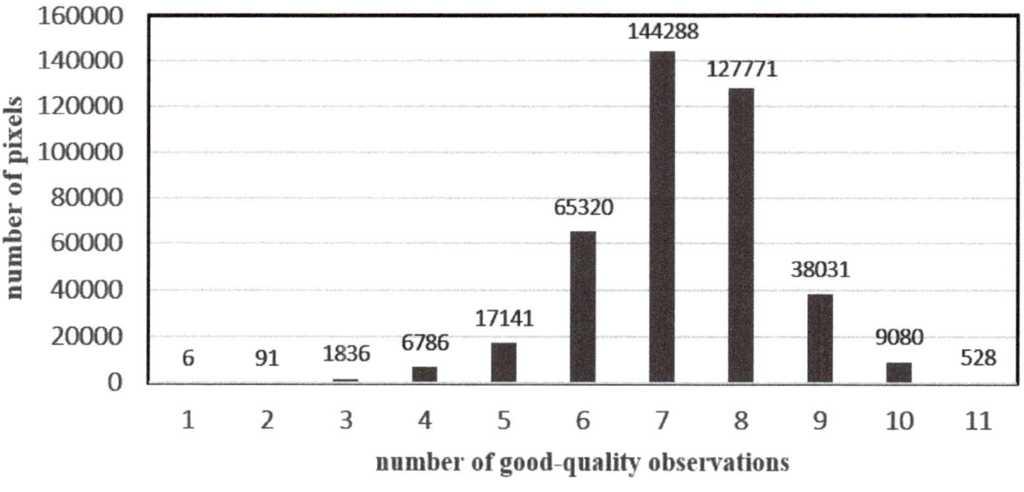

Figure A4. The number of pixels with good-quality observations in 2020.

References

1. Saintilan, N.; Khan, N.S.; Ashe, E.; Kelleway, J.J.; Rogers, K.; Woodroffe, C.D.; Horton, B.P. Thresholds of mangrove survival under rapid sea level rise. *Science* **2020**, *368*, 1118. [CrossRef] [PubMed]
2. Friess, D.A.; Rogers, K.; Lovelock, C.E.; Krauss, K.W.; Hamilton, S.E.; Lee, S.Y.; Lucas, R.; Primavera, J.; Rajkaran, A.; Shi, S. The State of the World's Mangrove Forests: Past, Present, and Future. *Annu. Rev. Environ. Resour.* **2019**, *44*, 89–115. [CrossRef]
3. Goldberg, L.; Lagomasino, D.; Thomas, N.; Fatoyinbo, T. Global declines in human-driven mangrove loss. *Glob. Change Biol.* **2020**, *26*, 5844–5855. [CrossRef] [PubMed]
4. Saintilan, N.; Asbridge, E.; Lucas, R.; Rogers, K.; Wen, L.; Powell, M.; Colloff, M.J.; Rodriguez, J.F.; Saco, P.M.; Sandi, S.; et al. Australian forested wetlands under climate change: Collapse or proliferation? *Mar. Freshw. Res.* **2021**, *73*, 1255–1262. [CrossRef]
5. Saintilan, N.; Wilson, N.C.; Rogers, K.; Rajkaran, A.; Krauss, K.W. Mangrove expansion and salt marsh decline at mangrove poleward limits. *Glob. Change Biol.* **2014**, *20*, 147–157. [CrossRef]
6. Hong, P.N.; San, H.T. *Mangroves of Vietnam*; IUCN: Bangkok, Thailand, 1993; p. 173.
7. Warner, R.; Kaidonis, M.; Dun, O.; Rogers, K.; Shi, Y.; Nguyen, T.T.X.; Woodroffe, C.D. Opportunities and challenges for mangrove carbon sequestration in the Mekong River Delta in Vietnam. *Sustain. Sci.* **2016**, *11*, 661–677. [CrossRef]

8. Veettil, B.K.; Ward, R.D.; Quang, N.X.; Trang, N.T.T.; Giang, T.H. Mangroves of Vietnam: Historical development, current state of research and future threats. *Estuar. Coast. Shelf Sci.* **2019**, *218*, 212–236. [CrossRef]
9. Hong, H.T.C.; Avtar, R.; Fujii, M. Monitoring changes in land use and distribution of mangroves in the southeastern part of the Mekong River Delta, Vietnam. *Trop. Ecol.* **2019**, *60*, 552–565. [CrossRef]
10. Pham, T.D.; Yoshino, K. Mangrove Mapping and Change Detection Using Multi-temporal Landsat imagery in Hai Phong city, Vietnam. In Proceedings of the International Symposium on Cartography in Internet and Ubiquitous Environments 2015, Tokyo, Japan, 17–19 March 2015.
11. Tran, L.X.; Fischer, A. Spatiotemporal changes and fragmentation of mangroves and its effects on fish diversity in Ca Mau Province (Vietnam). *J. Coast. Conserv.* **2017**, *21*, 355–368. [CrossRef]
12. Nguyen-Thanh, S.; Chen, C.-F.; Chang, N.-B.; Chen, C.-R.; Chang, L.-Y.; Bui, X.-T. Mangrove Mapping and Change Detection in Ca Mau Peninsula, Vietnam, Using Landsat Data and Object-Based Image Analysis. *IEEE J. Sel. Top. Appl. Earth Obs. Remote Sens.* **2015**, *8*, 503–510. [CrossRef]
13. Pham, L.T.H.; Brabyn, L. Monitoring mangrove biomass change in Vietnam using SPOT images and an object-based approach combined with machine learning algorithms. *ISPRS J. Photogramm. Remote Sens.* **2017**, *128*, 86–97. [CrossRef]
14. Pham, T.D.; Yoshino, K. Monitoring Mangrove Forest using Multi-temporal Satellite Data in the Northern Coast of Vietnam. In Proceedings of the 32nd Asian Conference on Remote Sensing, Taipei, Taiwan, 3–7 October 2011.
15. Lymburner, L.; Bunting, P.; Lucas, R.; Scarth, P.; Alam, I.; Phillips, C.; Ticehurst, C.; Held, A. Mapping the multi-decadal mangrove dynamics of the Australian coastline. *Remote Sens. Environ.* **2020**, *238*, 111185. [CrossRef]
16. Long, J.B.; Giri, C. Mapping the Philippines' Mangrove Forests Using Landsat Imagery. *Sensors* **2011**, *11*, 2972–2981. [CrossRef]
17. Giri, C.; Ochieng, E.; Tieszen, L.L.; Zhu, Z.; Shingh, A.; Loveland, T. Status and distribution of mangrove forests of the world using earth observation satellite data. *Glob. Ecol. Biogeogr.* **2011**, *20*, 154–159. [CrossRef]
18. Giri, C.; Pengra, B.; Zhu, Z.; Singh, A.; Tieszen, A.L.L. Monitoring mangrove forest dynamics of the Sundarbans in Bangladesh and India using multi-temporal satellite data from 1973 to 2000. *Estuar. Coast. Shelf Sci.* **2007**, *73*, 91–100. [CrossRef]
19. Pham, T.D.; Yokoya, N.; Bui, D.T.; Yoshino, K.; Friess, D.A. Remote Sensing Approaches for Monitoring Mangrove Species, Structure, and Biomass: Opportunities and Challenges. *Remote Sens.* **2019**, *11*, 230. [CrossRef]
20. Pham, T.D.; Xia, J.; Ha, N.T.; Bui, D.T.; Le, N.N.; Takeuchi, W. A Review of Remote Sensing Approaches for Monitoring Blue Carbon Ecosystems: Mangroves, Seagrassesand Salt Marshes during 2010–2018. *Sensors* **2019**, *19*, 1933. [CrossRef]
21. Yancho, J.M.M.; Jones, T.G.; Gandhi, S.R.; Ferster, C.; Lin, A.; Glass, L. The Google Earth Engine Mangrove Mapping Methodology (GEEMMM). *Remote Sens.* **2020**, *12*, 3758. [CrossRef]
22. Wang, X.; Xiao, X.; Zou, Z.; Hou, L.; Qin, Y.; Dong, J.; Doughty, R.B.; Chen, B.; Zhang, X.; Chen, Y.; et al. Mapping coastal wetlands of China using time series Landsat images in 2018 and Google Earth Engine. *ISPRS J. Photogramm. Remote Sens.* **2020**, *163*, 312–326. [CrossRef]
23. Mahdianpari, M.; Brisco, B.; Granger, J.E.; Mohammadimanesh, F.; Salehi, B.; Banks, S.; Homayouni, S.; Bourgeau-Chavez, L.; Weng, Q. The Second Generation Canadian Wetland Inventory Map at 10 Meters Resolution Using Google Earth Engine. *Can. J. Remote Sens.* **2020**, *46*, 360–375. [CrossRef]
24. Chen, B.; Xiao, X.; Li, X.; Pan, L.; Doughty, R.; Ma, J.; Dong, J.; Qin, Y.; Zhao, B.; Wu, Z.; et al. A mangrove forest map of China in 2015: Analysis of time series Landsat 7/8 and Sentinel-1A imagery in Google Earth Engine cloud computing platform. *ISPRS J. Photogramm. Remote Sens.* **2017**, *131*, 104–120. [CrossRef]
25. Masek, J.G.; Wulder, M.A.; Markham, B.; McCorkel, J.; Crawford, C.J.; Storey, J.; Jenstrom, D.T. Landsat 9: Empowering open science and applications through continuity. *Remote Sens. Environ.* **2020**, *248*, 111968. [CrossRef]
26. Pham, T.D.; Yokoya, N.; Nguyen, T.T.T.; Le, N.N.; Ha, N.T.; Xia, J.; Takeuchi, W.; Pham, T.D. Improvement of Mangrove Soil Carbon Stocks Estimation in North Vietnam Using Sentinel-2 Data and Machine Learning Approach. *GISci. Remote Sens.* **2021**, *58*, 68–87. [CrossRef]
27. Pham, T.D.; Yokoya, N.; Xia, J.; Ha, N.T.; Le, N.N.; Nguyen, T.T.T.; Dao, T.H.; Vu, T.T.P.; Pham, T.D.; Takeuchi, W. Comparison of Machine Learning Methods for Estimating Mangrove Above-Ground Biomass Using Multiple Source Remote Sensing Data in the Red River Delta Biosphere Reserve, Vietnam. *Remote Sens.* **2020**, *12*, 1334. [CrossRef]
28. Seto, K.C.; Fragkias, M. Mangrove conversion and aquaculture development in Vietnam: A remote sensing-based approach for evaluating the Ramsar Convention on Wetlands. *Glob. Environ. Change* **2007**, *17*, 486–500. [CrossRef]
29. Halls, A. *Wetlands, Biodiversity and the Ramsar Convention: The Role of the Convention on Wetlands in the Conservation and Wise Use of Biodiversity*; Ramsar Convention Bureau: Gland, Switzerland, 1997.
30. Xia, J.; Yokoya, N.; Pham, T.D. Probabilistic Mangrove Species Mapping with Multiple-Source Remote-Sensing Datasets Using Label Distribution Learning in Xuan Thuy National Park, Vietnam. *Remote Sens.* **2020**, *12*, 3834. [CrossRef]
31. Mai, C.V.; Stive, M.J.F.; Gelder, P.H. Coastal Protection Strategies for the Red River Delta. *J. Coast. Res.* **2009**, *25*, 105–116. [CrossRef]
32. Quartel, S.; Kroon, A.; Augustinus, P.G.E.F.; Van Santen, P.; Tri, N.H. Wave attenuation in coastal mangroves in the Red River Delta, Vietnam. *J. Asian Earth Sci.* **2007**, *29*, 576–584. [CrossRef]
33. Thomas, N.; Bunting, P.; Lucas, R.; Hardy, A.; Rosenqvist, A.; Fatoyinbo, T. Mapping Mangrove Extent and Change: A Globally Applicable Approach. *Remote Sens.* **2018**, *10*, 1466. [CrossRef]

34. Bunting, P.; Rosenqvist, A.; Hilarides, L.; Lucas, R.M.; Thomas, N. Global Mangrove Watch: Updated 2010 Mangrove Forest Extent (v2.5). *Remote Sens.* **2022**, *14*, 1034. [CrossRef]
35. Pham, T.D.; Xia, J.; Baier, G.; Le, N.N.; Yokoya, N. Mangrove Species Mapping Using Sentinel-1 and Sentinel-2 Data in North Vietnam. In Proceedings of IGARSS 2019—2019 IEEE International Geoscience and Remote Sensing Symposium, Yokohama, Japan, 28 July–2 August 2019; pp. 6102–6105.
36. Rouse, J.W., Jr.; Haas, R.; Schell, J.; Deering, D. *Monitoring Vegetation Systems in the Great Plains with ERTS*; NASA Special Publication: Washington, DC, USA, 1974.
37. Liu, H.Q.; Huete, A. A feedback based modification of the NDVI to minimize canopy background and atmospheric noise. *IEEE Trans. Geosci. Remote Sens.* **1995**, *33*, 457–465. [CrossRef]
38. Xiao, X.; Boles, S.; Liu, J.; Zhuang, D.; Liu, M. Characterization of forest types in Northeastern China, using multi-temporal SPOT-4 VEGETATION sensor data. *Remote Sens. Environ.* **2002**, *82*, 335–348. [CrossRef]
39. Xu, H. Modification of normalised difference water index (NDWI) to enhance open water features in remotely sensed imagery. *Int. J. Remote Sens.* **2006**, *27*, 3025–3033. [CrossRef]
40. Ha, N.T.; Manley-Harris, M.; Pham, T.D.; Hawes, I. A Comparative Assessment of Ensemble-Based Machine Learning and Maximum Likelihood Methods for Mapping Seagrass Using Sentinel-2 Imagery in Tauranga Harbor, New Zealand. *Remote Sens.* **2020**, *12*, 355. [CrossRef]
41. Quang, N.H.; Quinn, C.H.; Stringer, L.C.; Carrie, R.; Hackney, C.R.; Van Hue, L.T.; Van Tan, D.; Nga, P.T.T. Multi-Decadal Changes in Mangrove Extent, Age and Species in the Red River Estuaries of Viet Nam. *Remote Sens.* **2020**, *12*, 2289. [CrossRef]
42. Nguyen, H.-H.; Tran, L.T.N.; Le, A.T.; Nghia, N.H.; Duong, L.V.K.; Nguyen, H.T.T.; Bohm, S.; Premnath, C.F.S. Monitoring Changes in Coastal Mangrove Extents Using Multi-Temporal Satellite Data in Selected Communes, Hai Phong City, Vietnam. *For. Soc.* **2020**, *4*, 256–270. [CrossRef]
43. Hauser, L.T.; An Binh, N.; Viet Hoa, P.; Hong Quan, N.; Timmermans, J. Gap-Free Monitoring of Annual Mangrove Forest Dynamics in Ca Mau Province, Vietnamese Mekong Delta, Using the Landsat-7-8 Archives and Post-Classification Temporal Optimization. *Remote Sens.* **2020**, *12*, 3729. [CrossRef]
44. Phan, D.C.; Trung, T.H.; Truong, V.T.; Sasagawa, T.; Vu, T.P.T.; Bui, D.T.; Hayashi, M.; Tadono, T.; Nasahara, K.N. First comprehensive quantification of annual land use/cover from 1990 to 2020 across mainland Vietnam. *Sci. Rep.* **2021**, *11*, 9979. [CrossRef]
45. Truong, V.T.; Hoang, T.T.; Cao, D.P.; Hayashi, M.; Tadono, T.; Nasahara, K.N. JAXA Annual Forest Cover Maps for Vietnam during 2015–2018 Using ALOS-2/PALSAR-2 and Auxiliary Data. *Remote Sens.* **2019**, *11*, 2412. [CrossRef]
46. Pham, T.D.; Yoshino, K. Impacts of mangrove management systems on mangrove changes in the Northern Coast of Vietnam. *Tropics* **2016**, *24*, 141–151. [CrossRef]
47. Phuong, N.C.; Richard, J. Economic Transition and Accounting System Reform in Vietnam. *Eur. Account. Rev.* **2011**, *20*, 693–725. [CrossRef]
48. Leslie, M.; Nguyen, S.T.; Nguyen, T.K.D.; Pham, T.T.; Cao, T.T.N.; Le, T.Q.; Dang, T.T.; Nguyen, T.H.T.; Nguyen, T.B.N.; Le, H.N.; et al. Bringing social and cultural considerations into environmental management for vulnerable coastal communities: Responses to environmental change in Xuan Thuy National Park, Nam Dinh Province, Vietnam. *Ocean. Coast. Manag.* **2018**, *158*, 32–44. [CrossRef]
49. Hai, N.T.; Dell, B.; Phuong, V.T.; Harper, R.J. Towards a more robust approach for the restoration of mangroves in Vietnam. *Ann. For. Sci.* **2020**, *77*, 18. [CrossRef]
50. Nguyen, H.T.; Yoneda, R.; Ninomiya, I.; Harada, K.; Dao, T.V.; Sy, T.M.; Phan, H.N. The effects of stand-age and inundation on carbon accumulation in mangrove plantation soil in Namdinh, Northern Vietnam. *Tropics* **2004**, *14*, 21–37. [CrossRef]
51. Powell, N.; Osbeck, M.; Tan, S.B.; Toan, V.C. Mangrove restoration and rehabilitation for climate change adaptation in Vietnam. In *World Resources Report Case Study*; World Resources Report; World Resources: Washington, DC, USA, 2011. Available online: http://www.worldresourcesreport.org (accessed on 25 November 2021).
52. Nguyen, T.P.; Van Tam, N.; Quoi, L.P.; Parnell, K.E. Community perspectives on an internationally funded mangrove restoration project: Kien Giang province, Vietnam. *Ocean. Coast. Manag.* **2016**, *119*, 146–154. [CrossRef]
53. Pham, T.T.; Vu, T.P.; Hoang, T.L.; Dao, T.L.C.; Nguyen, D.T.; Pham, D.C.; Dao, L.H.T.; Nguyen, V.T.; Hoang, N.V.H. The Effectiveness of Financial Incentives for Addressing Mangrove Loss in Northern Vietnam. *Front. For. Glob. Change* **2022**, *4*, 709073. [CrossRef]

remote sensing

MDPI

Review

A Review of Spectral Indices for Mangrove Remote Sensing

Thuong V. Tran [1,2,*], Ruth Reef [1] and Xuan Zhu [1]

1 School of Earth, Atmosphere and Environment, Monash University, Clayton, VIC 3800, Australia
2 Institute of Engineering and Technology, Thu Dau Mot University, Thu Dau Mot City 75000, Vietnam
* Correspondence: thuong.tran@monash.edu

Abstract: Mangrove ecosystems provide critical goods and ecosystem services to coastal communities and contribute to climate change mitigation. Over four decades, remote sensing has proved its usefulness in monitoring mangrove ecosystems on a broad scale, over time, and at a lower cost than field observation. The increasing use of spectral indices has led to an expansion of the geographical context of mangrove studies from local-scale studies to intercontinental and global analyses over the past 20 years. In remote sensing, numerous spectral indices derived from multiple spectral bands of remotely sensed data have been developed and used for multiple studies on mangroves. In this paper, we review the range of spectral indices produced and utilised in mangrove remote sensing between 1996 and 2021. Our findings reveal that spectral indices have been used for a variety of mangrove aspects but excluded identification of mangrove species. The included aspects are mangrove extent, distribution, mangrove above ground parameters (e.g., carbon density, biomass, canopy height, and estimations of LAI), and changes to the aforementioned aspects over time. Normalised Difference Vegetation Index (NDVI) was found to be the most widely applied index in mangroves, used in 82% of the studies reviewed, followed by the Enhanced Vegetation Index (EVI) used in 28% of the studies. Development and application of potential indices for mangrove cover characterisation has increased (currently 6 indices are published), but NDVI remains the most popular index for mangrove remote sensing. Ultimately, we identify the limitations and gaps of current studies and suggest some future directions under the topic of spectral index application in connection to time series imagery and the fusion of optical sensors for mangrove studies in the digital era.

Keywords: vegetation index; mangrove index; mangrove forest; mangrove above ground; biomass; carbon sink; bibliometric analysis

Citation: Tran, T.V.; Reef, R.; Zhu, X. A Review of Spectral Indices for Mangrove Remote Sensing. *Remote Sens.* **2022**, *14*, 4868. https://doi.org/10.3390/rs14194868

Academic Editor: Chandra Giri

Received: 12 August 2022
Accepted: 26 September 2022
Published: 29 September 2022

1. Introduction

Mangrove is a term which corresponds to intertidal ecosystems or lignified plant communities that grow in coastal environments between 40°S and 30°N throughout the world (Figure 1). The mangrove boundary is extended to the south of Japan (30.4°N) and Bermuda (32.4°N); to the south of New Zealand (38.05°S), Australia (38.85°S), and the east coast of South Africa (32.98°S) [1,2]. Mangrove distribution is restricted generally to areas where the mean temperature ranges 20–35°C, annual rainfall is between 1500–2500 mm, and there is a substantial riverine input of freshwater discharge [2]. Actually, the number of frozen days in the year may play on mangrove presence at high latitudes [3]. Decreases in the frequency of extreme cold occurrences could lead to considerable increases in mangrove cover near the current poleward limits of mangrove forests. The global mangrove distribution is classified into two groups, including the Indo-West Pacific (IWP) and the Atlantic East Pacific (AEP). Mangroves initially developed on the Tethys Sea's coastlines in the late Cretaceous-early Tertiary period [1,2,4]. Three million years ago, modern mangrove taxa emerged on the eastern borders of Tethys, diversified into present-day IWP regions, and subsequently spread into AEP regions [5,6]. The richness in the distribution of mangrove species reduces from the IWP to AEP. Globally, there are approximately 77 mangrove species, but about 54 species in 20 genera from 16 families constitute the group of "true mangroves" occurring only in mangrove habitats. Among these 77 species, 65 species in 32 genera and 24 families are recorded from IWP, while only 15 species in 10 genera and 8 families are in AEP [1,2]. The most commonly found genera in both IWP and AEP are *Rhizophora* and *Avicennia* [2].

The biophysical variables of mangroves (i.e., leaf area, basal area, tree height, percent canopy closure, diameter at breast height, carbon stock, and biomass) mostly depend on climatic conditions, while sea level rise has an influence on the structure and spatial distribution of mangroves [7,8]. Temperature, precipitation, and storminess explain 74% of the global trends in the maximum values of canopy height and above-ground biomass [7]. Globally, 75% of mangroves are distributed in tropical regions. The largest cover and highest mangrove diversity are found in Asia (39%), followed by Africa (21%, but mostly on the eastern side), North and Central America (15%), South America (12.6%) and Oceania (Australia, Papua New Guinea, New Zealand, south Pacific islands) (12.4%) [9,10]. The mangroves in the equatorial regions have the maximum biomass and tree canopies can reach an average height of 30–40 m. The highest mangrove forests are found in Gabon, an equatorial African nation, where heights reach 62.8 m. A quarter of the estimated 11.7 Pg C global mangrove carbon stock, which includes soil, above- and below-ground biomass, is stored in Indonesia [7,8]. These biophysical parameters gradually decrease with increasing latitude due to varied temperature and environmental conditions.

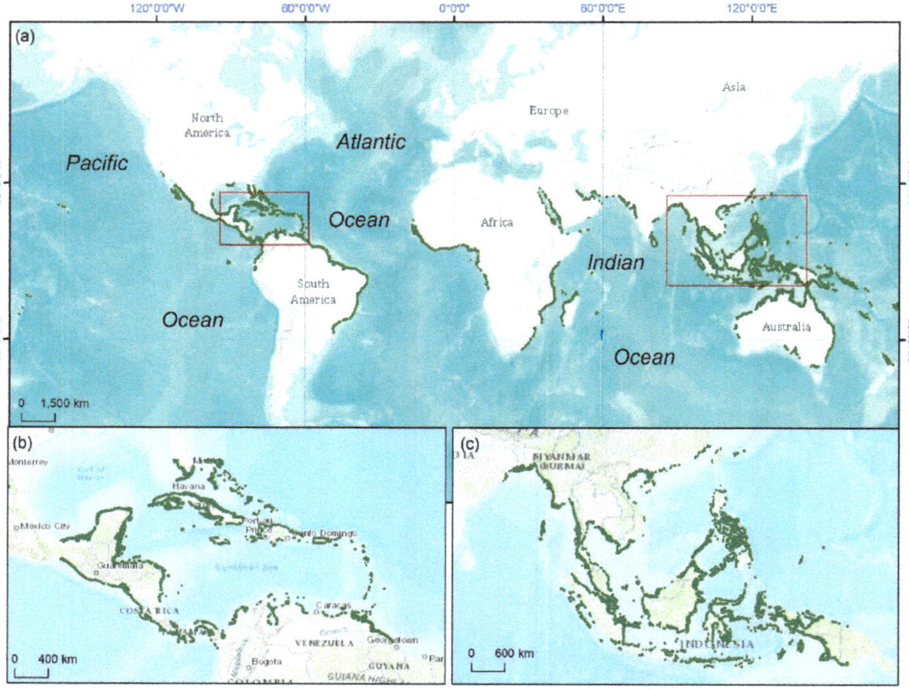

Figure 1. (**a**) The distribution of mangroves (in pale green) in the world, (**b**) in the Caribbean, and (**c**) in South and Southeast Asia (Source: adapted from Global Mangrove Watch, [11,12]).

Mangrove ecosystems produce valuable goods and services, including regulation (e.g., coastal protection, water filtration), supply (e.g., fisheries, aquaculture, timber, fuel, honey, construction materials, medicine), culture (e.g., tourism), and support (e.g., nursery habitats, climate mitigation) [13,14]. In some areas, mangroves have been proposed to provide a natural barrier to coastal erosion process, defending inland areas home to 120 million people from natural hazards (e.g., typhoon, cyclones, tsunamis) [7,15]. Mangrove restoration for coastal defence is expected to be up to five times more cost-effective than "grey infrastructure" such as breakwaters [16]. Mangroves also regulate water quality, and it is estimated that 2–5 ha of mangroves can treat the effluent from 1 ha of some aquaculture practices [17]. The carbon storage potential of mangroves is 3–5 times higher than that of tropical upland forests due to strong carbon storage in the soil [18,19]. Mangroves are also a valued source of timber, fuel, and tourism. There are over 2000 mangrove related attractions globally, such as boat tours, boardwalks, kayaking, and fishing [20]. Together, the economic value from mangrove ecosystem services has been estimated to exceed 800 billion per year [15,21]. However, approximately 35% of the global mangrove forests has been lost over the past 50 years due to both anthropogenic activities and physical stressors [22,23]. While restoration of mangrove forests has been increasing over the past 40 years, the net reduction in mangrove cover area and species richness is still a high 1–2% per year [22,24].

Traditionally, monitoring mangrove ecosystems used field observation and survey methods [25–27]. However, these approaches are difficult to monitor and measure mangroves in situ due to their dense understory and geographical location in intertidal zones [21,25]. Additionally, field observation and survey methods are labour-intensive, costly, and frequently limited in extent. Many surveys are qualitative and difficult to reproduce or revisit over time. Remote sensing (RS) has overcome the drawbacks of traditional field surveys and is has been continuously improving in terms of spatial resolution, revisit time and user costs over the past four decades [25,28]. RS is acknowledged in this context as the science and technology of acquisition of information about Earth's surface materials from a distance, typically from aircrafts or satellites [29]. The two types of remote sensing we refer to are (i) optical and (ii) radar sensors, which are classified according to the energy source of the signal used to identify the object. The remotely sensed data, acquired from these sensors, allows us to gather accurate information about the geographical distribution of mangrove ecosystems and biophysical properties at the pixel level [13,27].

In remote sensing, mangroves can be identified based on the textural and spectral properties of the canopy and leaves [13,30]. Their structural appearance, which can be either homogenous across the forest or heterogeneous, is affected by factors including species composition, growth form, density, and stand height. Almost all mangrove species can be discriminated within the visible and near infrared (NIR) region because of scattering in the spongy mesophyll cells in plants [31,32]. Using structural information extracted from several remotely sensed products regional and global estimation can be made of mangrove height, canopy, species succession, biomass, and carbon stocks [26]. The highest spectral reflectance of mangroves was observed in the NIR region for both Landsat 8 and Sentinel-2A surface reflectance sources (Figure 2). With the Sentinel-2A in particular, mangrove reflectance was observed to rise rapidly at the red-edge. Therefore, mangrove ecosystems can be observed using indices computed from spectral bands in the visible and NIR regions of optical remote sensing.

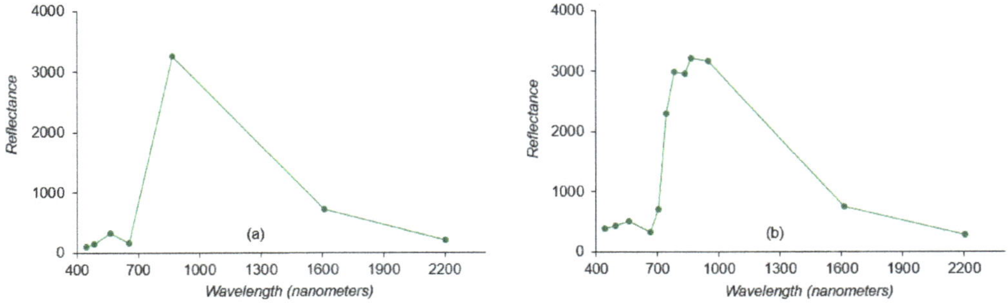

Figure 2. Spectral signatures of mangroves (*Rhizophora*) in Can Gio Mangrove Biosphere Reserve, Vietnam, derived from median values of (**a**) Landsat 8 and (**b**) Sentinel-2A surface reflectance in February 2021 (Source: obtained from Google Earth Engine).

The importance of remote sensing in mangrove studies has been recognised in many review studies [25,26,28,30,33–35]. These publications serve as a good starting point for researchers who are interested in mangrove remote sensing. However, the application of different spectral indices in mangroves has not been reviewed extensively in most of these studies. For instance, Green et al. (1998) [33] considered the significance of remote sensing for mangrove mapping from 1972 to 1996. While the study is the first paper that mentioned applying NDVI to mangrove classification, it only focused on NDVI even though at the time there were over 40 vegetation indices that could have been relevant to mangrove ecosystems. During the 1998–2018 period, most review papers highlighted remote sensing as a technique or approach for mangrove studies, while remote sensing has been defined as the science of acquiring information from distance [29,36,37]. Recently, Wang et al. (2019) [25] revealed common gaps in previous publications (i.e., research topics, key milestones, and mangrove driving forces) in mangrove remote sensing and investigated the importance of remote sensing for mangrove studies from 1956 to 2018. However, Wang et al. (2019) did not clearly state what kinds of spectral indices are specifically applied to mangrove remote sensing.

The present study intends to address the aforementioned knowledge gaps, by answering the following research question: what spectral indices have been applied and have been proven effective for mangrove remote sensing? Our objectives are to (i) examine and categorise spectral indices used in publications related to mangrove remote sensing; (ii) assess their applications in the study of mangrove ecosystems; and (iii) propose future directions for the application of additional spectral indices in mangrove remote sensing.

2. Search Strategy and Data Analysis

Mangrove scholars used various qualitative and quantitative approaches to understand and organise earlier findings of mangrove studies. Among these, a quantitative analysis of academic literature, defined as bibliometrics, was investigated as a potential tool to introduce a systematic, transparent, and reproducible review process [38–41]. Compared to other literature review techniques, bibliometric analysis of the published literature is effective to identify research gaps and direct future avenues of research [41]. For bibliographic citations, Web of Science (WoS) and the Scopus platforms are the most extensive databases which are widely used to obtain metadata for bibliometric analysis [42,43]. Scopus was launched in 2004, but WoS launched in 1997 and is considered the earliest international bibliographic database [42,44]. WoS comprises four citation databases with more than 10,000 journals [45]. Journals indexed in the WoS must meet 28 criteria (i.e., 24 quality criteria and 4 impact criteria) [44,46]. The fulfilment of 28 criteria contributes to enhancing academic quality and minimising the influence of multiple predatory journals. The journal listed in the WoS database primarily provides impact factor (a ratio between citations and citable items published the previous year) and h-index (an index based on a list of publications ranked in descending order by Times Cited count) [46]. Therefore, journals with high impact factor or h-index are cited more often than journals with lower impact factor or h-index.

Various keywords were entered in the searching process associated with global mangrove ecosystems based on spectral indices application (Table 1). We used Thomson Data Analyzer (TDA) integrated in the Web of Science (WoS) Core Collection to retrieve annual publications and their citations [47]. The keyword search resulted in 293 papers published between 1992 and 2021. We then reviewed the abstract and content of the 293 papers and removed from our study the papers that did not relate to the application of spectral indices in mangroves. This left 195 publications (including 90 journal papers, 14 conference papers and one book chapter) for our review.

Table 1. The predefined keywords of the searching process.

No.	Keywords
1	vegetation index and mangrove
2	comprehensive mangrove quality index
3	a mangrove recognition index
4	mangrove and an index analysis approach
5	leaf area index and mangrove
6	spectral mangrove index

Full records (i.e., author, title, source, abstract) and cited references of the search results were downloaded in the BibTex format in several batches, each comprising no more than 500 data entries. For further processing, all of the obtained result files were zipped together and imported into the R-statistical and VOSviewer software packages. The bibliometric analysis was carried out with the help of Bibliometrix package in R [38]. The annual cited times of the gathered articles were calculated using TDA's citation report tool, while the top journal sources for publication and citation were determined using Bibliometrix package. Finally, using VOSViewer software [48], we performed co-word analysis to visualise density networks of author keywords for trend analysis in mangrove remote sensing. The co-word method is a technique to analyse the co-occurrences of key words and identify relationships and interactions between the topics researched and emerging research trends [48]. Details regarding the theory and practical function of the co-word approach utilising VOSViewer software may be found at [45,48]. However, the record of the online bibliographic database prior to 1990 may be incomplete [42,44,49] because the internet-based Web of Science was firstly launched in 1997. Therefore, several publications from print only sources may be missing. We thus supplemented our database with metadata from outcomes of Green et al. (1998) [33] and Bannari et al. (1995) [50] to increase the number of studies applying spectral indices in mangrove studies prior to 1996.

Publications and annual citations in this field of research, retrieved from citation report analytics of the Web of Science Core Collection, significantly increased between 1996 and 2021 (Figure 3). The number of publications dropped in 2021, which may be related to COVID-19 pandemic because most field trips were delayed or cancelled and this impacted field data collection for validation and other known impacts on academic workloads [51]. Most of the publications (21) were published in the journal of Remote Sensing (IF: 4.848, Open access), others were mainly from the International Journal of Applied Earth Observation and Geoinformation (8, IF: 5.993, Open access), International Journal of Remote Sensing (8, IF: 3.362), Estuarine, Coastal and Shelf Science (7, IF: 2.904), Modeling Earth Systems and Environment (6, IF: N/A) and ISPRS Journal of Photogrammetry and Remote Sensing (5, IF: 10.565). The journals Ecological Indicators, Regional Studies in Marine Science, and Wetlands had four papers each. The remaining journals had published less than three papers involving mangrove remote sensing since 1996. In-text citations of work that applied mangrove remote sensing indices were found in Remote Sensing of Environment (881, h-index: 281), Remote sensing (445, h-index: 124), International Journal of Remote Sensing (437, h-index: 174), Estuarine, Coastal and Shelf Science (249, h-index: 134), Aquatic Botany (170, h-index: 94). The publications most cited were those in higher ranked (higher IF) journals, regardless of the open access status of the journal.

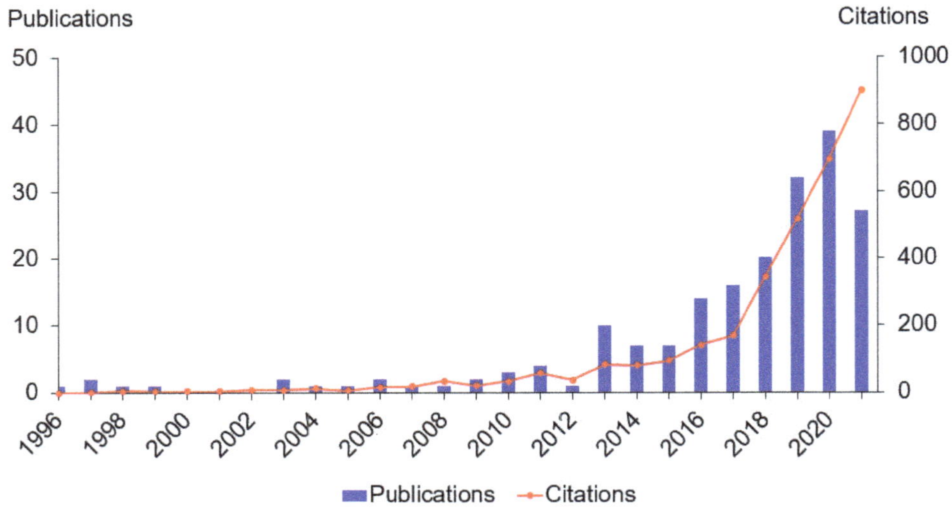

Figure 3. Change in 195 peer-reviewed publications and their citations per year between 1996 and 2021 (based on the citation report tool in TDA using mentioned keywords).

3. Overview of Spectral Indices Used in Mangrove Remote Sensing

A spectral index is an equation that combines pixel values from two or more spectral bands in a multispectral image using various algorithms, mainly focused on band ratio or feature scaling (e.g., normalised or standardised algorithms) [52,53]. Spectral indices are calculated to highlight pixels in an image that not only show the relative abundance of a land cover of interest, but also emphasise an ecosystem function [52–54]. They show better sensitivity than individual spectral bands for spectral signature detection. Throughout the mission of Earth surface observation, spectral indices have significantly contributed a more thorough understanding of environments and ecosystems across space and time [28]. The geographical extent of mangrove studies using spectral indices has also seen a significant change over time (Table 2). Most studies (92%) were carried out at the national level, while a few publications implemented research on intercontinental/global scale during the 2000–2015 period and this trend is increasing (16% in 2021 vs. 8% in 2016). The study areas were mainly in India (12.8%), Mainland China (12.3%), Indonesia (9.7%), the US (9.7%), and Mexico (8.2%).

Table 2. The mangrove study area of spectral indices application, obtained from our searched publications.

Country/Region	Before 2000s	2001–2015	2016–2021	%
US	2 (33%)	6 (15%)	11 (7%)	9.7
Mexico	1 (17%)	6 (15%)	9 (6%)	8.2
India	0	6 (15%)	19 (13%)	12.8
China	0	3 (8%)	21(14%)	12.3
Australia	0	1 (3%)	6 (4%)	3.5
Malaysia	0	5 (13%)	7 (5%)	6.2
Indonesia	0	1 (3%)	18 (12%)	9.7
Vietnam	0	1 (3%)	7 (5%)	4.1
Others	3 (50%)	7 (18%)	29 (19%)	20
Intercontinental regions and globe	0	3 (8%)	25 (16%)	13.5

Spectral indices can be categorised as either satellite or airborne system indices based on the platforms used for data acquisition. Depending on the spectral bands of passive satellite remote sensing, spectral indices may be further grouped into indices with (i) simple ratio, (ii) visible and near-infrared (VNIR) bands, (iii) visible and red edge bands, (iv) visible and mid-infrared bands, and (v) visible and shortwave infrared (SWIR) bands. In addition, following the applications of spectral indices in mangrove remote sensing, we separate spectral vegetation indices and spectral mangrove-specific indices. The key differences between the two types of indices are their applications and the spectral bands used in the indices. Vegetation indices are spectral indices computed using spectral bands in the visible, red edge, and near-infrared regions [55]. They have been widely applied to mangrove ecosystems previously [13,30]. However, as mangroves have common spectral characteristics as other vegetation, separating mangroves from other types of vegetation using a single vegetation index is challenging [56]. To address this issue, some mangrove-specific spectral indices have been proposed for separating mangroves from terrestrial vegetation [56–60]. These spectral mangrove indices include spectral bands from VNIR to SWIR regions. In this review, based on publications that used spectral indices for mangrove ecosystems and the approaches they applied, we classified indices into four categories: (i) visible and near-infrared bands; (ii) visible and red edge bands; (iii) visible bands of airborne systems; and (iv) spectral mangrove specific- indices.

3.1. Spectral Indices with Visible and Near-Infrared Bands (VNIR)

Indices measured from spectral bands in the VNIR regions are acknowledged as vegetation indices [50], except Normalized Difference Water Index (NDWI) [61]. The history of spectral vegetation indices development is associated with the Landsat mission in 1972. Particularly, Pearson and Miller (1972) developed the first vegetation indices, i.e., Ratio Vegetation Index (RVI) and Vegetation Index Number (VIN), to estimate and monitor vegetative cover [50,62]. Following Pearson and Miller (1972), Rouse et al. (1973) introduced the Normalized Difference Vegetation Index (NDVI) that is now widely applied for land cover and environmental studies [63,64]. Over the last four decades, more than 40 vegetation indices have been developed, but 28 of these were used in mangrove investigations (Table 3). Two categories of vegetation indices can be separated: ones that include only spectral bands and others that include spectral bands that are adjusted by non-spectral factors [e.g., soil adjustment factor (L), soil line factors (a, b), and coefficients of atmosphere resistance (c_1 and c_2)]. The first category (e.g., RVI, VIN, NDVI–Table 3) is based on linear combinations (difference or sum) of spectral bands or raw band ratios without considering environmental interactions. The second group (e.g., SAVI, TSAVI, EVI–Table 3) is based on the knowledge of physical phenomena which explains interactions between electromagnetic radiation, the atmosphere, the vegetative cover, and the soil background.

In the first group without adjustment factors, NDVI is the first index to show the highest correlation with field measured mangrove canopy cover (r = 0.91), higher than PVI, GVI, and RVI [65]. NDVI (No.3–Table 3) is a simple indicator that is acquired in red (visible) and near-infrared (NIR) regions, based on a normalised algorithm. The NIR band in the NDVI equation is useful for vegetation detection because healthy vegetation (which contains chlorophyll) reflects more NIR compared to other wavelengths [63]. The normalised algorithm mitigates the atmospheric effects and the impacts of sensor calibration degradation in the red and NIR bands [64,66]. Mathematically, NDVI also forms the basis for other indices. For example, a bijective relationship between NDVI and VIN is demonstrated by Equation (1). However, the NDVI values are affected by soil background when the green leaf area is small or the majority of the scene is soil [67,68]. Therefore, a number of vegetation indices in the second category have been developed for taking into account environmental conditions. For example, to adjust soil background, soil factors (e.g., L, a, and b) were included in SAVI and TSAVI (No.8–9, Table 3). Both indices SAVI and TSAVI are equal to NDVI, if the value of the soil adjustment factor in SAVI (No. 8, Table 3) is zero (L = 0) or the slope (a) and ordinate (b) at the origin of bare soil line parameters in TSAVI (No.9, Table 3) are one and zero (a = 1 and b = 0), respectively.

$$NDVI = \frac{NIR - R}{NIR + R} = \frac{NIR/R - 1}{NIR/R + 1} = \frac{VIN - 1}{VIN + 1} \tag{1}$$

Some studies show that the indices that include adjustment factors are less sensitive to atmospheric and soil background effects than the first-generation spectral indices [50,64,65]. Enhanced Vegetation Index (EVI, No.4–Table 3) [69], for example, which corrects for some atmospheric conditions and canopy background noise, and provides a better estimation of mangrove biophysical properties in high density forests than does NDVI [70]. However, the blue band requirement may be a disadvantage of EVI, which cannot be generated from optical sensors that do not have a blue band, such as the Advanced Very High-Resolution Radiometer (AVHRR) and the Advanced Spaceborne Thermal Emission and Reflection Radiometer (ASTER). Alternatively, the EVI-2 (No.28–Table 3) with two spectral bands in the red and NIR regions was proposed, achieving similar performance to EVI, which can be produced by almost all optical sensors [71]. However, the EVI-2 is sensitive to the impact of the bidirectional reflectance distribution function (BRDF). Therefore, each index has advantages and disadvantages in relation to vegetation characteristics with overlapping spectral features due to background signals from soil and confounding factors (e.g., sensor and calibration effects, quality assurance and quality control, BRDF, and atmospheric and topographic effects). Hence, there are no perfect vegetation indices for all aspects of mangrove studies under all conditions.

Table 3. The 28 indices in the spectral wavelength 400–1000 nm used in mangrove research (R: Red; G: Green; B: Blue; and NIR: Near Infrared). For whole algorithms, L is a soil adjustment factor that addresses nonlinear model, differential NIR and red radiant transfer through a canopy [72]; a and b are soil line factors [73]; c_1 and c_2 are coefficients of atmosphere resistance term which use blue band to correct for aerosol influences in red band [69].

No.	Spectral Index	Formula	Reference
1	Vegetation Index Number [62]	$VIN = \frac{R}{NIR}$	[74,75]
2	Ratio Vegetation Index [62]	$RVI = \frac{NIR}{R}$	[58,59,65,76–91]
3	Normalized Difference Vegetation Index [63]	$NDVI = \frac{NIR-R}{NIR+R}$	From 1991 to 2011: [65,76–78,92–105] Since 2012: [58–60,74,75,80–87,89–91,106–209]
4	Enhanced Vegetation Index [69]	$EVI = \frac{2.5(NIR-R)}{NIR+c_1R-c_2B+L}$	[3,59,74,75,80,82,84,90,123,144,146,149, 153,160,169,172,176,186,191,196,197,201, 206,207,210–225]
5	Perpendicular Vegetation Index [73]	$PVI = \frac{NIR-aR-b}{\sqrt{a^2+1}}$	[76,138]
6	Normalized Green-red Difference Index [226]	$NGRDI = \frac{G-R}{G+R} + 0.08$	[227]
7	Difference Vegetation Index [228]	$DVI = NIR - R$	[74–76,84,97]
8	Soil Adjusted Vegetation Index [67]	$SAVI = (1+L)\frac{NIR-R}{NIR+R+L}$	[58,76,83–85,89–91,108,110,120,123,132, 138,139,145,146,150,169,170,201,207,225]
9	Transformed soil adjusted vegetation index [68]	$TSAVI = \frac{a(NIR-aR-b)}{R+aNIR-ab}$	[76,196]
10	Soil adjusted ratio vegetation index 2 [229]	$SAVI_2 = \frac{NIR}{R+a \div b}$	[76]
11	Global environment monitoring index [230]	$GEMI = n(1-0.25n) - \frac{R-0.125}{1-R}$ at $n = \frac{2(NIR^2-R^2)+1.5NIR+0.5R}{NIR+R+0.5}$	[74,75,84,109,139,196]
12	Atmospherically Resistant Vegetation Index [231]	$ARVI = \frac{NIR-2R+B}{NIR+2R-B}$	[74,84,196,232]
13	Non-linear vegetation index [233]	$NLI = \frac{NIR^2-R}{NIR^2+R}$	[83]
14	Modified soil-adjusted vegetation index [234]	$MSAVI = \frac{2NIR+1-\sqrt{(2NIR+1)^2-8(NIR-R)}}{2}$	[81,85,86,108,132,145,169,191,207]
15	Renormalized difference vegetation index [235]	$RDVI = \frac{NIR-R}{\sqrt{NIR+R}}$	[81,85,108]
16	Modified Simple Ratio [236]	$MSR = \frac{NIR/R-1}{\sqrt{NIR/R+1}}$	[85,145]
17	Green Normalized Difference Vegetation Index [237]	$GARI = \frac{NIR-[G-1.7*(B-R)]}{NIR+[G-1.7*(B-R)]}$	[130,196]

Table 3. *Cont.*

No.	Spectral Index	Formula	Reference
18	Optimized Soil Adjusted Vegetation Index [238]	$OSAVI = \frac{NIR-R}{NIR+R+0.16}$	[82,89,109,110,115,130,177,199]
19	Green Normalized Difference Vegetation Index [239]	$GNDVI = \frac{NIR-G}{NIR+G}$	[81,85,87,89,119,139,145,193,207]
20	Green Leaf Index [240]	$GLI = \frac{(G-R)+(G-B)}{(G+R)+(G+B)} + 0.07$	[227]
21	Triangular Vegetation Index [241]	$TVI = 0.5 \times [120(NIR-G) - 200(R-G)]$	[74,82,84,85,108,207]
22	Transformed Difference Vegetation Index [242]	$TDVI = 1.5\left[\frac{NIR-R}{\sqrt{NIR^2+R+0.5}}\right]$	[109,115]
23	Modified Non-Linear Vegetation Index [243]	$MNLI = \frac{1.5(NIR^2-R)}{NIR^2+R+0.5}$	[83]
24	Modified Chlorophyll Absorption Ratio Index 1 [244]	$MCARI_1 = 1.2 \times [2.5(NIR-R) - 1.3(NIR-G)]$	[85,89,199,232]
25	Modified Chlorophyll Absorption Ratio Index 2 [244]	$MCARI_2 = MTVI_2 = \frac{1.5[1.2(NIR-G)-2.5(R-G)]}{\sqrt{(2NIR+1)^2-(6NIR-5\sqrt{R})-0.5}}$	
26	Modified Triangular Vegetation Index [244]		
27	Chlorophyll Vegetation Index [245]	$CVI = \frac{NIR*R}{G^2}$	[119,145,196]
28	Enhanced Vegetation Index 2 [71]	$EVI_2 = 2.5\left(\frac{NIR-R}{NIR+2.4R+L}\right)$	[59,74,75,84,89,132,193,196,207,246]

3.2. Spectral Indices in Visible and Red-Edge Bands

The red edge is a band in the red-NIR transition zone that indicates the transition between red visible absorption by chlorophyll and NIR scattering due to leaf internal structure [25]. This transition zone is used in various vegetation indices, the most important of which is the normalised difference between red visible (0.6 nm) and NIR (0.8 nm) reflectance. Spectral indices obtained from the red edge region (Table 4) were used in several investigations for mangrove chlorophyll and biophysical parameters [81,119], mangrove density and carbon analyses [85,247], and mangrove biomass [207]. These investigations found strong correlation ($r > 0.95$) between ground truth data and red edge vegetation indices. Normally, red edge bands are available from hyperspectral datasets that are difficult for large-scale collection. Alternatively, since 2015, three red-edge bands have been incorporated in the Sentinel-2A sensor, which could be combined with other visible bands to provide essential information about mangrove ecosystems at a spatial resolution of 10 m. Therefore, the use of red-edge bands promises the potential to benefit mangrove ecosystems associated with mangrove health monitoring in the future.

Table 4. The used indices with VNIR and red-edge bands (R: Red; B: Blue; and NIR: Near Infrared). With three Red-edge bands of Sentinel-2A, each formula can compute three sub-equations.

No.	Spectral Index	Formula	Reference
1	Normalised Difference Index [248]	$NDI = \frac{R_{edge}-R}{R_{edge}+R}$	[81,85,119,207,247]
2	Red edge NDVI [249]	$NDVI_{\text{Red-edge}} = \frac{NIR-R_{edge}}{NIR+R_{edge}}$	
3	Plant Senescence Reflectance Index [250]	$PSRI = \frac{R-B}{R_{edge}}$	
4	Normalized Difference Red edge Index [251]	$NDRE = \frac{NIR-R_{edge}}{NIR+R_{edge}}$	
5	Red-edge Chlorophyll Index [252]	$CI_{\text{Red-edge}} = \frac{NIR}{R_{edge}} - 1$	
6	MERIS Terrestrial Chlorophyll Index [253]	$MTCI = \frac{NIR-R_{edge}}{R_{edge}-R}$	

3.3. Spectral Indices with Visible Bands of Airborne Systems

Traditionally, aerial photographs (AP) were widely used for mangrove mapping and assessment [94]. Nowadays, with the support of Unmanned Aerial Vehicles (UAVs) or drones, the application of aerial photographs has become more convenient. Some spectral indices for mangrove classification have been proposed using visible bands of aerial photographs that can achieve an overall accuracy of over 95% in mangrove cover mapping [157,254] (Table 5). Notably, integration of Red-edge, NIR, and shortwave infrared (SWIR) bands in the optical sensors of UAVs promises an advantage for monitoring mangrove ecosystems at a spatial resolution of centimetres. This drone-based multispectral remote sensing can be as the future for mangrove remote sensing. However, these applications are only suitable for a small area and can be employed in a short time due to cost and energy limitation. More importantly, the use of UAVs for field data collection is regulated by the government in most countries throughout the world.

Table 5. The used UAV spectral indices for mangrove ecosystems.

No.	Spectral Index	Formula	Reference
1	Excess Green Vegetation Index [248]	$ExG = 2 \times G - R - B + 50$	[196,227,254]
2	Normalized Difference Index [248]	$NDI = \frac{G-R}{G+R}$	
3	Negative Excess Red Vegetation Index [255]	$NegExR = G - 1.4 \times R$	
4	Visible Atmospheric Resistant Index [256]	$VARI = \frac{G-R}{G+R-B}$	
5	Colour Index of Vegetation Extraction [257]	$CIVE = 0.441R - 0.881G + 0.385B + 18.78745$	
6	Vegetative Index [258]	$VEG = \frac{G}{R^{0.667}} \times B^{0.333}$	
7	Excess Green minus Excess Red [259]	$ExG - ExR$	
8	Triangular Greenness Index [260]	$TGI = G - 0.39R - 0.61B$	
9	Combined Index [261]	$CI = 0.25ExG + 0.3ExGR + 0.33CIVE + 0.12VEG$	
10	Visible-band Difference Vegetation Index [262]	$VDVI = \frac{2G-B-R}{2G+B+R}$	

3.4. Mangrove-Specific Spectral Indices

Mangroves are found in coastal wetlands where they are regularly submerged by tides [55,60,263]. As a result, fluctuations in tide levels, and thus the presence of water, results in tide-dependent variation in spectral signatures for mangrove forests, leading to inaccurate mapping results, particularly in locations with large tidal ranges [57,135]. Recently, several studies have developed specific spectral indices that can adapt to changes in tide conditions and applied them in order to separate mangroves from non-mangroves [57,58,60] and from other land cover types [55]. These are the mangrove-specific indices. During the 2013–2021 period, six mangrove-specific indices were proposed to improve the accuracy of single image remotely sensed data during high tide (Table 6).

Zhang and Tian (2013) [57] proposed a mangrove recognition index (MRI–Equation (2)) for mangrove detection using multi-temporal Landsat TM images that is insensitive to the stage of the tide. Winarso et al. (2014) [56] further developed this method to create a mangrove discrimination index (MDI–Equation (3)) for estimate mangrove density from Landsat 8 images. Kumar et al. (2017) [59] proposed two new vegetation indices (Normalised Difference Wetland Vegetation Index and Shortwave Infrared Absorption Index) and combined them with two previously published indices (Normalised Difference Infrared Index and Atmospherically Corrected Vegetation Index) to integrate the Mangrove Probability Vegetation Index (MPVI–Equation (4)) for mangrove classification. Gupta et al. (2018) [58] published a Combined Mangrove Recognition Index (CMRI–Equation (5)) to distinguish mangrove from non-mangrove with an accuracy of more than 60%. Jia et al. (2019) [60] generated the Mangrove Forest Index (MFI–Equation (6)) using spectral bands from Sentinel-2 data to detect submerged mangrove forests at high tide.

Table 6. The proposed mangrove indices for mangrove classification (*EQN: equation number*). In the Equation (2), *GVI* and *WI* are green vegetation index and wetness index at low (*L*) tide and high (*H*) tide, respectively. In Equation (4), n is the total number of bands in the image, R_i is the reflectance value at band i for a pixel of the reflectance image, and r_i is the reflectance value at band i for candidate spectrum of mangrove forest. In Equation (6), the ρ_λ is the reflectance of the band centre of λ, and i ranged from 1 to 4; λ_1, λ_2, λ_3, λ_4 represent the centre wavelengths at 705, 740, 783 and 865 nm, respectively. λ_i is the baseline reflectance in λ_i. ρ665 and ρ2190 are the reflectance of band 4 (centred at 665 nm) and 12 (centred at 2190 nm), respectively. In Equations (3), (5), and (7), G, R, NIR, and SWIR are green, red, near-infrared, and shortwave infrared bands, respectively.

Index	Formula	EQN	Reference
Mangrove Recognition Index	$MRI = \lvert GVI_L - GVI_H \rvert \times GVI_L \times (WI_L + WI_H)$	(2)	[57]
Mangrove Damage Index	$MDI = \frac{NIR-SWIR}{NIR \times SWIR} \times 10{,}000$	(3)	[56]
Mangrove Probability Vegetation Index	$MPVI = \frac{n\sum_{i=1}^{n} R_i r_i - \sum_{i=1}^{n} R_i \sum_{i=1}^{n} r_i}{\sqrt{n\sum_{i=1}^{n} R_i^2 - \left(\sum_{i=1}^{n} R_i\right)^2}\sqrt{n\sum_{i=1}^{n} r_i^2 - \left(\sum_{i=1}^{n} r_i\right)^2}}$	(4)	[59]
Combined Mangrove Recognition Index	$CMRI = \frac{NIR-R}{NIR+R} - \frac{G-NIR}{G+NIR}$	(5)	[58]
Mangrove Forest Index	$MFI = [(\rho_{\lambda1} - \rho_{B\lambda1}) + (\rho_{\lambda2} - \rho_{B\lambda2}) + (\rho_{\lambda3} - \rho_{B\lambda3}) + (\rho_{\lambda4} - \rho_{B\lambda4})]/4$ $\rho_{B\lambda i} = \rho_{2190} + (\rho_{665} - \rho_{2190}) \times (2190 - \lambda_i)/(2190 - 665)$	(6)	[60]
Mangrove Vegetation Index	$MVI = \frac{NIR-G}{SWIR_1-G}$	(7)	[55]

These proposed mangrove indices involve the signature of mangroves in the context of tidal fluctuation, which is sensitive to greenness and wetness patterns. Thus, MRI and MPVI are sensitive to tidal extent and period, and cannot be used in site comparisons where sites differ in hydrology. Additionally, the number of spectral bands required for the MFI calculation is only available using Sentinel-2 or hyperspectral sensors. Baloloy et al. (2020) [55] recently analysed the shortcomings of earlier integrated mangrove forest indices (i.e., MRI, CMRI, MPVI, NDI, and MFI) and developed a new index: the mangrove vegetation index (MVI–Equation (7)), for enhancing the accuracy of mangrove forest extent mapping. MVI is a single index that classifies mangroves, terrestrial vegetation (forest and non-forest), bare soil, built-up areas, water, and clouds using reflectance data from Sentinel-2A and Landsat-8 in the NIR, Green, and SWIR bands. MVI validation was initially used at an intercontinental scale and demonstrated an accuracy of more than 80% for the entire set of geographical research locations. The high index accuracy of MVI can provide a possibility for global mangrove studies, although MVI is limited by biophysical and environmental parameters due to its reliance on SWIR. Using the SWIR spectrum, in particular, has been a problem for sensor systems constructed with solely visible and NIR wavelengths (e.g., Landsat-1,4 and Planetscope). Additionally, SWIR reflectance value is frequently mixed with built-up land noise, water bodies, and vegetation surrounding [61,264,265]. Neri et al. (2021) [266] investigated misclassification of mangroves from other land cover types in aquaculture zones, irrigated croplands, and palm tree sites when applying MVI due to spectral similarity between mangroves and vegetation in these areas. Notably, the significant drawback of MVI is that it does not have a specific optimal threshold for mangroves, which differs from ranges of vegetation and other mangrove-specific indices for mangrove separation.

In summary, there are several newly improved spectral indices for mangrove classification, but none of these can completely reduce the impact of environmental factors (e.g., tidal influences, land cover mixture).

4. Evaluation of Spectral Indices Applications in Mangrove Remote Sensing

As mentioned in Section 3, a variety of spectral indices have been used for (i) mapping mangrove extents and distributions; (ii) measuring above-ground properties of mangroves; and (iii) detecting mangrove changes. The co-word map of keywords that resulted from the co-word analysis of the literature on spectral indices application for mangroves is presented in Figure 4. The mangrove and forest were core words of the network, and NDVI and remote sensing were prominent terms in a field study. This showed that NDVI has a strong relationship with aspects of mangrove remote sensing in terms of spectral index application. Also, the result predicted that NDVI was the most common index applied to mangrove ecosystems. In total, 82% of our reviewed publications used NDVI and 28% used EVI.

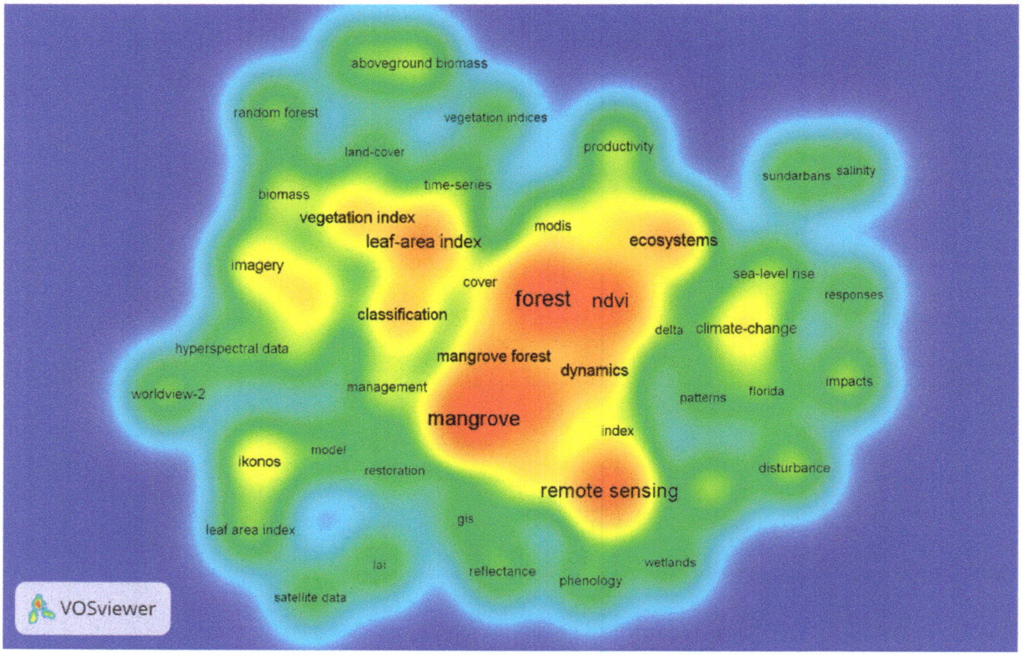

Figure 4. Co-word keywords map of spectral indices' application in mangrove research. Colours indicated the density of keywords or terms, ranging from blue (lowest density) to red (highest density). The larger the number of items in the vicinity of a point and the higher the weights of the neighbouring items, the closer the point is to being red.

4.1. Mangrove Extent and Distribution

Mapping mangroves is important for recording the present mangrove area and species distribution, provides information to support management decisions around potential threats of degradation due to uncontrolled development, and enables modelling change in relation to driving factors. Mangrove mapping can be binary, which classifies an image into two classes: mangroves and non-mangroves, or into a number of land covers based on a particular land over classification scheme with mangrove as one of the land cover types. Binary mapping normally uses single spectral indices (e.g., NDVI, EVI, SAVI, or mangrove indices) to highlight the pixels of mangrove vegetation and separate them from non-mangrove pixels [58]. Mapping can also include continuous classifiers such as height and canopy area.

Image classification for mangrove mapping can be supervised or unsupervised, pixel-based or object-based. Almost all previous research applied composited bands of satellite images based on classification approaches such as unsupervised algorithms (e.g., ISODATA or K-means) and parametric supervised algorithms (e.g., maximum likelihood) to detect mangrove extents without spectral indices application [25,26,267]. The overall accuracy of post-classification from these approaches ranges from 42% to 68% for most optical sensors and methodologies to measure accuracy [26]. Since spectral indices were used for mangrove ecosystems under pixel-based or object-based categories based on machine learning algorithms (e.g., artificial neural network, support vector machine, and random forest), an improvement of post-classification overall accuracy has been achieved to more than 80% [26]. More recent studies have used mangrove classification using spectral indices based on random forest, one of the machine learning algorithms, and offer the highest post-classification overall accuracy (>92%) [27,196,268].

From using single spectral indices for mangrove separation, some studies found that in dense mangrove areas, the NDVI or EVI value threshold for discriminating mangroves was 0.3 and above [104,269]. Le et al. (2020) applied NDVI, derived from Sentinel-2, to detect mangrove cover in the Can Gio Mangrove Biosphere Reserve, Vietnam [183]. This study considered that the NDVI mangrove value as NDVI > 0.3 with an overall classification accuracy of 83%. In addition to NDVI/EVI, six mangrove spectral indices (Table 6) have been developed and applied for mangrove separation with an overall classification accuracy of 80% above [55].

Vegetation, soil, and water are the three principal factors that contribute to the pixel composition of remotely sensed data in mangroves (Figure 5). In addition, seasonal and diurnal intertidal interactions influence the surface appearance [13,263]. These factors have a significant impact on the spectral characterisation of picture components. Therefore, depending on the effects of natural surroundings and mangrove density, the NDVI/EVI thresholds can be adjusted. In addition to physical influences, mangroves and other types of vegetation may generate similar signals from the vegetation index [56]. As a result, utilising a single NDVI or EVI threshold to distinguish mangroves from other types of vegetation may lead to an inaccurate outcome. Therefore, several soil and water spectral indices (Table 7) have been used concurrent with vegetation indices to improve mangrove detection [150,198].

Figure 5. Mangroves in Victoria State, Australia (Source: photo taken by author).

Table 7. The indices in the Visible, Near Infrared (NIR), and Shortwave Infrared (SWIR) bands.

No.	Spectral Index	Formula	Reference
1	Normalized Difference Moisture Index or Land Surface Water Index [264,270,271]	$NDMI = LSWI = \frac{NIR-SWIR}{NIR+SWIR}$	[59,60,91,123,134,135,152,191, 198,201]
2	Normalised Difference Water Index [264]	$NDWI = \frac{R-SWIR}{R+SWIR}$	[97,105,138,150,195,206]
3	Modified Normalized Difference Water Index [272]	$MNDWI = \frac{G-SWIR}{G+SWIR}$	[60,134,201]
4	Normalised Difference Soil Index [105]	$NDSI = \frac{SWIR-NIR}{SWIR+NIR}$	[105]
	Normalized Difference Water Index [61]	$NDWI = \frac{G-NIR}{G+NIR}$	[58,130,195,196,198,201]

The reflected signals from the SWIR regions capture information on radiation absorption by water, cellulose and lignin, and a variety of other biological elements. Nevertheless, collecting satellite imagery at SWIR wavelengths has distinct advantages, such as better atmospheric penetration and better contrasts among different vegetation types. The utilisation of the SWIR band in conjunction with the visible and NIR bands, in particular, aids in enhancing the presence of water in plant leaves [264] or urban characteristics [265]. For example, NDWI and NDBI threshold values more than 0 visualise water bodies and impervious surface, respectively [61,264,265]. Besides, several studies used elevation data and tasselled cap transformation to further improve the classification accuracy [55,57,134]. Consequently, an improved performance ($\geq$90% of overall accuracy) was investigated when apply multiple spectral indices for detecting the mangrove cover [27,267].

Our search did not reveal a spectral index that has been applied for mangrove species separation. Previous studies applied spectral bands to separate mangrove species based on maximum likelihood classification or machine learning algorithms (e.g., random forest and support vector machine) because each mangrove specie reflects a particular wavelength of the spectrum [25–27,164,273]. These studies revealed that spectral reflectance properties of some mangrove species are similar, making a challenge for identification. Hirata et al. (2014) [273] proved that the spectral reflectance properties for *A. alba* and *S. alba* were clearly distinct in three of four VNIR bands (i.e., Green, Red, and NIR), whereas those for the *Rhizophora* and *Bruguiera* species were similar in most spectral bands. A similarity of spectral bands among mangrove species leads to a uniformity/resemblance as using spectral indices because spectral index is computed from spectral bands ratio. It shows that almost all of mangrove species also have the same threshold value in spectral index. For instance, red mangroves (*Rhizophora*), black mangroves (*Avicennia*), and white mangroves (*Laguncularia racemosa*) may have particular spectral reflectance in a single spectral band, but the signals of three species in NDVI/EVI are normally more than 0.3.

4.2. *Above-Ground Properties of Mangroves Estimation*

The term "above-ground mangroves properties" in our study refers to the estimation in aspects of mangrove ecosystems above ground such as leaf area index (LAI), biomass, carbon, vertical structure, and mangrove health. Understanding these variables is beneficial in detecting the interaction of vegetation, the stability of that interaction, and the change in mangrove population [7,263,274]. Historically, the majority of research on these parameters' estimation employed ground-based approaches that were time-consuming, costly, and distribute sparsely across space, making regional mangrove monitoring challenging [25]. Alternatively, a number of papers predicted mangrove above-ground properties using vegetation indicators [177,209,220]. These research applied regression analyses to establish empirical relationships between remotely sensed vegetation indices and measured-above ground mangrove (AGM) data (e.g., leaf area index, height canopy, carbon sink, and biomass) [207,209,275,276]. These studies considered that indices derived from satellite data successfully modelled and estimated the mangrove above ground features. However, above ground mangrove ecosystems today have not been compared with the patterns of 30–50 years ago.

In these studies, NDVI and EVI demonstrate the most explanatory curvilinear relationships with AGM [25,65,93]. In fact, there is a saturation issue with NDVI that is mainly due to the red band, the energy in which is strongly absorbed by pigments. When a leaf contains a certain number of pigments, the reflectance remains low and practically constant with more pigment (e.g., increased leaf area) [70,214]. As a result, where forests are high in biomass, NDVI struggles to differentiate moderately high plant cover from very high plant cover [153,214]. Since 2012, most studies confirmed that EVI indicated a higher correlation coefficient with mangrove in field measurements than NDVI to significant extents [59,153,169]. Meanwhile, a few publications found NDVI to be the best mangrove predictor, relative to the performance of EVI and other vegetation indices [80,144,160]. The scale of the analysis also affects the suitability of the index used, with a few studies finding that at a higher resolution (smaller scale) NDVI is the preferred index [84]. In fact, each ecosystem has its unique characteristics, and each index is a separate indication for green vegetation. The best vegetation index to use for AGM estimation varies and thus its selection requires substantial field measurements to validate the results.

4.3. Mangrove Changes

Understanding of variations in mangrove patterns is critical in providing fundamental source for proposing appropriate strategies in mangrove ecosystem management and serving as a reference for broader worldwide applications. The changes in mangrove ecosystems are acknowledged as a result of natural influences and human activities. To investigate mangrove changes, two methodologies are usually applied (i) bi-temporal analysis and (ii) long-term monitoring. Bi-temporal analysis uses two images per 5 or 10 years to assess the changes in mangrove cover. A bi-temporal technique is common and easy to apply, and it calculates the differences of mangrove cover at two times in the context of land use and land cover change. However, there are some limitations in regard to environmental factors (e.g., tide variations, terrain, and atmospheric conditions) if images are only obtained on a single day of the year. This is because we can utilise the method outlined in Section 4.1 to retrieve information about land cover for each year before employing an intersection of two scenes. Besides, the significance of physical factors (e.g., erosion, typhon) as a source of mangrove loss and the trend in mangrove cover may be underestimated.

In contrast, long-term monitoring normally applies time series of spectral indices to understand mangrove dynamics through space and time. Most of studies applied NDVI or EVI and linear regression algorithms to analyse spatiotemporal change and anticipate trends in mangrove distribution at a local scale [172,183,214,277]. These analyses concluded that the loss of mangrove ecosystems is mainly caused by conversion in land use and land cover, compared to natural factors. The aquaculture ponds and impervious surface expansion in the coastal area are a threat to mangrove ecosystems [28]. Globally, Hansen et al. (2013) [277] first used NDVI and ordinary least squares slope of the regression to examine forest loss and gain from 2000 to 2012. The study revealed that a decreased trend in mangrove cover occurred in Asian and Caribbean countries. However, the global analysis overlooked driving factors (e.g., land use and land cover transformation and physical hazards) because policies for land use and land cover changes are different among nations in the world.

Overall, the application of spectral indices for mangrove remote sensing provides several advantages in relation to mapping spatial distribution, above-ground mangrove properties, and mangrove changes. However, examples of knowledge gaps from previous studies should be included (i) visualising mangrove changes from the past to the present; (ii) identifying the driving elements impacting mangroves; and (iii) evaluating effects of environmental factors on satellite images.

5. Discussion and Future Directions

5.1. The Potential Indices for Mangrove Remote Sensing

Over the past 50 years, the importance of spectral indices in mangrove remote sensing has been recognised, but several knowledge gaps still exist in relation to the best index selection for mangrove characterisation. A perfect spectral index for an ensemble of mangrove biophysical parameters is yet to be developed. Our study explored that NDVI accounted for the highest proportion (82%) of the applied spectral indices for mangrove ecosystems, followed by EVI (28%). These normalised algorithms mitigate the atmospheric effects and the impacts of sensor calibration degradation in the red and NIR bands [64,66]. The widespread adoption of remote sensing has resulted in the creation of low-cost image data that may be used to broaden NDVI applications. Hence, NDVI will continue to be a dominant vegetation index used for mangrove remote sensing. However, this does not mean that NDVI is always effective because of its limitations in relation to soil background and vegetation density.

Each index has its own advantages and disadvantages and can be affected by the impacts of the soil background and confounding factors such as sensor and calibration effects, bidirectional reflectance distribution function, atmospheric and topographic effects, or other local environmental conditions (e.g., tide). Hence, for future applications, instead of constructing or discovering a prospective mangrove index, we should examine local conditions and the factors influencing the effectiveness of spectral indices before deciding on the use of them for analysis. Additionally, the number of spectral bands available on optical sensors influenced the indices chosen for mangrove remote sensing. For instance, the four-band version of the Planetscope instrument with no SWIR band is only able to produce spectral indices in the visible and NIR regions. In the case that one index cannot meet the needs of mangrove assessment or other purposes, another index should be applied.

5.2. Long-Term Mangrove Monitoring with Time Series-Based Approaches in Relation to Driving Factors

Monitoring mangrove dynamics normally includes seasonal and annual changes that require a series of historical and regular imagery. In fact, there are many factors, including tide conditions, atmospheric factors, or missing or mis-registered data, that can cause errors in image acquisitions. Therefore, using single-date images to calculate spectral index has shown significant limitations on a large scale because environmental conditions vary from day to day and across sites. Alternatively, generating optical images using averaging is less susceptible to high resolution noise and are thus capable of characterising both long-term and abrupt mangrove changes. For example, using annual mean/median spectral indices that are derived from daily/5-days/8-days/16-days timeseries data enables us to reduce the environmental factors' influence on the image of interest. The study of multiple remotely sensed data has been widely employed in phenological investigations of mangrove ecosystems [121,134,153].

In addition to data, time series analysis provides pieces of information on the timing of mangrove change, as well as improving the quality and accuracy of information being derived using remotely sensed data [167,172,174,217,219]. Also, the time series analysis of spectral indices data evaluates trends and predicts the persistence of mangrove trends under spatial regression application. A variety of time series analysis techniques have been produced [(e.g., National Forest Trend [278], Recurrent Neural Network [279] to analyse and monitor spatiotemporal changes in mangrove ecosystems [191,221,280]. The digital number (DN) value of each pixel from time series images gives more sensitivity than single composited spectral band so that it can easily compare with natural factors (e.g., rainfall, temperature, and ocean dynamics) to certain significance of physical influences. This method holds significant promise for studying the long-term dynamics of environmental variations, and it can monitor future mangrove regeneration.

5.3. Fusion of Images from Multiple Sensors

In the process of the Earth's surface observation and particularly in mangrove remote sensing, selecting a potential optical sensor to calculate spectral index is crucial for assessment with high accuracy. However, there are several factors that can influence the choice of optical remote sensing platforms, such as the purpose of the research, the data availability, the national context, the budget constraints, the scale of the study, and the location of the study area. For instance, to explore spatiotemporal changes in annual mangrove patterns, a long-term time series analysis from Landsat imagery should be preferred because the data is available from 1972 to date. Additionally, to understand mangrove quality or seasonal changes, a variety of MODerate Resolution Imaging Spectroradiometer (MODIS) products with a high temporal resolution (1 day) may be the best choice. Recently, several studies fused multi-sensor images to have more information about mangrove ecosystems. For example, Kanniah et al. (2021) [281] used three optical sensors (i.e., Landsat, MODIS, and Sentinel-2A) to study mangrove fragmentation and health conditions. Guo et al. (2021) [209] used UAV and WorldView-2 datasets to validate the Sentinel-2 imagery for LAI estimation. Besides, several publications fused passive and active sensor images to understand mangrove structure or biomass. Pham et al. (2020) [89] combined optical bands (Sentinel-2A) with active sensors (i.e., Sentinel-1 and ALOS-2 PALSAR-2) to calculate some vegetation indices for mangrove above-ground biomass. These studies concluded that fusing multiple remote sensing sources helps to provide a large amount of information about mangrove ecosystems, compared to single sensor applications.

Fusion of multiple sensors can enhance the accuracy of the data. Integration of NDVI from Advanced Very High-Resolution Radiometer (AVHRR–launched in 1979) and MODIS (launched in the 2000s) enabled a long-term dataset from 1979 to date. AVHRR NDVI composites at 1 km spatial resolution [92,93] was used for mangrove monitoring prior to the 2000s. However, the AVHRR satellite system has degraded in orbit to the point that it is advised that NDVI MODIS products should be used for longer periods in the future [282]. Additionally, combination of ASTER (launched 1999) and Landsat 4,5, 8, and 9 is an alternative approach for line correction of Landsat 7, allowing us to obtain a set of data at 30 m spatial resolution from 1988 to the present. Notably, using multiple sensors enables improving the re-visit days of satellite data, which is better for smoothing data [25,27]. For example, when Landsat 8 and Landsat 9 are combined, the re-visit days are reduced from 16 to 8 days. Hence, fusion of multiple sensors (i.e., passive, and active sensors) is a recommendable approach to compute the spectral index for mangrove studies in the future.

6. Conclusions

Land use and land cover transformation in relation to natural hazards are the primary factors threatening mangroves in the future. Spectral indices have been applied to mangroves and demonstrated their effectiveness in various studies over 50 years. Each spectral index has its own strength and limitation in mapping mangrove distributions and measuring their above-ground biophysical properties in various environments. Therefore, to select a potential index, we should understand the interaction between the local conditions and mangrove ecosystems. NDVI is the most popular index that can be applied for mangrove ecosystems, followed by EVI, although both are sensitive to environmental conditions.

Long-term mangrove monitoring is crucial for identifying the trend in mangrove pattern changes in connection to driving variables. Using time series analysis of spectral indexes helps to reduce the effect of external influences. Using multiple sensors enables obtaining a set of databases for long-term monitoring associated with natural hazards and human activities. Nowadays, accessing big data has become easier with the help of technology and digital cloud platforms (e.g., Google Earth Engine). These technological advancements will shift mangrove studies from a local to a global scale and imply the necessity to learn programming skills.

In the context of the digital era, mangrove scholars should apply the advantages of cloud computing platforms for spectral index computation. These approaches assist image processing quickly and enable analysis of mangrove ecosystems at global scale. However, approaching these techniques requests users to have some knowledge about information technology and quite understand about coding. Therefore, developing a tool or application on cloud storage for mangrove monitoring based on vegetation index should be taken into account for new users or scholars who do not have good information technology skills. In addition to tools, a guideline for algorithm selection (e.g., machine learning, deep learning) should be developed to save time for spectral index computation.

Author Contributions: Conceptualization, methodology, formal analysis, investigation, writing—original draft preparation: T.V.T.; writing—review and editing: T.V.T., R.R. and X.Z.; supervision: R.R. and X.Z. All authors have read and agreed to the published version of the manuscript.

Funding: This study was funded by an Australian Research Council Discovery Award DP180103444 to R.R.

Data Availability Statement: The data presented in this study are available on request from the corresponding author.

Acknowledgments: Acknowledgement is given to Monash University for supporting this research. The authors would like to acknowledge the valuable comments of anonymous reviewers and editors that assisted with the finalisation of this manuscript.

Conflicts of Interest: The authors declare no conflict of interest.

Acronyms

EVI	Enhanced Vegetation Index
LAI	Leaf Area Index
NDVI	Normalised Difference Vegetation Index
PVI	Perpendicular Vegetation Index
RVI	Ratio Vegetation Index
SAVI	Soil Adjusted Vegetation Index
TSAVI	Transformed Soil Adjusted Vegetation Index
TVI	Triangular Vegetation Index
VIN	Vegetation Index Number

References

1. Tomlinson, P.B. *The Botany of Mangroves*; Cambridge University Press: Cambridge, UK, 2016; ISBN 1-316-79065-7.
2. Mandal, R.N.; Bar, R. *Mangroves for Building Resilience to Climate Change*; Apple Academic Press: Waretown, NJ, USA, 2018; ISBN 0-429-48778-9.
3. Cavanaugh, K.C.; Osland, M.J.; Bardou, R.; Hinojosa-Arango, G.; López-Vivas, J.M.; Parker, J.D.; Rovai, A.S. Sensitivity of Mangrove Range Limits to Climate Variability. *Glob. Ecol. Biogeogr.* **2018**, *27*, 925–935. [CrossRef]
4. Ellison, A.M.; Farnsworth, E.J.; Merkt, R.E. Origins of Mangrove Ecosystems and the Mangrove Biodiversity Anomaly. *Glob. Ecol. Biogeogr.* **1999**, *8*, 95–115.
5. Steenis, C.G.G.J. The Distribution of Mangrove Plant Genera and Its Significance for Palaeogeography. *Proc. Kon. Net. Amst.* **1962**, *65*, 164–169.
6. Friess, D.A.; Rogers, K.; Lovelock, C.E.; Krauss, K.W.; Hamilton, S.E.; Lee, S.Y.; Lucas, R.; Primavera, J.; Rajkaran, A.; Shi, S. The State of the World's Mangrove Forests: Past, Present, and Future. *Annu. Rev. Environ. Resour.* **2019**, *44*, 89–115. [CrossRef]
7. Simard, M.; Fatoyinbo, L.; Smetanka, C.; Rivera-Monroy, V.H.; Castañeda-Moya, E.; Thomas, N.; Van der Stocken, T. Mangrove Canopy Height Globally Related to Precipitation, Temperature and Cyclone Frequency. *Nat. Geosci* **2019**, *12*, 40–45. [CrossRef]
8. Kauffman, J.B.; Adame, M.F.; Arifanti, V.B.; Schile-Beers, L.M.; Bernardino, A.F.; Bhomia, R.K.; Donato, D.C.; Feller, I.C.; Ferreira, T.O.; del Carmen Jesus Garcia, M.; et al. Total Ecosystem Carbon Stocks of Mangroves across Broad Global Environmental and Physical Gradients. *Ecol. Monogr.* **2020**, *90*, e01405. [CrossRef]
9. Sandilyan, S.; Kathiresan, K. Mangrove Conservation: A Global Perspective. *Biodivers. Conserv.* **2012**, *21*, 3523–3542. [CrossRef]
10. FAO of the UN. *The World's Mangroves 1980–2005: A Thematic Study Prepared in the Framework of the Global Forest Resources Assessment 2005*; FAO: Rome, Italy, 2007; p. 153.
11. Bunting, P.; Rosenqvist, A.; Lucas, R.; Rebelo, L.-M.; Hilarides, L.; Thomas, N.; Hardy, A.; Itoh, T.; Shimada, M.; Finlayson, C. The Global Mangrove Watch—A New 2010 Global Baseline of Mangrove Extent. *Remote Sens.* **2018**, *10*, 1669. [CrossRef]
12. Bunting, P.; Rosenqvist, A.; Hilarides, L.; Lucas, R.M.; Thomas, N. Global Mangrove Watch: Updated 2010 Mangrove Forest Extent (v2. 5). *Remote Sens.* **2022**, *14*, 1034. [CrossRef]

13. Kuenzer, C.; Bluemel, A.; Gebhardt, S.; Quoc, T.V.; Dech, S. Remote Sensing of Mangrove Ecosystems: A Review. *Remote Sens.* **2011**, *3*, 878–928. [CrossRef]
14. Lee, S.Y.; Primavera, J.H.; Dahdouh-Guebas, F.; McKee, K.; Bosire, J.O.; Cannicci, S.; Diele, K.; Fromard, F.; Koedam, N.; Marchand, C. Ecological Role and Services of Tropical Mangrove Ecosystems: A Reassessment. *Glob. Ecol. Biogeogr.* **2014**, *23*, 726–743. [CrossRef]
15. Duke, N.; Nagelkerken, I.; Agardy, T.; Wells, S.; Van Lavieren, H. *The Importance of Mangroves to People: A Call to Action*; United Nations Environment Programme World Conservation Monitoring Centre: Cambridge, UK, 2014; ISBN 92-807-3397-4.
16. Narayan, S.; Beck, M.W.; Reguero, B.G.; Losada, I.J.; Van Wesenbeeck, B.; Pontee, N.; Sanchirico, J.N.; Ingram, J.C.; Lange, G.-M.; Burks-Copes, K.A. The Effectiveness, Costs and Coastal Protection Benefits of Natural and Nature-Based Defences. *PLoS ONE* **2016**, *11*, e0154735. [CrossRef]
17. Primavera, J.H. Overcoming the Impacts of Aquaculture on the Coastal Zone. *Ocean Coast. Manag.* **2006**, *49*, 531–545. [CrossRef]
18. Donato, D.C.; Kauffman, J.B.; Mackenzie, R.A.; Ainsworth, A.; Pfleeger, A.Z. Whole-Island Carbon Stocks in the Tropical Pacific: Implications for Mangrove Conservation and Upland Restoration. *J. Environ. Manag.* **2012**, *97*, 89–96. [CrossRef] [PubMed]
19. Donato, D.C.; Kauffman, J.B.; Murdiyarso, D.; Kurnianto, S.; Stidham, M.; Kanninen, M. Mangroves among the Most Carbon-Rich Forests in the Tropics. *Nat. Geosci.* **2011**, *4*, 293–297. [CrossRef]
20. Spalding, M.; Parrett, C.L. Global Patterns in Mangrove Recreation and Tourism. *Mar. Policy* **2019**, *110*, 103540. [CrossRef]
21. Giri, C.; Ochieng, E.; Tieszen, L.L.; Zhu, Z.; Singh, A.; Loveland, T.; Masek, J.; Duke, N. Status and Distribution of Mangrove Forests of the World Using Earth Observation Satellite Data. *Glob. Ecol. Biogeogr.* **2011**, *20*, 154–159. [CrossRef]
22. Carugati, L.; Gatto, B.; Rastelli, E.; Lo Martire, M.; Coral, C.; Greco, S.; Danovaro, R. Impact of Mangrove Forests Degradation on Biodiversity and Ecosystem Functioning. *Sci. Rep.* **2018**, *8*, 13298. [CrossRef]
23. Goldberg, L.; Lagomasino, D.; Thomas, N.; Fatoyinbo, T. Global Declines in Human-Driven Mangrove Loss. *Glob. Change Biol.* **2020**, *26*, 5844–5855. [CrossRef]
24. Su, J.; Friess, D.A.; Gasparatos, A. A Meta-Analysis of the Ecological and Economic Outcomes of Mangrove Restoration. *Nat. Commun.* **2021**, *12*, 5050. [CrossRef]
25. Wang, L.; Jia, M.; Yin, D.; Tian, J. A Review of Remote Sensing for Mangrove Forests: 1956–2018. *Remote Sens. Environ.* **2019**, *231*, 111223. [CrossRef]
26. Pham, T.D.; Yokoya, N.; Bui, D.T.; Yoshino, K.; Friess, D.A. Remote Sensing Approaches for Monitoring Mangrove Species, Structure, and Biomass: Opportunities and Challenges. *Remote Sens.* **2019**, *11*, 230. [CrossRef]
27. Maurya, K.; Mahajan, S.; Chaube, N. Remote Sensing Techniques: Mapping and Monitoring of Mangrove Ecosystem—A Review. *Complex Intell. Syst.* **2021**, *7*, 2797–2818. [CrossRef]
28. Cardenas, N.Y.; Joyce, K.E.; Maier, S.W. Monitoring Mangrove Forests: Are We Taking Full Advantage of Technology? *Int. J. Appl. Earth Obs. Geoinf.* **2017**, *63*, 1–14.
29. Campbell, J.B.; Wynne, R.H. *Introduction to Remote Sensing*; Guilford Press: New York, NY, USA, 2011; ISBN 1-60918-177-8.
30. Thakur, S.; Mondal, I.; Ghosh, P.B.; Das, P.; De, T.K. A Review of the Application of Multispectral Remote Sensing in the Study of Mangrove Ecosystems with Special Emphasis on Image Processing Techniques. *Spat. Inf. Res.* **2020**, *28*, 39–51. [CrossRef]
31. Zulfa, A.W.; Norizah, K.; Hamdan, O.; Faridah-Hanum, I.; Rhyma, P.P.; Fitrianto, A. Spectral Signature Analysis to Determine Mangrove Species Delineation Structured by Anthropogenic Effects. *Ecol. Indic.* **2021**, *130*, 108148. [CrossRef]
32. Zulfa, A.W.; Norizah, K.; Hamdan, O.; Zulkifly, S.; Faridah-Hanum, I.; Rhyma, P.P. Discriminating Trees Species from the Relationship between Spectral Reflectance and Chlorophyll Contents of Mangrove Forest in Malaysia. *Ecol. Indic.* **2020**, *111*, 106024. [CrossRef]
33. Green, E.P.; Clark, C.D.; Mumby, P.J.; Edwards, A.J.; Ellis, A.C. Remote Sensing Techniques for Mangrove Mapping. *Int. J. Remote Sens.* **1998**, *19*, 935–956. [CrossRef]
34. Heumann, B.W. Satellite Remote Sensing of Mangrove Forests: Recent Advances and Future Opportunities. *Prog. Phys. Geogr.* **2011**, *35*, 87–108. [CrossRef]
35. Purnamasayangsukasih, P.R.; Norizah, K.; Ismail, A.A.; Shamsudin, I. A Review of Uses of Satellite Imagery in Monitoring Mangrove Forests. In Proceedings of the IOP Conference Series: Earth and Environmental Science; IOP Publishing: Bristol, UK, 2016; Volume 37, p. 012034.
36. Estes, J.; Kline, K.; Collins, E. Remote Sensing. In *International Encyclopedia of the Social & Behavioral Sciences*; Smelser, N.J., Baltes, P.B., Eds.; Pergamon: Oxford, UK, 2001; pp. 13144–13150. ISBN 978-0-08-043076-8.
37. Pricope, N.G.; Mapes, K.L.; Woodward, K.D. Remote Sensing of Human–Environment Interactions in Global Change Research: A Review of Advances, Challenges and Future Directions. *Remote Sens.* **2019**, *11*, 2783. [CrossRef]
38. Aria, M.; Cuccurullo, C. Bibliometrix: An R-Tool for Comprehensive Science Mapping Analysis. *J. Informetr.* **2017**, *11*, 959–975. [CrossRef]
39. Derviş, H. Bibliometric Analysis Using Bibliometrix an R Package. *J. Scientometr. Res.* **2019**, *8*, 156–160. [CrossRef]
40. Aria, M.; Misuraca, M.; Spano, M. Mapping the Evolution of Social Research and Data Science on 30 Years of Social Indicators Research. *Soc. Indic. Res.* **2020**, *149*, 803–831. [CrossRef]
41. Duan, P.; Wang, Y.; Yin, P. Remote Sensing Applications in Monitoring of Protected Areas: A Bibliometric Analysis. *Remote Sens.* **2020**, *12*, 772. [CrossRef]

42. Pranckutė, R. Web of Science (WoS) and Scopus: The Titans of Bibliographic Information in Today's Academic World. *Publications* **2021**, *9*, 12. [CrossRef]
43. Zhu, J.; Liu, W. A Tale of Two Databases: The Use of Web of Science and Scopus in Academic Papers. *Scientometrics* **2020**, *123*, 321–335. [CrossRef]
44. Aghaei Chadegani, A.; Salehi, H.; Yunus, M.; Farhadi, H.; Fooladi, M.; Farhadi, M.; Ale Ebrahim, N. *A Comparison between Two Main Academic Literature Collections: Web of Science and Scopus Databases*; Social Science Research Network: Rochester, NY, USA, 2013.
45. Moral-Muñoz, J.A.; Herrera-Viedma, E.; Santisteban-Espejo, A.; Cobo, M.J. Software Tools for Conducting Bibliometric Analysis in Science: An up-to-Date Review. *Prof. Inf.* **2020**, *29*, e290103. [CrossRef]
46. Clarivate, A. Web of Science Journal Evaluation Process and Selection Criteria. Available online: https://clarivate.com/webofsciencegroup/journal-evaluation-process-and-selection-criteria/ (accessed on 7 February 2022).
47. Shu, F. Research on the Application of Thomson Data Analyzer to Analyses the Patent Intelligence of Scientific Institutions. *Inf. Sci.* **2008**, *26*, 1833–1843.
48. Perianes-Rodriguez, A.; Waltman, L.; van Eck, N.J. Constructing Bibliometric Networks: A Comparison between Full and Fractional Counting. *J. Informetr.* **2016**, *10*, 1178–1195. [CrossRef]
49. Liu, W. The Data Source of This Study Is Web of Science Core Collection? Not Enough. *Scientometrics* **2019**, *121*, 1815–1824. [CrossRef]
50. Bannari, A.; Morin, D.; Bonn, F.; Huete, A.R. A Review of Vegetation Indices. *Remote Sens. Rev.* **1995**, *13*, 95–120. [CrossRef]
51. Raynaud, M.; Goutaudier, V.; Louis, K.; Al-Awadhi, S.; Dubourg, Q.; Truchot, A.; Brousse, R.; Saleh, N.; Giarraputo, A.; Debiais, C.; et al. Impact of the COVID-19 Pandemic on Publication Dynamics and Non-COVID-19 Research Production. *BMC Med. Res. Methodol.* **2021**, *21*, 255. [CrossRef] [PubMed]
52. Vinay, V.; Julia, L. Introducing the Spectral Index Library in ArcGIS. Available online: https://www.esri.com/about/newsroom/arcuser/spectral-library/ (accessed on 6 February 2022).
53. Xue, J.; Su, B. Significant Remote Sensing Vegetation Indices: A Review of Developments and Applications. *J. Sens.* **2017**, *2017*, 1353691. [CrossRef]
54. Deng, Y.; Wu, C.; Li, M.; Chen, R. RNDSI: A Ratio Normalized Difference Soil Index for Remote Sensing of Urban/Suburban Environments. *Int. J. Appl. Earth Obs. Geoinf.* **2015**, *39*, 40–48. [CrossRef]
55. Baloloy, A.B.; Blanco, A.C.; Sta, A.R.R.C.; Nadaoka, K. Development and Application of a New Mangrove Vegetation Index (MVI) for Rapid and Accurate Mangrove Mapping. *ISPRS J. Photogramm. Remote Sens.* **2020**, *166*, 95–117. [CrossRef]
56. Winarso, G.; Purwanto, A.; Yuwono, D.; Center, R.S.A. New Mangrove Index as Degradation Health Indicator Using Remote Sensing Data: Segara Anakan and Alas Purwo Case Study. In Proceedings of the 12th Biennial Conference of Pan Ocean Remote Sensing Conference (PORSEC 2014), Bali, Indonesia, 4–7 November 2014; pp. 4–7.
57. Zhang, X.; Tian, Q. A Mangrove Recognition Index for Remote Sensing of Mangrove Forest from Space. *Curr. Sci.* **2013**, *105*, 1149–1154.
58. Gupta, K.; Mukhopadhyay, A.; Giri, S.; Chanda, A.; Majumdar, S.D.; Samanta, S.; Mitra, D.; Samal, R.N.; Pattnaik, A.K.; Hazra, S. An Index for Discrimination of Mangroves from Non-Mangroves Using LANDSAT 8 OLI Imagery. *MethodsX* **2018**, *5*, 1129–1139. [CrossRef]
59. Kumar, A.; Stupp, P.; Dahal, S.; Remillard, C.; Bledsoe, R.; Stone, A.; Cameron, C.; Rastogi, G.; Samal, R.; Mishra, D.R. A Multi-Sensor Approach for Assessing Mangrove Biophysical Characteristics in Coastal Odisha, India. *Proc. Natl. Acad. Sci. USA India Sect. A Phys. Sci.* **2017**, *87*, 679–700. [CrossRef]
60. Jia, M.; Wang, Z.; Wang, C.; Mao, D.; Zhang, Y. A New Vegetation Index to Detect Periodically Submerged Mangrove Forest Using Single-Tide Sentinel-2 Imagery. *Remote Sens.* **2019**, *11*, 2043. [CrossRef]
61. McFeeters, S.K. The Use of the Normalized Difference Water Index (NDWI) in the Delineation of Open Water Features. *Int. J. Remote Sens.* **1996**, *17*, 1425–1432. [CrossRef]
62. Pearson, R.L.; Miller, L.D. Remote Mapping of Standing Crop Biomass for Estimation of the Productivity of the Shortgrass Prairie. *Remote Sens. Environ.* **1972**, *8*, 1355–1379.
63. Rouse, J.W.; Haas, R.H.; Schell, J.A.; Deering, D.W. Monitoring Vegetation Systems in the Great Plains with ERTS. In *Third Earth Reserves Technology Satellite Symposium*; Greenbelt: NASA SP-351; NASA: Washington, DC, USA, 1973; Volume 30103017, pp. 309–317.
64. Huang, S.; Tang, L.; Hupy, J.P.; Wang, Y.; Shao, G. A Commentary Review on the Use of Normalized Difference Vegetation Index (NDVI) in the Era of Popular Remote Sensing. *J. For. Res.* **2021**, *32*, 1–6. [CrossRef]
65. Jensen, J.R.; Lin, H.; Yang, X.; Ramsey, E., III; Davis, B.A.; Thoemke, C.W. The Measurement of Mangrove Characteristics in Southwest Florida Using SPOT Multispectral Data. *Geocarto Int.* **1991**, *6*, 13–21. [CrossRef]
66. Forkel, M.; Carvalhais, N.; Verbesselt, J.; Mahecha, M.D.; Neigh, C.S.R.; Reichstein, M. Trend Change Detection in NDVI Time Series: Effects of Inter-Annual Variability and Methodology. *Remote Sens.* **2013**, *5*, 2113–2144. [CrossRef]
67. Huete, A.R. A Soil-Adjusted Vegetation Index (SAVI). *Remote Sens. Environ.* **1988**, *25*, 295–309. [CrossRef]
68. Baret, F.; Guyot, G.; Major, D.J. TSAVI: A Vegetation Index Which Minimizes Soil Brightness Effects on LAI and APAR Estimation. In Proceedings of the 12th Canadian Symposium on Remote Sensing Geoscience and Remote Sensing Symposium, Vancouver, BC, Canada, 10–14 July 1989; Volume 3, pp. 1355–1358.

69. Huete, A.; Didan, K.; Miura, T.; Rodriguez, E.P.; Gao, X.; Ferreira, L.G. Overview of the Radiometric and Biophysical Performance of the MODIS Vegetation Indices. *Remote Sens. Environ.* **2002**, *83*, 195–213. [CrossRef]
70. Tran, T.V.; Tran, D.X.; Nguyen, H.; Latorre-Carmona, P.; Myint, S.W. Characterising Spatiotemporal Vegetation Variations Using LANDSAT Time-Series and Hurst Exponent Index in the Mekong River Delta. *Land Degrad. Dev.* **2021**, *32*, 3507–3523. [CrossRef]
71. Jiang, Z.; Huete, A.R.; Didan, K.; Miura, T. Development of a Two-Band Enhanced Vegetation Index without a Blue Band. *Remote Sens. Environ.* **2008**, *112*, 3833–3845. [CrossRef]
72. Liu, H.Q.; Huete, A. A Feedback Based Modification of the NDVI to Minimize Canopy Background and Atmospheric Noise. *IEEE Trans. Geosci. Remote Sens.* **1995**, *33*, 457–465. [CrossRef]
73. Richardson, A.J.; Wiegand, C.L. Distinguishing Vegetation from Soil Background Information. *Photogramm. Eng. Remote Sens.* **1977**, *43*, 1541–1552.
74. Wicaksono, P.; Danoedoro, P.; Hartono; Nehren, U. Mangrove Biomass Carbon Stock Mapping of the Karimunjawa Islands Using Multispectral Remote Sensing. *Int. J. Remote Sens.* **2015**, *37*, 26–52. [CrossRef]
75. Wicaksono, P. Mangrove Above-Ground Carbon Stock Mapping of Multi-Resolution Passive Remote-Sensing Systems. *Int. J. Remote Sens.* **2017**, *38*, 1551–1578. [CrossRef]
76. Díaz, B.M.; Blackburn, G.A. Remote Sensing of Mangrove Biophysical Properties: Evidence from a Laboratory Simulation of the Possible Effects of Background Variation on Spectral Vegetation Indices. *Int. J. Remote Sens.* **2003**, *24*, 53–73. [CrossRef]
77. Kovacs, J.M.; Flores-Verdugo, F.; Wang, J.; Aspden, L.P. Estimating Leaf Area Index of a Degraded Mangrove Forest Using High Spatial Resolution Satellite Data. *Aquat. Bot.* **2004**, *80*, 13–22. [CrossRef]
78. Kovacs, J.M.; King, J.M.L.; Flores de Santiago, F.; Flores-Verdugo, F. Evaluating the Condition of a Mangrove Forest of the Mexican Pacific Based on an Estimated Leaf Area Index Mapping Approach. *Environ. Monit Assess* **2009**, *157*, 137–149. [CrossRef]
79. Kovacs, J.M.; Liu, Y.; Zhang, C.; Flores-Verdugo, F.; de Santiago, F.F. A Field Based Statistical Approach for Validating a Remotely Sensed Mangrove Forest Classification Scheme. *Wetl. Ecol. Manag.* **2011**, *19*, 409. [CrossRef]
80. Kamal, M.; Phinn, S.; Johansen, K. Assessment of Multi-Resolution Image Data for Mangrove Leaf Area Index Mapping. *Remote Sens. Environ.* **2016**, *176*, 242–254. [CrossRef]
81. Heenkenda, M.K.; Maier, S.W.; Joyce, K.E. Estimating Mangrove Biophysical Variables Using WorldView-2 Satellite Data: Rapid Creek, Northern Territory, Australia. *J. Imaging* **2016**, *2*, 24. [CrossRef]
82. Zhu, Y.; Liu, K.; Liu, L.; Myint, S.W.; Wang, S.; Liu, H.; He, Z. Exploring the Potential of WorldView-2 Red-Edge Band-Based Vegetation Indices for Estimation of Mangrove Leaf Area Index with Machine Learning Algorithms. *Remote Sens.* **2017**, *9*, 1060. [CrossRef]
83. George, R.; Padalia, H.; Sinha, S.K.; Kumar, A.S. Evaluation of the Use of Hyperspectral Vegetation Indices for Estimating Mangrove Leaf Area Index in Middle Andaman Island, India. *Remote Sens. Lett.* **2018**, *9*, 1099–1108. [CrossRef]
84. Wicaksono, P.; Hafizt, M. Dark Target Effectiveness for Dark-Object Subtraction Atmospheric Correction Method on Mangrove above-Ground Carbon Stock Mapping. *IET Image Processing* **2018**, *12*, 582–587. [CrossRef]
85. Muhsoni, F.F.; Sambah, A.B.; Mahmudi, M.; Wiadnya, D.G.R. Estimation of Mangrove Carbon Stock with Hybrid Method Using Image Sentinel-2. *GEOMATE J.* **2018**, *15*, 185–192. [CrossRef]
86. Oostdijk, M.; Santos, M.J.; Whigham, D.; Verhoeven, J.; Silvestri, S. Assessing Rehabilitation of Managed Mangrove Ecosystems Using High Resolution Remote Sensing. *Estuar. Coast. Shelf Sci.* **2018**, *211*, 238–247. [CrossRef]
87. Razali, S.M.; Nuruddin, A.A.; Kamarudin, N. Mapping Mangrove Density for Conservation of the RAMSAR Site in Peninsular Malaysia. *Int. J. Conserv. Sci.* **2020**, *11*, 153–164.
88. Ávila-Flores, G.; Juárez-Mancilla, J.; Hinojosa-Arango, G.; Cruz-Chávez, P.; López-Vivas, J.M.; Arizpe-Covarrubias, O. A Practical Index to Estimate Mangrove Conservation Status: The Forests from La Paz Bay, Mexico as a Case Study. *Sustainability* **2020**, *12*, 858. [CrossRef]
89. Pham, T.D.; Yokoya, N.; Xia, J.; Ha, N.T.; Le, N.N.; Nguyen, T.T.T.; Dao, T.H.; Vu, T.T.P.; Pham, T.D.; Takeuchi, W. Comparison of Machine Learning Methods for Estimating Mangrove Above-Ground Biomass Using Multiple Source Remote Sensing Data in the Red River Delta Biosphere Reserve, Vietnam. *Remote Sens.* **2020**, *12*, 1334. [CrossRef]
90. Xia, Q.; Qin, C.-Z.; Li, H.; Huang, C.; Su, F.-Z.; Jia, M.-M. Evaluation of Submerged Mangrove Recognition Index Using Multi-Tidal Remote Sensing Data. *Ecol. Indic.* **2020**, *113*, 106196. [CrossRef]
91. Ali, A.; Nayyar, Z.A. Extraction of Mangrove Forest through Landsat 8 Mangrove Index (L8MI). *Arab. J. Geosci.* **2020**, *13*, 1132. [CrossRef]
92. Ramsey, E.W.R., III; Jensen, J.R. Remote Sensing of Mangrove Wetlands: Relating Canopy Spectra to Site-Specific Data. *Photogramm. Eng. Remote Sens.* **1996**, *62*, 939–948.
93. Green, E.P.; Mumby, P.J.; Edwards, A.J.; Clark, C.D.; Ellis, A.C. Estimating Leaf Area Index of Mangroves from Satellite Data. *Aquat. Bot.* **1997**, *58*, 11–19. [CrossRef]
94. Green, E.; Mumby, P.; Edwards, A.; Clark, C.; Ellis, A. The Assessment of Mangrove Areas Using High Resolution Multispectral Airborne Imagery. *J. Coast. Res.* **1998**, *14*, 433–443.
95. Ruiz-Luna, A.; Berlanga-Robles, C. Modifications in Coverage Patterns and Land Use around the Huizache-Caimanero Lagoon System, Sinaloa, Mexico: A Multi-Temporal Analysis Using LANDSAT Images. *Estuar. Coast. Shelf Sci.* **1999**, *49*, 37–44. [CrossRef]
96. Kovacs, J.M.; Wang, J.; Flores-Verdugo, F. Mapping Mangrove Leaf Area Index at the Species Level Using IKONOS and LAI-2000 Sensors for the Agua Brava Lagoon, Mexican Pacific. *Estuar. Coast. Shelf Sci.* **2005**, *62*, 377–384. [CrossRef]

97. Mantri, V.A.; Mishra, A.K. On Monitoring Mangrove Vegetation of Sagar Island by Remote Sensing. *Natl. Acad. Sci. Lett.* **2006**, *29*, 45–48. [CrossRef]
98. Nichol, C.J.; Rascher, U.; Matsubara, S.; Osmond, B. Assessing Photosynthetic Efficiency in an Experimental Mangrove Canopy Using Remote Sensing and Chlorophyll Fluorescence. *Trees* **2005**, *20*, 9. [CrossRef]
99. Li, X.; Gar-On Yeh, A.; Wang, S.; Liu, K.; Liu, X.; Qian, J.; Chen, X. Regression and Analytical Models for Estimating Mangrove Wetland Biomass in South China Using Radarsat Images. *Int. J. Remote Sens.* **2007**, *28*, 5567–5582. [CrossRef]
100. Kovacs, J.M.; Zhang, C.; Flores-Verdugo, F.J. Mapping the Condition of Mangroves of the Mexican Pacific Using C-Band ENVISAT ASAR and Landsat Optical Data. *Cienc. Mar.* **2008**, *34*, 407–418. [CrossRef]
101. Lee, T.-M.; Yeh, H.-C. Applying Remote Sensing Techniques to Monitor Shifting Wetland Vegetation: A Case Study of Danshui River Estuary Mangrove Communities, Taiwan. *Ecol. Eng.* **2009**, *35*, 487–496. [CrossRef]
102. Rajitha, K.; Mukherjee, C.K.; Vinu Chandran, R.; Prakash Mohan, M.M. Land-Cover Change Dynamics and Coastal Aquaculture Development: A Case Study in the East Godavari Delta, Andhra Pradesh, India Using Multi-Temporal Satellite Data. *Int. J. Remote Sens.* **2010**, *31*, 4423–4442. [CrossRef]
103. Ruiz-Luna, A.; Cervantes Escobar, A.; Berlanga-Robles, C. Assessing Distribution Patterns, Extent, and Current Condition of Northwest Mexico Mangroves. *Wetlands* **2010**, *30*, 717–723. [CrossRef]
104. Satyanarayana, B.; Mohamad, K.A.; Idris, I.F.; Husain, M.-L.; Dahdouh-Guebas, F. Assessment of Mangrove Vegetation Based on Remote Sensing and Ground-Truth Measurements at Tumpat, Kelantan Delta, East Coast of Peninsular Malaysia. *Int. J. Remote Sens.* **2011**, *32*, 1635–1650. [CrossRef]
105. Kamthonkiat, D.; Rodfai, C.; Saiwanrungkul, A.; Koshimura, S.; Matsuoka, M. Geoinformatics in Mangrove Monitoring: Damage and Recovery after the 2004 Indian Ocean Tsunami in Phang Nga, Thailand. *Nat. Hazards Earth Syst. Sci.* **2011**, *11*, 1851–1862. [CrossRef]
106. Pujiono, E.; Kwak, D.-A.; Lee, W.-K.; Sulistyanto; Kim, S.-R.; Lee, J.Y.; Lee, S.-H.; Park, T.; Kim, M.-I. RGB-NDVI Color Composites for Monitoring the Change in Mangrove Area at the Maubesi Nature Reserve, Indonesia. *For. Sci. Technol.* **2013**, *9*, 171–179. [CrossRef]
107. Vo, Q.T.; Oppelt, N.; Leinenkugel, P.; Kuenzer, C. Remote Sensing in Mapping Mangrove Ecosystems—An Object-Based Approach. *Remote Sens.* **2013**, *5*, 183–201. [CrossRef]
108. Wong, F.K.K.; Fung, T. Combining Hyperspectral and Radar Imagery for Mangrove Leaf Area Index Modeling. *Photogramm. Eng. Remote Sens.* **2013**, *79*, 479–490. [CrossRef]
109. Manna, S.; Mondal, P.P.; Mukhopadhyay, A.; Akhand, A.; Hazra, S.; Mitra, D. Vegetation Cover Change Analysis from Multi-Temporal Satellite Data in Jharkhali Island, Sundarbans, India. *IJMS* **2013**, *42*, 331–342.
110. Hamdan, O.; Khairunnisa, M.; Ammar, A.; Hasmadi, I.M.; Aziz, H.K. Mangrove Carbon Stock Assessment by Optical Satellite Imagery. *J. Trop. For. Sci.* **2013**, *25*, 554–565.
111. Wohlfart, C.; Wegmann, M.; Leimgruber, P. Mapping Threatened Dry Deciduous Dipterocarp Forest in South-East Asia for Conservation Management. *Trop. Conserv. Sci.* **2014**, *7*, 597–613. [CrossRef]
112. Almeida, P.; Altobelli, A.; D'Aietti, L.; Feoli, E.; Ganis, P.; Giordano, F.; Napolitano, R.; Simonetti, C. The Role of Vegetation Analysis by Remote Sensing and GIS Technology for Planning Sustainable Development: A Case Study for the Santos Estuary Drainage Basin (Brazil). *Plant Biosyst.-Int. J. Deal. All Asp. Plant Biol.* **2014**, *148*, 540–546. [CrossRef]
113. Fuller, D.; Wang, Y. Recent Trends in Satellite Vegetation Index Observations Indicate Decreasing Vegetation Biomass in the Southeastern Saline Everglades Wetlands. *Wetlands* **2014**, *34*, 67–77. [CrossRef]
114. Chellamani, P.; Singh, C.P.; Panigrahy, S. Assessment of the Health Status of Indian Mangrove Ecosystems Using Multi Temporal Remote Sensing Data. *Trop. Ecol.* **2014**, *55*, 245–253.
115. Manna, S.; Nandy, S.; Chanda, A.; Akhand, A.; Hazra, S.; Dadhwal, V.K. Estimating Aboveground Biomass in Avicennia Marina Plantation in Indian Sundarbans Using High-Resolution Satellite Data. *J. Appl. Remote Sens.* **2014**, *8*, 083638. [CrossRef]
116. Anwar, M.S.; Takewaka, S. Analyses on Phenological and Morphological Variations of Mangrove Forests along the Southwest Coast of Bangladesh. *J. Coast. Conserv.* **2014**, *18*, 339–357. [CrossRef]
117. Patil, V.; Singh, A.; Naik, N.; Unnikrishnan, S. Estimation of Mangrove Carbon Stocks by Applying Remote Sensing and GIS Techniques. *Wetlands* **2015**, *35*, 695–707. [CrossRef]
118. Ibharim, N.A.; Mustapha, M.A.; Lihan, T.; Mazlan, A.G. Mapping Mangrove Changes in the Matang Mangrove Forest Using Multi Temporal Satellite Imageries. *Ocean Coast. Manag.* **2015**, *114*, 64–76. [CrossRef]
119. Heenkenda, M.K.; Joyce, K.E.; Maier, S.W.; de Bruin, S. Quantifying Mangrove Chlorophyll from High Spatial Resolution Imagery. *ISPRS J. Photogramm. Remote Sens.* **2015**, *108*, 234–244. [CrossRef]
120. Lagomasino, D.; Price, R.M.; Whitman, D.; Melesse, A.; Oberbauer, S.F. Spatial and Temporal Variability in Spectral-Based Surface Energy Evapotranspiration Measured from Landsat 5TM across Two Mangrove Ecotones. *Agric. For. Meteorol.* **2015**, *213*, 304–316. [CrossRef]
121. Alatorre, L.C.; Sánchez-Carrillo, S.; Miramontes-Beltrán, S.; Medina, R.J.; Torres-Olave, M.E.; Bravo, L.C.; Wiebe, L.C.; Granados, A.; Adams, D.K.; Sánchez, E.; et al. Temporal Changes of NDVI for Qualitative Environmental Assessment of Mangroves: Shrimp Farming Impact on the Health Decline of the Arid Mangroves in the Gulf of California (1990–2010). *J. Arid Environ.* **2016**, *125*, 98–109. [CrossRef]

122. Jana, A.; Maiti, S.; Biswas, A. Seasonal Change Monitoring and Mapping of Coastal Vegetation Types along Midnapur-Balasore Coast, Bay of Bengal Using Multi-Temporal Landsat Data. *Model. Earth Syst. Environ.* **2015**, *2*, 7. [CrossRef]
123. Zhang, K.; Thapa, B.; Ross, M.; Gann, D. Remote Sensing of Seasonal Changes and Disturbances in Mangrove Forest: A Case Study from South Florida. *Ecosphere* **2016**, *7*, e01366. [CrossRef]
124. Conti, L.A.; de Araújo, C.A.S.; Cunha-Lignon, M. Spatial Database Modeling for Mangrove Forests Mapping; Example of Two Estuarine Systems in Brazil. *Model. Earth Syst. Environ.* **2016**, *2*, 73. [CrossRef]
125. Rodriguez, W.; Feller, I.C.; Cavanaugh, K.C. Spatio-Temporal Changes of a Mangrove–Saltmarsh Ecotone in the Northeastern Coast of Florida, USA. *Glob. Ecol. Conserv.* **2016**, *7*, 245–261. [CrossRef]
126. Malone, S.L.; Barr, J.; Fuentes, J.D.; Oberbauer, S.F.; Staudhammer, C.L.; Gaiser, E.E.; Starr, G. Sensitivity to Low-Temperature Events: Implications for CO2 Dynamics in Subtropical Coastal Ecosystems. *Wetlands* **2016**, *36*, 957–967. [CrossRef]
127. Nardin, W.; Locatelli, S.; Pasquarella, V.; Rulli, M.C.; Woodcock, C.E.; Fagherazzi, S. Dynamics of a Fringe Mangrove Forest Detected by Landsat Images in the Mekong River Delta, Vietnam. *Earth Surf. Process. Landf.* **2016**, *41*, 2024–2037. [CrossRef]
128. Son, N.-T.; Chen, C.-F.; Chen, C.-R. Mapping Mangrove Density from Rapideye Data in Central America. *Open Geosci.* **2017**, *9*, 211–220. [CrossRef]
129. Yagci, A.L.; Santanello, J.A.; Jones, J.W.; Barr, J. Estimating Evaporative Fraction from Readily Obtainable Variables in Mangrove Forests of the Everglades, U.S.A. *Int. J. Remote Sens.* **2017**, *38*, 3981–4007. [CrossRef]
130. Abd-El Monsef, H.; Smith, S.E. A New Approach for Estimating Mangrove Canopy Cover Using Landsat 8 Imagery. *Comput. Electron. Agric.* **2017**, *135*, 183–194. [CrossRef]
131. Lovelock, C.E.; Feller, I.C.; Reef, R.; Hickey, S.; Ball, M.C. Mangrove Dieback during Fluctuating Sea Levels. *Sci. Rep.* **2017**, *7*, 1680. [CrossRef]
132. Pham, L.T.; Brabyn, L. Monitoring Mangrove Biomass Change in Vietnam Using SPOT Images and an Object-Based Approach Combined with Machine Learning Algorithms. *ISPRS J. Photogramm. Remote Sens.* **2017**, *128*, 86–97. [CrossRef]
133. Tian, J.; Wang, L.; Li, X.; Gong, H.; Shi, C.; Zhong, R.; Liu, X. Comparison of UAV and WorldView-2 Imagery for Mapping Leaf Area Index of Mangrove Forest. *Int. J. Appl. Earth Obs. Geoinf.* **2017**, *61*, 22–31. [CrossRef]
134. Chen, B.; Xiao, X.; Li, X.; Pan, L.; Doughty, R.; Ma, J.; Dong, J.; Qin, Y.; Zhao, B.; Wu, Z. A Mangrove Forest Map of China in 2015: Analysis of Time Series Landsat 7/8 and Sentinel-1A Imagery in Google Earth Engine Cloud Computing Platform. *ISPRS J. Photogramm. Remote Sens.* **2017**, *131*, 104–120. [CrossRef]
135. Zhang, X.; Treitz, P.M.; Chen, D.; Quan, C.; Shi, L.; Li, X. Mapping Mangrove Forests Using Multi-Tidal Remotely-Sensed Data and a Decision-Tree-Based Procedure. *Int. J. Appl. Earth Obs. Geoinf.* **2017**, *62*, 201–214. [CrossRef]
136. Castillo, J.A.A.; Apan, A.A.; Maraseni, T.N.; Salmo, S.G. Estimation and Mapping of Above-Ground Biomass of Mangrove Forests and Their Replacement Land Uses in the Philippines Using Sentinel Imagery. *ISPRS J. Photogramm. Remote Sens.* **2017**, *134*, 70–85. [CrossRef]
137. Galeano, A.; Urrego, L.E.; Botero, V.; Bernal, G. Mangrove Resilience to Climate Extreme Events in a Colombian Caribbean Island. *Wetl. Ecol. Manag.* **2017**, *25*, 743–760. [CrossRef]
138. Milani, A.S. Mangrove Forests of the Persian Gulf and the Gulf of Oman. In *Threats to Mangrove Forests: Hazards, Vulnerability, and Management*; Makowski, C., Finkl, C.W., Eds.; Coastal Research Library; Springer International Publishing: Cham, Switzerland, 2018; pp. 53–75. ISBN 978-3-319-73016-5.
139. Muhd-Ekhzarizal, M.; Mohd-Hasmadi, I.; Hamdan, O.; Mohamad-Roslan, M.; Noor-Shaila, S. Estimation of Aboveground Biomass in Mangrove Forests Using Vegetation Indices from SPOT-5 Image. *J. Trop. For. Sci.* **2018**, *30*, 224–233.
140. Valderrama-Landeros, L.; Flores-de-Santiago, F.; Kovacs, J.M.; Flores-Verdugo, F. An Assessment of Commonly Employed Satellite-Based Remote Sensors for Mapping Mangrove Species in Mexico Using an NDVI-Based Classification Scheme. *Environ. Monit. Assess* **2018**, *190*, 23. [CrossRef]
141. Flores-Cárdenas, F.; Millán-Aguilar, O.; Díaz-Lara, L.; Rodríguez-Arredondo, L.; Hurtado-Oliva, M.Á.; Manzano-Sarabia, M. Trends in the Normalized Difference Vegetation Index for Mangrove Areas in Northwestern Mexico. *J. Coast. Res.* **2018**, *34*, 877–882. [CrossRef]
142. Wang, M.; Cao, W.; Guan, Q.; Wu, G.; Wang, F. Assessing Changes of Mangrove Forest in a Coastal Region of Southeast China Using Multi-Temporal Satellite Images. *Estuar. Coast. Shelf Sci.* **2018**, *207*, 283–292. [CrossRef]
143. Marshall, A.; Schulte to Bühne, H.; Bland, L.; Pettorelli, N. Assessing Ecosystem Collapse Risk in Ecosystems Dominated by Foundation Species: The Case of Fringe Mangroves. *Ecol. Indic.* **2018**, *91*, 128–137. [CrossRef]
144. Chen, Y.-C.; Chu, T.-J.; Wei, J.-D.; Shih, C.-H. Effects of Mangrove Removal on Benthic Organisms in the Siangshan Wetland in Hsinchu, Taiwan. *PeerJ.* **2018**, *6*, e5670. [CrossRef]
145. Staben, G.; Lucieer, A.; Scarth, P. Modelling LiDAR Derived Tree Canopy Height from Landsat TM, ETM+ and OLI Satellite Imagery—A Machine Learning Approach. *Int. J. Appl. Earth Obs. Geoinf.* **2018**, *73*, 666–681. [CrossRef]
146. Wan, R.; Wang, P.; Wang, X.; Yao, X.; Dai, X. Modeling Wetland Aboveground Biomass in the Poyang Lake National Nature Reserve Using Machine Learning Algorithms and Landsat-8 Imagery. *J. Appl. Remote Sens.* **2018**, *12*, 046029. [CrossRef]
147. Selvam, P.P.; Ramesh, R.; Purvaja, R.; Srinivasalu, S. Temporal Changes in Mangrove Forest Coverage and Seasonal Influence on NDVI in Pichavaram Mangrove Forest, India. *Int. J. Ecol. Dev.* **2019**, *34*, 49–61.
148. Taureau, F.; Robin, M.; Proisy, C.; Fromard, F.; Imbert, D.; Debaine, F. Mapping the Mangrove Forest Canopy Using Spectral Unmixing of Very High Spatial Resolution Satellite Images. *Remote Sens.* **2019**, *11*, 367. [CrossRef]

149. Shrestha, S.; Miranda, I.; Kumar, A.; Pardo, M.L.E.; Dahal, S.; Rashid, T.; Remillard, C.; Mishra, D.R. Identifying and Forecasting Potential Biophysical Risk Areas within a Tropical Mangrove Ecosystem Using Multi-Sensor Data. *Int. J. Appl. Earth Obs. Geoinf.* **2019**, *74*, 281–294. [CrossRef]
150. Roy, S.; Mahapatra, M.; Chakraborty, A. Mapping and Monitoring of Mangrove along the Odisha Coast Based on Remote Sensing and GIS Techniques. *Modeling Earth Syst. Environ.* **2019**, *5*, 217–226. [CrossRef]
151. Mafi-Gholami, D.; Zenner, E.K.; Jaafari, A.; Ward, R.D. Modeling Multi-Decadal Mangrove Leaf Area Index in Response to Drought along the Semi-Arid Southern Coasts of Iran. *Sci. Total Environ.* **2019**, *656*, 1326–1336. [CrossRef]
152. Bera, R.; Maiti, R. Quantitative Analysis of Erosion and Accretion (1975–2017) Using DSAS—A Study on Indian Sundarbans. *Reg. Stud. Mar. Sci.* **2019**, *28*, 100583. [CrossRef]
153. Alejandro Berlanga-Robles, C.; Ruiz-Luna, A.; Nepita Villanueva, M.R. Seasonal Trend Analysis (STA) of MODIS Vegetation Index Time Series for the Mangrove Canopy of the Teacapan-Agua Brava Lagoon System, Mexico. *GIScience Remote Sens.* **2019**, *56*, 338–361. [CrossRef]
154. Chuai, X.; Yuan, Y.; Zhang, X.; Guo, X.; Zhang, X.; Xie, F.; Zhao, R.; Li, J. Multiangle Land Use-Linked Carbon Balance Examination in Nanjing City, China. *Land Use Policy* **2019**, *84*, 305–315. [CrossRef]
155. Calva, L.G.; Golubov, J.; Mandujano, M.D.C.; Lara-Domínguez, A.L.; López-Portillo, J. Assessing Google Earth Pro Images for Detailed Conservation Diagnostics of Mangrove Communities. *Coas* **2019**, *92*, 33–43. [CrossRef]
156. Zhu, X.; Song, L.; Weng, Q.; Huang, G. Linking in Situ Photochemical Reflectance Index Measurements with Mangrove Carbon Dynamics in a Subtropical Coastal Wetland. *J. Geophys. Res. Biogeosci.* **2019**, *124*, 1714–1730. [CrossRef]
157. Yaney-Keller, A.; Tomillo, P.S.; Marshall, J.M.; Paladino, F.V. Using Unmanned Aerial Systems (UAS) to Assay Mangrove Estuaries on the Pacific Coast of Costa Rica. *PLoS ONE* **2019**, *14*, e0217310. [CrossRef]
158. Santos, R.D.O.; Delgado, R.C.; Pereira, M.G.; de Souza, L.P.; Tedoro, P.E.; da Silva Junior, C.A.; Costa, G.D.O. Space-Time Variability of the Roncador River Basin in the Change of Land Use and Cover and Its Correlation with Climatic Variables. *Biosci. J.* **2019**, *35*, 1033–1042. [CrossRef]
159. Vázquez-Lule, A.; Colditz, R.; Herrera-Silveira, J.; Guevara, M.; Rodríguez-Zúñiga, M.T.; Cruz, I.; Ressl, R.; Vargas, R. Greenness Trends and Carbon Stocks of Mangroves across Mexico. *Environ. Res. Lett.* **2019**, *14*, 075010. [CrossRef]
160. Pandey, P.C.; Anand, A.; Srivastava, P.K. Spatial Distribution of Mangrove Forest Species and Biomass Assessment Using Field Inventory and Earth Observation Hyperspectral Data. *Biodivers. Conserv.* **2019**, *28*, 2143–2162. [CrossRef]
161. Faridah-Hanum, I.; Yusoff, F.M.; Fitrianto, A.; Ainuddin, N.A.; Gandaseca, S.; Zaiton, S.; Norizah, K.; Nurhidayu, S.; Roslan, M.K.; Hakeem, K.R.; et al. Development of a Comprehensive Mangrove Quality Index (MQI) in Matang Mangrove: Assessing Mangrove Ecosystem Health. *Ecol. Indic.* **2019**, *102*, 103–117. [CrossRef]
162. Li, W.; El-Askary, H.; Qurban, M.A.; Li, J.; ManiKandan, K.P.; Piechota, T. Using Multi-Indices Approach to Quantify Mangrove Changes over the Western Arabian Gulf along Saudi Arabia Coast. *Ecol. Indic.* **2019**, *102*, 734–745. [CrossRef]
163. Zhang, C.; Durgan, S.D.; Lagomasino, D. Modeling Risk of Mangroves to Tropical Cyclones: A Case Study of Hurricane Irma. *Estuar. Coast. Shelf Sci.* **2019**, *224*, 108–116. [CrossRef]
164. Li, Q.; Wong, F.K.K.; Fung, T. Classification of Mangrove Species Using Combined WordView-3 and LiDAR Data in Mai Po Nature Reserve, Hong Kong. *Remote Sens.* **2019**, *11*, 2114. [CrossRef]
165. Ashournejad, Q.; Amiraslani, F.; Moghadam, M.K.; Toomanian, A. Assessing the Changes of Mangrove Ecosystem Services Value in the Pars Special Economic Energy Zone. *Ocean Coast. Manag.* **2019**, *179*, 104838. [CrossRef]
166. Rayegani, B.; Barati, S.; Goshtasb, H.; Sarkheil, H.; Ramezani, J. An Effective Approach to Selecting the Appropriate Pan-Sharpening Method in Digital Change Detection of Natural Ecosystems. *Ecol. Inform.* **2019**, *53*, 100984. [CrossRef]
167. Li, H.; Jia, M.; Zhang, R.; Ren, Y.; Wen, X. Incorporating the Plant Phenological Trajectory into Mangrove Species Mapping with Dense Time Series Sentinel-2 Imagery and the Google Earth Engine Platform. *Remote Sens.* **2019**, *11*, 2479. [CrossRef]
168. Garcia del Toro, E.M.; Mas-Lopez, M.I. Changes in Land Cover in Cacheu River Mangroves Natural Park, Guinea-Bissau: The Need for a More Sustainable Management. *Sustainability* **2019**, *11*, 6247. [CrossRef]
169. Younes, N.; Joyce, K.E.; Northfield, T.D.; Maier, S.W. The Effects of Water Depth on Estimating Fractional Vegetation Cover in Mangrove Forests. *Int. J. Appl. Earth Obs. Geoinf.* **2019**, *83*, 101924. [CrossRef]
170. Rhyma, P.P.; Norizah, K.; Hamdan, O.; Faridah-Hanum, I.; Zulfa, A.W. Integration of Normalised Different Vegetation Index and Soil-Adjusted Vegetation Index for Mangrove Vegetation Delineation. *Remote Sens. Appl. Soc. Environ.* **2020**, *17*, 100280. [CrossRef]
171. Marins, R.V.; Lacerda, L.D.; Araujo, I.C.S.; Fonseca, L.V.; Silva, F.A. Phosphorus and Suspended Matter Retention in Mangroves Affected by Shrimp Farm Effluents in NE Brazil. *An. Acad. Bras. Ciênc.* **2020**, *92*, e20200758. [CrossRef]
172. Mandal, M.S.H.; Kamruzzaman, M.; Hosaka, T. Elucidating the Phenology of the Sundarbans Mangrove Forest Using 18-Year Time Series of MODIS Vegetation Indices. *Tropics* **2020**, *29*, 41–55. [CrossRef]
173. Xiao, H.; Su, F.; Fu, D.; Wang, Q.; Huang, C. Coastal Mangrove Response to Marine Erosion: Evaluating the Impacts of Spatial Distribution and Vegetation Growth in Bangkok Bay from 1987 to 2017. *Remote Sens.* **2020**, *12*, 220. [CrossRef]
174. Svejkovsky, J.; Ogurcak, D.E.; Ross, M.S.; Arkowitz, A. Satellite Image-Based Time Series Observations of Vegetation Response to Hurricane Irma in the Lower Florida Keys. *Estuaries Coasts* **2020**, *43*, 1058–1069. [CrossRef]
175. Chen, N. Mapping Mangrove in Dongzhaigang, China Using Sentinel-2 Imagery. *J. Appl. Remote Sens.* **2020**, *14*, 014508. [CrossRef]

176. Anand, A.; Pandey, P.C.; Petropoulos, G.P.; Pavlides, A.; Srivastava, P.K.; Sharma, J.K.; Malhi, R.K.M. Use of Hyperion for Mangrove Forest Carbon Stock Assessment in Bhitarkanika Forest Reserve: A Contribution towards Blue Carbon Initiative. *Remote Sens.* **2020**, *12*, 597. [CrossRef]
177. Arshad, M.; Eid, E.M.; Hasan, M. Mangrove Health along the Hyper-Arid Southern Red Sea Coast of Saudi Arabia. *Environ. Monit. Assess.* **2020**, *192*, 189. [CrossRef] [PubMed]
178. Thakur, S.; Maity, D.; Mondal, I.; Basumatary, G.; Ghosh, P.B.; Das, P.; De, T.K. Assessment of Changes in Land Use, Land Cover, and Land Surface Temperature in the Mangrove Forest of Sundarbans, Northeast Coast of India. *Environ. Dev. Sustain.* **2021**, *23*, 1917–1943. [CrossRef]
179. Bindu, G.; Rajan, P.; Jishnu, E.S.; Joseph, K.A. Carbon Stock Assessment of Mangroves Using Remote Sensing and Geographic Information System. *Egypt. J. Remote Sens. Space Sci.* **2020**, *23*, 1–9. [CrossRef]
180. Nguyen, H.-H.; Tran, L.T.N.; Le, A.T.; Nghia, N.H.; Duong, L.V.K.; Nguyen, H.T.T.; Bohm, S.; Premnath, C.F.S. Monitoring Changes in Coastal Mangrove Extents Using Multi-Temporal Satellite Data in Selected Communes, Hai Phong City, Vietnam. *For. Soc.* **2020**, *4*, 256–270. [CrossRef]
181. Mandal, M.S.H.; Hosaka, T. Assessing Cyclone Disturbances (1988–2016) in the Sundarbans Mangrove Forests Using Landsat and Google Earth Engine. *Nat. Hazards* **2020**, *102*, 133–150. [CrossRef]
182. Rossi, R.E.; Archer, S.K.; Giri, C.; Layman, C.A. The Role of Multiple Stressors in a Dwarf Red Mangrove (Rhizophora Mangle) Dieback. *Estuar. Coast. Shelf Sci.* **2020**, *237*, 106660. [CrossRef]
183. Le, H.T.; Tran, T.V.; Gyeltshen, S.; Nguyen, C.P.T.; Tran, D.X.; Luu, T.H.; Duong, M.B. Characterizing Spatiotemporal Patterns of Mangrove Forests in Can Gio Biosphere Reserve Using Sentinel-2 Imagery. *Appl. Sci.* **2020**, *10*, 4058. [CrossRef]
184. Taillie, P.J.; Roman-Cuesta, R.; Lagomasino, D.; Cifuentes-Jara, M.; Fatoyinbo, T.; Ott, L.E.; Poulter, B. Widespread Mangrove Damage Resulting from the 2017 Atlantic Mega Hurricane Season. *Environ. Res. Lett.* **2020**, *15*, 064010. [CrossRef]
185. Nur, H.; Islam, M.N. Hot Spot (Gi*) Model for Forest Vulnerability Assessment: A Remote Sensing-Based Geo-Statistical Investigation of the Sundarbans Mangrove Forest, Bangladesh. *Modeling Earth Syst. Environ.* **2020**, *6*, 2141–2151.
186. Sakti, A.D.; Fauzi, A.I.; Wilwatikta, F.N.; Rajagukguk, Y.S.; Sudhana, S.A.; Yayusman, L.F.; Syahid, L.N.; Sritarapipat, T.; Principe, J.A.; Trang, N.T.Q. Multi-Source Remote Sensing Data Product Analysis: Investigating Anthropogenic and Naturogenic Impacts on Mangroves in Southeast Asia. *Remote Sens.* **2020**, *12*, 2720. [CrossRef]
187. Dayathilake, D.D.T.L.; Lokupitiya, E.; Wijeratne, V.P.I.S. Estimation of Aboveground and Belowground Carbon Stocks in Urban Freshwater Wetlands of Sri Lanka. *Carbon Balance Manag.* **2020**, *15*, 17. [CrossRef] [PubMed]
188. Yu, M.; Gao, Q. Topography, Drainage Capability, and Legacy of Drought Differentiate Tropical Ecosystem Response to and Recovery from Major Hurricanes. *Environ. Res. Lett.* **2020**, *15*, 104046. [CrossRef]
189. Castillo, Y.B.; Kim, K.; Kim, H.S. Thirty-Two Years of Mangrove Forest Land Cover Change in Parita Bay, Panama. *For. Sci. Technol.* **2021**, *17*, 67–79. [CrossRef]
190. Etemadi, H.; Smoak, J.M.; Abbasi, E. Spatiotemporal Pattern of Degradation in Arid Mangrove Forests of the Northern Persian Gulf. *Oceanologia* **2021**, *63*, 99–114. [CrossRef]
191. Aljahdali, M.O.; Munawar, S.; Khan, W.R. Monitoring Mangrove Forest Degradation and Regeneration: Landsat Time Series Analysis of Moisture and Vegetation Indices at Rabigh Lagoon, Red Sea. *Forests* **2021**, *12*, 52. [CrossRef]
192. Nardin, W.; Vona, I.; Fagherazzi, S. Sediment Deposition Affects Mangrove Forests in the Mekong Delta, Vietnam. *Cont. Shelf Res.* **2021**, *213*, 104319. [CrossRef]
193. Valderrama-Landeros, L.; Flores-Verdugo, F.; Rodríguez-Sobreyra, R.; Kovacs, J.M.; Flores-de-Santiago, F. Extrapolating Canopy Phenology Information Using Sentinel-2 Data and the Google Earth Engine Platform to Identify the Optimal Dates for Remotely Sensed Image Acquisition of Semiarid Mangroves. *J. Environ. Manag.* **2021**, *279*, 111617. [CrossRef]
194. Thakur, S.; Mondal, I.; Bar, S.; Nandi, S.; Ghosh, P.B.; Das, P.; De, T.K. Shoreline Changes and Its Impact on the Mangrove Ecosystems of Some Islands of Indian Sundarbans, North-East Coast of India. *J. Clean. Prod.* **2021**, *284*, 124764. [CrossRef]
195. Zhao, C.-P.; Qin, C.-Z. A Detailed Mangrove Map of China for 2019 Derived from Sentinel-1 and -2 Images and Google Earth Images. *Geosci. Data J.* **2022**, *9*, 74–88. [CrossRef]
196. Jiang, Y.; Zhang, L.; Yan, M.; Qi, J.; Fu, T.; Fan, S.; Chen, B. High-Resolution Mangrove Forests Classification with Machine Learning Using Worldview and Uav Hyperspectral Data. *Remote Sens.* **2021**, *13*, 1529. [CrossRef]
197. Purnamasari, E.; Kamal, M.; Wicaksono, P. Comparison of Vegetation Indices for Estimating Above-Ground Mangrove Carbon Stocks Using PlanetScope Image. *Reg. Stud. Mar. Sci.* **2021**, *44*, 101730. [CrossRef]
198. Lee, C.K.; Duncan, C.; Nicholson, E.; Fatoyinbo, T.E.; Lagomasino, D.; Thomas, N.; Worthington, T.A.; Murray, N.J. Mapping the Extent of Mangrove Ecosystem Degradation by Integrating an Ecological Conceptual Model with Satellite Data. *Remote Sens.* **2021**, *13*, 2047. [CrossRef]
199. Cui, L.; Zuo, X.; Dou, Z.; Huang, Y.; Zhao, X.; Zhai, X.; Lei, Y.; Li, J.; Pan, X.; Li, W. Plant Identification of Beijing Hanshiqiao Wetland Based on Hyperspectral Data. *Spectrosc. Lett.* **2021**, *54*, 381–394. [CrossRef]
200. Obida, C.B.; Blackburn, G.A.; Whyatt, J.D.; Semple, K.T. Counting the Cost of the Niger Delta's Largest Oil Spills: Satellite Remote Sensing Reveals Extensive Environmental Damage with >1million People in the Impact Zone. *Sci. Total Environ.* **2021**, *775*, 145854. [CrossRef]
201. Niu, C.; Phinn, S.; Roelfsema, C. Global Sensitivity Analysis for Canopy Reflectance and Vegetation Indices of Mangroves. *Remote Sens.* **2021**, *13*, 2617. [CrossRef]

202. Singgalen, Y.A.; Gudiato, C.; Prasetyo, S.Y.J.; Fibriani, C. Mangrove Monitoring Using Normalized Difference Vegetation Index (NDVI): Case Study In North Halmahera, Indonesia. *J. Teknol. Kelaut. Trop.* **2021**, *13*, 219–239. [CrossRef]
203. Kamal, M.; Sidik, F.; Prananda, A.R.A.; Mahardhika, S.A. Mapping Leaf Area Index of Restored Mangroves Using WorldView-2 Imagery in Perancak Estuary, Bali, Indonesia. *Remote Sens. Appl. Soc. Environ.* **2021**, *23*, 100567. [CrossRef]
204. Maina, J.M.; Bosire, J.O.; Kairo, J.G.; Bandeira, S.O.; Mangora, M.M.; Macamo, C.; Ralison, H.; Majambo, G. Identifying Global and Local Drivers of Change in Mangrove Cover and the Implications for Management. *Glob. Ecol. Biogeogr.* **2021**, *30*, 2057–2069. [CrossRef]
205. Idris, N.S.; Mustapha, M.A.; Sulaiman, N.; Khamis, S.; Husin, S.M.; Darbis, N.D.A. The Dynamics of Landscape Changes Surrounding a Firefly Ecotourism Area. *Glob. Ecol. Conserv.* **2021**, *29*, e01741. [CrossRef]
206. Mishra, M.; Acharyya, T.; Santos, C.A.G.; da Silva, R.M.; Kar, D.; Mustafa Kamal, A.H.; Raulo, S. Geo-Ecological Impact Assessment of Severe Cyclonic Storm Amphan on Sundarban Mangrove Forest Using Geospatial Technology. *Estuar. Coast. Shelf Sci.* **2021**, *260*, 107486. [CrossRef]
207. Nguyen, H.-H.; Nguyen, T.T.H. Above-Ground Biomass Estimation Models of Mangrove Forests Based on Remote Sensing and Field-Surveyed Data: Implications for C-PFES Implementation in Quang Ninh Province, Vietnam. *Reg. Stud. Mar. Sci.* **2021**, *48*, 101985. [CrossRef]
208. Meijer, K.J.; El-Hacen, E.-H.M.; Govers, L.L.; Lavaleye, M.; Piersma, T.; Olff, H. Mangrove-Mudflat Connectivity Shapes Benthic Communities in a Tropical Intertidal System. *Ecol. Indic.* **2021**, *130*, 108030. [CrossRef]
209. Guo, X.; Wang, M.; Jia, M.; Wang, W. Estimating Mangrove Leaf Area Index Based on Red-Edge Vegetation Indices: A Comparison among UAV, WorldView-2 and Sentinel-2 Imagery. *Int. J. Appl. Earth Obs. Geoinf.* **2021**, *103*, 102493. [CrossRef]
210. Barr, J.G.; Engel, V.; Smith, T.J.; Fuentes, J.D. Hurricane Disturbance and Recovery of Energy Balance, CO2 Fluxes and Canopy Structure in a Mangrove Forest of the Florida Everglades. *Agric. For. Meteorol.* **2012**, *153*, 54–66. [CrossRef]
211. Barr, J.G.; Engel, V.; Fuentes, J.D.; Fuller, D.O.; Kwon, H. Modeling Light Use Efficiency in a Subtropical Mangrove Forest Equipped with CO_2 Eddy Covariance. *Biogeosciences* **2013**, *10*, 2145–2158. [CrossRef]
212. Shoemaker, W.B.; Anderson, F.; Barr, J.G.; Graham, S.L.; Botkin, D.B. Carbon Exchange between the Atmosphere and Subtropical Forested Cypress and Pine Wetlands. *Biogeosciences* **2015**, *12*, 2285–2300. [CrossRef]
213. Dutta, D.; Das, P.K.; Paul, S.; Sharma, J.R.; Dadhwal, V.K. Assessment of Ecological Disturbance in the Mangrove Forest of Sundarbans Caused by Cyclones Using MODIS Time-Series Data (2001–2011). *Nat. Hazards* **2015**, *79*, 775–790. [CrossRef]
214. Ishtiaque, A.; Myint, S.W.; Wang, C. Examining the Ecosystem Health and Sustainability of the World's Largest Mangrove Forest Using Multi-Temporal MODIS Products. *Sci. Total Environ.* **2016**, *569–570*, 1241–1254. [CrossRef]
215. Bolivar, J.M.; Gutierrez-Velez, V.H.; Sierra, C.A. Carbon Stocks in Aboveground Biomass for Colombian Mangroves with Associated Uncertainties. *Reg. Stud. Mar. Sci.* **2018**, *18*, 145–155. [CrossRef]
216. Songsom, V.; Koedsin, W.; Ritchie, R.J.; Huete, A. Mangrove Phenology and Environmental Drivers Derived from Remote Sensing in Southern Thailand. *Remote Sens.* **2019**, *11*, 955. [CrossRef]
217. Nepita-Villanueva, M.R.; Berlanga-Robles, C.A.; Ruiz-Luna, A.; Morales Barcenas, J.H. Spatio-Temporal Mangrove Canopy Variation (2001–2016) Assessed Using the MODIS Enhanced Vegetation Index (EVI). *J. Coast. Conserv.* **2019**, *23*, 589–597. [CrossRef]
218. Feagin, R.A.; Forbrich, I.; Huff, T.P.; Barr, J.G.; Ruiz-Plancarte, J.; Fuentes, J.D.; Najjar, R.G.; Vargas, R.; Vázquez-Lule, A.; Windham-Myers, L. Tidal Wetland Gross Primary Production across the Continental United States, 2000–2019. *Glob. Biogeochem. Cycles* **2020**, *34*, e2019GB006349. [CrossRef]
219. Berlanga-Robles, C.A.; Ruiz-Luna, A. Assessing Seasonal and Long-Term Mangrove Canopy Variations in Sinaloa, Northwest Mexico, Based on Time Series of Enhanced Vegetation Index (EVI) Data. *Wetl. Ecol. Manag.* **2020**, *28*, 229–249. [CrossRef]
220. Parida, B.R.; Kumari, A. Mapping and Modeling Mangrove Biophysical and Biochemical Parameters Using Sentinel-2A Satellite Data in Bhitarkanika National Park, Odisha. *Model. Earth Syst. Environ.* **2021**, *7*, 2463–2474. [CrossRef]
221. Younes, N.; Northfield, T.D.; Joyce, K.E.; Maier, S.W.; Duke, N.C.; Lymburner, L. A Novel Approach to Modelling Mangrove Phenology from Satellite Images: A Case Study from Northern Australia. *Remote Sens.* **2020**, *12*, 4008. [CrossRef]
222. Younes, N.; Joyce, K.E.; Maier, S.W. All Models of Satellite-Derived Phenology Are Wrong, but Some Are Useful: A Case Study from Northern Australia. *Int. J. Appl. Earth Obs. Geoinf.* **2021**, *97*, 102285. [CrossRef]
223. Peereman, J.; Hogan, J.A.; Lin, T.-C. Disturbance Frequency, Intensity and Forest Structure Modulate Cyclone-Induced Changes in Mangrove Forest Canopy Cover. *Glob. Ecol. Biogeogr.* **2022**, *31*, 37–50. [CrossRef]
224. Zhu, B.; Liao, J.; Shen, G. Spatio-Temporal Simulation of Mangrove Forests under Different Scenarios: A Case Study of Mangrove Protected Areas, Hainan Island, China. *Remote Sens.* **2021**, *13*, 4059. [CrossRef]
225. Poortinga, A.; Tenneson, K.; Shapiro, A.; Nquyen, Q.; San Aung, K.; Chishtie, F.; Saah, D. Mapping Plantations in Myanmar by Fusing Landsat-8, Sentinel-2 and Sentinel-1 Data along with Systematic Error Quantification. *Remote Sens.* **2019**, *11*, 831. [CrossRef]
226. Tucker, C.J. Red and Photographic Infrared Linear Combinations for Monitoring Vegetation. *Remote Sens. Environ.* **1979**, *8*, 127–150. [CrossRef]
227. Liu, X.; Wang, L. Feasibility of Using Consumer-Grade Unmanned Aerial Vehicles to Estimate Leaf Area Index in Mangrove Forest. *Remote Sens. Lett.* **2018**, *9*, 1040–1049. [CrossRef]

228. Clevers, J. The Application of a Vegetation Index in Correcting the Infrared Reflectance for Soil Background. In Proceedings of the Remote Sensing for Ressources Development and Environmental Management, International Symposium. 7, Enschede, The Netherlands, 25–29 August 1986; pp. 221–226.
229. Major, D.J.; Baret, F.; Guyot, G. A Ratio Vegetation Index Adjusted for Soil Brightness. *Int. J. Remote Sens.* **1990**, *11*, 727–740. [CrossRef]
230. Pinty, B.; Verstraete, M.M. GEMI: A Non-Linear Index to Monitor Global Vegetation from Satellites. *Vegetatio* **1992**, *101*, 15–20. [CrossRef]
231. Kaufman, Y.J.; Tanre, D. Atmospherically Resistant Vegetation Index (ARVI) for EOS-MODIS. *IEEE Trans. Geosci. Remote Sens.* **1992**, *30*, 261–270. [CrossRef]
232. Hati, J.P.; Goswami, S.; Samanta, S.; Pramanick, N.; Majumdar, S.D.; Chaube, N.R.; Misra, A.; Hazra, S. Estimation of Vegetation Stress in the Mangrove Forest Using AVIRIS-NG Airborne Hyperspectral Data. *Modeling Earth Syst. Environ.* **2021**, *7*, 1877–1889. [CrossRef]
233. Goel, N.S.; Qin, W. Influences of Canopy Architecture on Relationships between Various Vegetation Indices and LAI and Fpar: A Computer Simulation. *Remote Sens. Rev.* **1994**, *10*, 309–347. [CrossRef]
234. Qi, J.; Chehbouni, A.; Huete, A.R.; Kerr, Y.H.; Sorooshian, S. A Modified Soil Adjusted Vegetation Index. *Remote Sens. Environ.* **1994**, *48*, 119–126. [CrossRef]
235. Roujean, J.-L.; Breon, F.-M. Estimating PAR Absorbed by Vegetation from Bidirectional Reflectance Measurements. *Remote Sens. Environ.* **1995**, *51*, 375–384. [CrossRef]
236. Chen, J.M. Evaluation of Vegetation Indices and a Modified Simple Ratio for Boreal Applications. *Can. J. Remote Sens.* **1996**, *22*, 229–242. [CrossRef]
237. Gitelson, A.A.; Kaufman, Y.J.; Merzlyak, M.N. Use of a Green Channel in Remote Sensing of Global Vegetation from EOS-MODIS. *Remote Sens. Environ.* **1996**, *58*, 289–298. [CrossRef]
238. Rondeaux, G.; Steven, M.; Baret, F. Optimization of Soil-Adjusted Vegetation Indices. *Remote Sens. Environ.* **1996**, *55*, 95–107. [CrossRef]
239. Gitelson, A.A.; Merzlyak, M.N. Remote Sensing of Chlorophyll Concentration in Higher Plant Leaves. *Adv. Space Res.* **1998**, *22*, 689–692. [CrossRef]
240. Louhaichi, M.; Borman, M.M.; Johnson, D.E. Spatially Located Platform and Aerial Photography for Documentation of Grazing Impacts on Wheat. *Geocarto Int.* **2001**, *16*, 65–70. [CrossRef]
241. Broge, N.H.; Leblanc, E. Comparing Prediction Power and Stability of Broadband and Hyperspectral Vegetation Indices for Estimation of Green Leaf Area Index and Canopy Chlorophyll Density. *Remote Sens. Environ.* **2001**, *76*, 156–172. [CrossRef]
242. Bannari, A.; Asalhi, H.; Teillet, P.M. Transformed Difference Vegetation Index (TDVI) for Vegetation Cover Mapping. In Proceedings of the IEEE International geoscience and remote sensing symposium, Toronto, ON, Canada, 24–28 June 2002; Volume 5, pp. 3053–3055.
243. Gong, P.; Pu, R.; Biging, G.S.; Larrieu, M.R. Estimation of Forest Leaf Area Index Using Vegetation Indices Derived from Hyperion Hyperspectral Data. *IEEE Trans. Geosci. Remote Sens.* **2003**, *41*, 1355–1362. [CrossRef]
244. Haboudane, D.; Miller, J.R.; Pattey, E.; Zarco-Tejada, P.J.; Strachan, I.B. Hyperspectral Vegetation Indices and Novel Algorithms for Predicting Green LAI of Crop Canopies: Modeling and Validation in the Context of Precision Agriculture. *Remote Sens. Environ.* **2004**, *90*, 337–352. [CrossRef]
245. Vincini, M.; Frazzi, E.; D'alessio, P.; Stafford, J.V. Comparison of Narrow-Band and Broad-Band Vegetation Indexes for Canopy Chlorophyll Density Estimation in Sugar Beet. In Proceedings of the Precision agriculture '07: Proceedings of the 6th European Conference on Precision Agriculture, Skiathos, Greece, 3–6 June 2007; pp. 189–196.
246. Rahman, A.F.; Dragoni, D.; Didan, K.; Barreto-Munoz, A.; Hutabarat, J.A. Detecting Large Scale Conversion of Mangroves to Aquaculture with Change Point and Mixed-Pixel Analyses of High-Fidelity MODIS Data. *Remote Sens. Environ.* **2013**, *130*, 96–107. [CrossRef]
247. Muhsoni, F.F.; Sambah, A.B.; Mahmudi, M.; Wiadnya, D.G.R. Comparison of Different Vegetation Indices for Assessing Mangrove Density Using Sentinel-2 Imagery. *GEOMATE J.* **2018**, *14*, 42–51.
248. Woebbecke, D.; Meyer, G.; Bargen, K.V.; Mortensen, D. Color Indices for Weed Identification under Various Soil, Residue, and Lighting Conditions. *Trans. ASAE* **1995**, *38*, 259–269. [CrossRef]
249. Gitelson, A.; Merzlyak, M.N. Spectral Reflectance Changes Associated with Autumn Senescence of Aesculus Hippocastanum L. and Acer Platanoides L. Leaves. Spectral Features and Relation to Chlorophyll Estimation. *J. Plant Physiol.* **1994**, *143*, 286–292. [CrossRef]
250. Merzlyak, M.N.; Gitelson, A.A.; Chivkunova, O.B.; Rakitin, V.Y. Non-destructive Optical Detection of Pigment Changes during Leaf Senescence and Fruit Ripening. *Physiol. Plant.* **1999**, *106*, 135–141. [CrossRef]
251. Barnes, E.M.; Clarke, T.R.; Richards, S.E.; Colaizzi, P.D.; Haberland, J.; Kostrzewski, M.; Waller, P.; Choi, C.; Riley, E.; Thompson, T. Coincident Detection of Crop Water Stress, Nitrogen Status and Canopy Density Using Ground Based Multispectral Data. In Proceedings of the Fifth International Conference on Precision Agriculture, Bloomington, MN, USA, 16–19 July 2000; Volume 1619.
252. Gitelson, A.A.; Viña, A.; Arkebauer, T.J.; Rundquist, D.C.; Keydan, G.; Leavitt, B. Remote Estimation of Leaf Area Index and Green Leaf Biomass in Maize Canopies. *Geophys. Res. Lett.* **2003**, *30*, 1248. [CrossRef]
253. Dash, J.; Curran, P.J. The MERIS Terrestrial Chlorophyll Index. *Int. J. Remote Sens.* **2004**, *25*, 5403–5413. [CrossRef]

254. Biswas, H.; Zhang, K.; Ross, M.S.; Gann, D. Delineation of Tree Patches in a Mangrove-Marsh Transition Zone by Watershed Segmentation of Aerial Photographs. *Remote Sens.* **2020**, *12*, 2086. [CrossRef]
255. Meyer, G.E.; Hindman, T.W.; Laksmi, K. *Machine Vision Detection Parameters for Plant Species Identification*; Meyer, G.E., DeShazer, J.A., Eds.; SPIE: Boston, MA, USA, 1999; pp. 327–335.
256. Gitelson, A.A.; Kaufman, Y.J.; Stark, R.; Rundquist, D. Novel Algorithms for Remote Estimation of Vegetation Fraction. *Remote Sens. Environ.* **2002**, *80*, 76–87. [CrossRef]
257. Kataoka, T.; Kaneko, T.; Okamoto, H.; Hata, S. Crop Growth Estimation System Using Machine Vision. In Proceedings of the Proceedings 2003 IEEE/ASME International Conference on Advanced Intelligent Mechatronics (AIM 2003), Kobe, Japan, 20–24 July 2003; Volume 2, pp. b1079–b1083.
258. Hague, T.; Tillett, N.D.; Wheeler, H. Automated Crop and Weed Monitoring in Widely Spaced Cereals. *Precis. Agric.* **2006**, *7*, 21–32. [CrossRef]
259. Meyer, G.E.; Neto, J.C. Verification of Color Vegetation Indices for Automated Crop Imaging Applications. *Comput. Electron. Agric.* **2008**, *63*, 282–293. [CrossRef]
260. Hunt, E.R.; Daughtry, C.S.T.; Eitel, J.U.; Long, D.S. Remote Sensing Leaf Chlorophyll Content Using a Visible Band Index. *Agron. J.* **2011**, *103*, 1090–1099. [CrossRef]
261. Yang, W.; Wang, S.; Zhao, X.; Zhang, J.; Feng, J. Greenness Identification Based on HSV Decision Tree. *Inf. Processing Agric.* **2015**, *2*, 149–160. [CrossRef]
262. Xiaoqin, W.; Miaomiao, W.; Shaoqiang, W.; Yundong, W. Extraction of Vegetation Information from Visible Unmanned Aerial Vehicle Images. *Trans. Chin. Soc. Agric. Eng.* **2015**, *31*, 152–159.
263. Spencer, T.; Möller, I.; Reef, R. Mangrove Systems and Environments. *Ref. Modul. Earth Syst. Environ. Sci.* **2022**, 675–712. [CrossRef]
264. Gao, B.-C. NDWI—A Normalized Difference Water Index for Remote Sensing of Vegetation Liquid Water from Space. *Remote Sens. Environ.* **1996**, *58*, 257–266. [CrossRef]
265. Bhatti, S.S.; Tripathi, N.K. Built-up Area Extraction Using Landsat 8 OLI Imagery. *GIScience Remote Sens.* **2014**, *51*, 445–467. [CrossRef]
266. Neri, M.P.; Baloloy, A.B.; Blanco, A.C. Limitation Assessment and Workflow Refinement of the Mangrove Vegetation Index (MVI)-Based Mapping Methodology Using Sentinel-2 Imagery. *Int. Arch. Photogramm. Remote Sens. Spat. Inf. Sci.* **2021**, *46*, 235–242. [CrossRef]
267. Kamal, M.; Phinn, S. Hyperspectral Data for Mangrove Species Mapping: A Comparison of Pixel-Based and Object-Based Approach. *Remote Sens.* **2011**, *3*, 2222–2242. [CrossRef]
268. Toosi, N.B.; Soffianian, A.R.; Fakheran, S.; Pourmanafi, S.; Ginzler, C.; Waser, L.T. Comparing Different Classification Algorithms for Monitoring Mangrove Cover Changes in Southern Iran. *Glob. Ecol. Conserv.* **2019**, *19*, e00662. [CrossRef]
269. Akhrianti, I. Spatial Distribution of Mangrove in Kelapan Island, South Bangka Regency. *Red* **2019**, *665*, 10.
270. Hardisky, M.A.; Daiber, F.C.; Roman, C.T.; Klemas, V. Remote Sensing of Biomass and Annual Net Aerial Primary Productivity of a Salt Marsh. *Remote Sens. Environ.* **1984**, *16*, 91–106. [CrossRef]
271. Xiao, X.; Zhang, Q.; Braswell, B.; Urbanski, S.; Boles, S.; Wofsy, S.; Moore, B.; Ojima, D. Modeling Gross Primary Production of Temperate Deciduous Broadleaf Forest Using Satellite Images and Climate Data. *Remote Sens. Environ.* **2004**, *91*, 256–270. [CrossRef]
272. Xu, H. Modification of Normalised Difference Water Index (NDWI) to Enhance Open Water Features in Remotely Sensed Imagery. *Int. J. Remote Sens.* **2006**, *27*, 3025–3033. [CrossRef]
273. Hirata, Y.; Tabuchi, R.; Patanaponpaiboon, P.; Poungparn, S.; Yoneda, R.; Fujioka, Y. Estimation of Aboveground Biomass in Mangrove Forests Using High-Resolution Satellite Data. *J. Res.* **2014**, *19*, 34–41. [CrossRef]
274. Bathmann, J.; Peters, R.; Reef, R.; Berger, U.; Walther, M.; Lovelock, C.E. Modelling Mangrove Forest Structure and Species Composition over Tidal Inundation Gradients: The Feedback between Plant Water Use and Porewater Salinity in an Arid Mangrove Ecosystem. *Agric. For. Meteorol.* **2021**, *308–309*, 108547. [CrossRef]
275. Fang, H.; Liang, S. Leaf Area Index Models. In *Reference Module in Earth Systems and Environmental Sciences*; Elsevier: Amsterdam, The Netherlands, 2014; pp. 2139–2148. ISBN 978-0-12-409548-9.
276. Zhen, J.; Jiang, X.; Xu, Y.; Miao, J.; Zhao, D.; Wang, J.; Wang, J.; Wu, G. Mapping Leaf Chlorophyll Content of Mangrove Forests with Sentinel-2 Images of Four Periods. *Int. J. Appl. Earth Obs. Geoinf.* **2021**, *102*, 102387. [CrossRef]
277. Hansen, M.C.; Potapov, P.V.; Moore, R.; Hancher, M.; Turubanova, S.A.; Tyukavina, A.; Thau, D.; Stehman, S.V.; Goetz, S.J.; Loveland, T.R.; et al. High-Resolution Global Maps of 21st-Century Forest Cover Change. *Science* **2013**, *342*, 850–853. [CrossRef] [PubMed]
278. Lehmann, E.A.; Wallace, J.F.; Caccetta, P.A.; Furby, S.L.; Zdunic, K. Forest Cover Trends from Time Series Landsat Data for the Australian Continent. *Int. J. Appl. Earth Obs. Geoinf.* **2013**, *21*, 453–462. [CrossRef]
279. Reddy, D.S.; Prasad, P.R.C. Prediction of Vegetation Dynamics Using NDVI Time Series Data and LSTM. *Modeling Earth Syst. Environ.* **2018**, *4*, 409–419. [CrossRef]
280. Bryan-Brown, D.N.; Connolly, R.M.; Richards, D.R.; Adame, F.; Friess, D.A.; Brown, C.J. Global Trends in Mangrove Forest Fragmentation. *Sci Rep.* **2020**, *10*, 7117. [CrossRef]

281. Kanniah, K.D.; Kang, C.S.; Sharma, S.; Amir, A.A. Remote Sensing to Study Mangrove Fragmentation and Its Impacts on Leaf Area Index and Gross Primary Productivity in the South of Peninsular Malaysia. *Remote Sens.* **2021**, *13*, 1427. [CrossRef]
282. USGS. USGS EROS Archive—Advanced Very High Resolution Radiometer (AVHRR)—Sensor Characteristics. Available online: https://www.usgs.gov/centers/eros/science/usgs-eros-archive-advanced-very-high-resolution-radiometer-avhrr#web-tools (accessed on 25 April 2022).

MDPI
St. Alban-Anlage 66
4052 Basel
Switzerland
www.mdpi.com

Remote Sensing Editorial Office
E-mail: remotesensing@mdpi.com
www.mdpi.com/journal/remotesensing

www.ingramcontent.com/pod-product-compliance
Lightning Source LLC
Chambersburg PA
CBHW061839210926
51850CB00006B/931